Lasers
in
Polymer Science
and
Technology:
Applications

Volume III

Editors

Jean-Pierre Fouassier, Ph.D.
Professor
Laboratory of General Photochemistry
Ecole Nationale Superieure de Chimie
University of Haute-Alsace
Mulhouse, France

Jan F. Rabek, Ph.D.
Department of Polymer Technology
The Royal Institute of Technology
Stockholm, Sweden

CRC Press, Inc.
Boca Raton, Florida

CHEMISTRY

a 3825255 (handwritten)

Library of Congress Cataloging-in-Publication Data

Lasers in polymer science and technology:applications / editors, Jean
-Pierre Fouassier and Jan F. Rabek.
   p. cm.
  Bibliography: p.
  Includes index.
  ISBN 0-8493-4844-7 (v. 1)
  1. Polymers--Analysis. 2. Laser spectroscopy. I. Fouassier,
Jean-Pierre, 1947- II. Rabek, J. F.
  TP1140.L37 1990
  668.9--dc20   89-9822
                                                      CIP

DEDICATION

To our wives, partners through life
Geneviève — Ewelina
and our children
Patrick, Laurence, and Yann — Dominika
for their patience and understanding.

PREFACE

Laser spectroscopy and laser technology have been growing ever since the first laser was developed in 1960 and cover now a wide range of applications. Among them, three groups came into prominence as regards polymer science and technology: molecular gas lasers (notably CO_2 lasers) in the IR region, gas, solid, and dye lasers in the visible and near IR region, and the relatively new group of UV excimer lasers. Lasers are unique sources of light. Many recent advances in science are dependent on the application of their uniqueness to specific problems. Lasers can produce the most spectrally pure light available, enabling atomic and molecular energy levels to be studied in greater detail than ever before. Certain types of laser can give rise to the shortest pulses of light available from any light source, thus providing a means for measuring some of the fastest processes in nature.

Measurements of luminescence (fluorescence and phosphorescence) provide some of the most sensitive and selective methods of spectroscopy. In addition, luminescence measurements provide important information about the properties of excited states, because the emitted light originates from electronically excited states. The measurement of luminescence intensities makes it possible to monitor the changes in concentration of the emitting chemical species as a function of time, whereas the wavelength distribution of the luminescence provides information on the nature and energy of the emitting species.

Such areas as laser luminescence spectroscopy, pico- and nanosecond absorption spectroscopy, CIDNP and CIDEP laser flash photolysis, holographic spectroscopy, and time-resolved diffuse reflectance laser spectroscopy, have evolved from esoteric research specialities into standard procedures, and in some cases routinely applied in a number of laboratories all over the world.

Application of Rayleigh, Brillouin, and Raman laser spectroscopy in polymer science gives information about local polymer chain motion, large-scale diffusion, relaxation behavior, phase transitions, and ordered states of macromolecules.

During the last decade the photochemistry and photophysics of polymers have grown into an important and pervasive branch of polymer science. Great strides have been made in the theory of photoreactions, energy transfer processes, the utilization of photoreactions in polymerization, grafting, curing, degradation, and stabilization of polymers. The progress of powerful laser techniques has not been limited to spectroscopical studies in polymer matrix, colloids, dyed fabrics, photoinitiators, photosensitizers, photoresists, materials for solar energy conversion, or biological molecules and macromolecules; it has also found a number of practical and even industrial applications.

One of the most important applications of lasers is the use of a high intensity beam for material processing in polymers. In these materials, the laser beam can be employed for drilling, cutting, and welding. Lasers can produce holes at very high speeds and dimensions, unobtainable by other processing methods.

Lasers can be successfully used to study surface processes and surface modification of polymeric materials, such as molecular beam scattering, oxidation, etching, annealing, phase transitions, surface mobility, and thin films and vapor phase deposition.

UV laser radiation causes the breakup and spontaneous removal of material from the surface of organic polymers (ablative photodecomposition). The surface of the solid is etched away to a depth of a few tenths of a micron, and the products are expelled at supersonic velocity. This method has found practical applications in photolithography, optics, electronics, and the aerospace industries.

The newest process includes stereolithography, which involves building three-dimensional plastic prototypes (models) from computer-aided designs. Stereolithography is actually a combination of four technologies: photochemistry, computer-aided, laser light, and laser-image formation. The device (which consists in a mechanically scanned, computer driven three-dimensional solid pattern generator) builds parts by creating, under the laser exposure, cross sections of the part out of a liquid photopolymer, then "fusing" the sections together until a complete model is formed.

Another new development is technology of micromachines such as gears, turbines, and motors which are 100 to 200 μm in diameter which can be used in a space technology, microrobots, or missile-guidance systems. These micromachines are made by a process of etching patterns on silicon chips. Beside making such micromachines, microscopic tools on a catheter, inserted through a blood vessel, would enable surgeons to do "closed heart" surgery. Developing of micromachine technology would not be possible without photopolymers and UV lasers.

The editors went to great lengths in order to secure the cooperation of the most outstanding specialists to complete this monography. A number of invited authorities were not able to accept our invitation, due to other commitments, but all authors who presented their contributions "poured their hearts out" in this endeavour. We would like to thank them for their efforts and cooperation. This monography strongly favors the inclusion of experimental details, apparatus, and techniques, thus allowing the neophyte to learn the "tricks of the trade" from the experts. This is an effort to show, in compact form, the bulk of information available on applications of lasers to polymer science and technology. The editors are pleased to submit to the readers the state-of-the art in this field.

J.-P. Fouassier and J. R. Rabek

THE EDITORS

Dr. Jean-Pierre Fouassier is head of the Laboratoire de Photochimie Générale, Ecole Nationale Supérieure de Chimie de Mulhouse, and Centre National de la Recherche Scientifique, and Professor of Physical Chemistry at the University of Haute Alsace.

Prof. J. P. Fouassier graduated in 1970 from The National School of Chemistry, Mulhouse, with an Engineer degree and obtained his Ph.D. in 1975 at the University of Strasbourg. After doing postdoctoral work at the Institüt für Makromolekulare Chemie, Freiburg, (West Germany), he was appointed as lecturer. It was in 1980 that he assumed his present position.

Prof. J. P. Fouassier is a member of the Société Française de Chimie, the Groupe Français des Polymères, the European Photochemistry Association, the ACS Polymer Division, and Radtech Europe.

Prof. J. P. Fouassier has been the recipient of research grants from the Centre National de la Recherche Scientifique, the Ministère de la Recherche, the Association Nationale pour la Valorisation et l'Aide à la Recherche, and French and European private industries. He has published more than 100 research papers. His current major research interests include time-resolved laser spectroscopies, excited state processes in photoinitiators and photosensitizers, laser-induced photopolymerization reactions, development of photosensitive systems for holographic recording, and light radiation curing.

Dr. Jan F. Rabek is Professor of Polymer Chemistry in the Department of Polymer Technology, The Royal Institute of Technology, Stockholm, working in the field of polymer photochemistry and photophysics since 1960. His research interests lie in the photodegradation, photooxidation, and photostabilization of polymers, singlet oxygen photooxidation, spectroscopy of molecular complexes in polymers, and recently photoconducting polymers and polymeric photosensors.

Dr. Rabek obtained his D.Sc. in Polymer Technology at the Department of Polymer Technology, Technical University, Wroclaw, Poland (1965) and his Ph.D. in Polymer Photochemistry at the Department of Chemistry, Sileasian Technical University, Gliwice, Poland (1968). He has published more than 120 research papers, review papers, and books on the photochemistry of polymers.

CONTRIBUTORS

Jean-Claude Andre, Ph.D.
Director of Research
Department of Physical Chemistry
Centre National de la Recherche
 Scientifique
Nancy, France

H. Anneser, Dipl. Chem.
Department of Physical Chemistry
University of Munich
Munich, West Germany

B. Braren
Staff Engineer
Department of Physical Sciences
IBM Corporation
Yorktown Heights, New York

Christoph Bräuchle, Prof. Dr.
Department of Physical Chemistry
University of Munich
Munich, West Germany

Miguel Cabrera, Ph.D.
Department of Physical Chemistry
Centre National de la Recherche
 Scientifique
Nancy, France

Peter M. Castle, Ph.D.
Advisory Scientist
Westinghouse Idaho Nuclear Corporation
Lawrence Livermore Laboratories
Livermore, California

Christian Decker, Ph.D.
Director of Research
Department of Photochemistry
Centre National de la Recherche
 Scientifique
Mulhouse, France

D. Ehlich, Dr.
Institute of Physical Chemistry
University of Mainz
Mainz, West Germany

Jean-Pierre Fouassier, Ph.D.
Professor
Laboratory of General Photochemistry
Ecole Nationale Superieure de Chimie
University of Haute-Alsace
Mulhouse, France

Mark A. Iannone, Ph.D.
Department of Chemistry
University of Pennsylvania
Philadelphia, Pennsylvania

Jean-Yves Jezequel, Ph.D.
Department of Physical Chemistry
Centre National de la Recherche
 Scientifique
Nancy, France

Robert M. O'Connell, Ph.D.
Department of Electrical and Computer
 Engineering
University of Missouri
Columbia, Missouri

Rajender K. Sadhir, Ph.D.
Fellow Scientist
Materials Technology Division
Westinghouse Research and Development
 Center
Pittsburgh, Pennsylvania

Gary W. Scott, Ph.D.
Professor
Department of Chemistry
University of California
Riverside, California

H. Sillescu, Prof. Dr.
Institute of Physical Chemistry
University of Mainz
Mainz, West Germany

R. Srinivasan, Ph.D.
Manager
Department of Physical Sciences
IBM Corporation
Yorktown Heights, New York

SERIES TABLE OF CONTENTS

TABLE OF CONTENTS

Chapter 1

LASER-INDUCED PHOTOPOLYMERIZATION: A MECHANISTIC APPROACH

Christian Decker and Jean Pierre Fouassier

TABLE OF CONTENTS

I. INTRODUCTION

More than 25 years of research and development resulted in a rapid growth of the overall laser market, thus reflecting the requirements of the users in the field of laser technology. On one hand, large possibilities in average power, power density, focalization, wavelength capability, efficiency, pulse characteristics, according to the type of lasers (e.g., ion, CO_2, solid-state, dye, excimer, gas, metallic vapor lasers, etc.) have been turned to account in important scientific sectors such as medicine, graphic arts, communications, information processing, microelectronics, industrial cutting, and metrology. In a number of these applications, a field of research on laser-induced polymer chemistry was called into existence by the development of laser processing. On the other hand, lasers made vital contributions to spectroscopy, especially to time-resolved spectroscopy (in the nano- and picosecond time range), leading to the development of powerful techniques, which allow real-time probing of energy levels, as well as primary excited state processes.

The fast development during the last decade of high-performance UV-curable systems has induced a growing number of applications in various industrial areas. The most successful so far have been in the coating industry for the surface protection of a large variety of materials (metal, plastics, wood, paper, etc.) and in the microlithography area for the fabrication of high-resolution relief images.[1,2] Further improvements made it possible to achieve photosensitization under visible light irradiation.

The main advantage of using radiation to initiate the polymerization process is that it takes only a fraction of a second to achieve an extensive through-cure of the photosensitive resin. For some specific applications, such as the coating of optical fibers or photoimaging, these large but still restricted cure speeds appear yet to be one of the important limitations, so that there is an increasing demand to develop ever-faster systems. This can be best achieved by replacing the conventional mercury lamps by powerful lasers which appear today as the ultimate light sources for reaching instant polymerization.

In this type of application, lasers will be primarily used as a high-intensity light source to produce radicals by a photochemical process, much alike in conventional UV curing. In order to make the economics attractive, it is yet important to apply this technology to chain reactions where the overall quantum yield is the highest possible and to use very efficient lasers which are appropriate to the present case, i.e., which emit in the near-UV and visible range, like the continuous-wave (CW) argon-ion laser and the pulsed excimer lasers.

Several investigations on laser-induced polymerization have been reported in the last few years,[3-24] but the actual development of this technology has apparently not met the efficiency, reliability, and economic requirements essential for industrial applications. Still, laser-initiated radical production offers some remarkable advantages over conventional initiation that result mainly from the large power output available and the spatial coherence of the laser beam which can be finely focused. Ultrafast polymerization of surface-adsorbed acrylic monomers was achieved by means of a submicronic focused Ar[+] laser emitting in the UV range,[9] as well as the rapid curing of multiacrylic resins irradiated by UV[10-15,18] or visible[6,19] laser beams. Both the thiol-ene system[16,17] and lauryl methacrylate[24] were shown to polymerize readily by exposure to high-powered excimer lasers, the extent of the reaction being highly dependent on the pulse repetition rate. Full use of the high power of the laser beam can only be made if the irradiated system obeys the reciprocity law, i.e., if the product of the light intensity and the required exposure time is independent of the intensity.

Applications in polymer chemistry are known, for example, in the field of microelectronics (semiconductor microlithography), reprography (production of printing plates), and holography. One major problem in laser-induced photopolymerization is to match the photosensitizer absorption and the laser wavelength emission and to enhance the efficiency with which the photochemical system induces chain propagation.

This chapter deals mainly with the photophysical and photochemical processes involved in laser-induced polymerization reactions. Another chapter in this book covers the potential applications.[25]

II. BASIC DEFINITIONS

A. BASIC CHEMISTRY OF UV CURING

A polymer network can be formed either by bridging together existing polymer chains, e.g., in vulcanization of rubbers, or by promoting the polymerization of a monomer that contains at least two reactive double bonds in its molecule. The latter process being a chain reaction, it can develop very rapidly to a large extent, in particular when intense UV radiation is used to produce the initiating species. This UV curing reaction leads ultimately to a three-dimensional polymer network which usually exhibits a very high crosslink concentration, up to 10 mol 1^{-1}, depending on the length of the monomer unit.

Since most of the monomers commonly employed are not producing initiating species with a sufficiently high yield upon UV exposure, it is necessary to introduce a photoinitiator that will make the polymerization start. A typical UV-curable formulation will therefore contain two basic components: (1) a photoinitiator that must effectively absorb the incident light and produce initiating species with a high efficiency and (2) a monomer and/or an oligomer bearing at least two unsaturations, usually acrylates, that will generate the polymer network.

1. The Photoinitiator

The choice of the photoinitiator is of prime importance in light-induced polymerizations, since it directly governs the rate of cure. A suitable photoinitiator system must first present a high absorption in the emission range of the laser. In addition, the excited states thus formed must both have a short lifetime to avoid quenching by oxygen or by the monomer and split into reactive radicals or ionic species with the highest possible quantum yield. Figure 1 shows schematically the various deactivation pathways of an excited singlet molecule: internal conversion, fluorescence, intersystem crossing, phosphorescence, quenching, cleavage, hydrogen abstraction, and electron transfer. Besides the initiation efficiency, other factors have to be considered in selecting the proper photoinitiator such as the solubility in the monomer, the storage stability, and the nature of the photoproducts, which should not be colored or toxic or induce some degradation of the polymer upon aging.

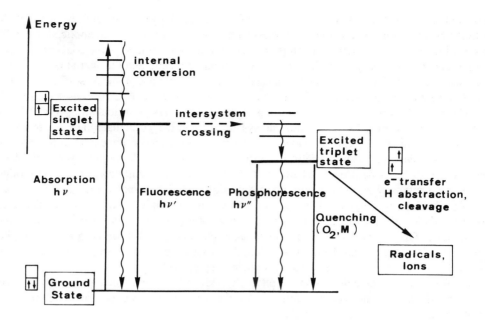

FIGURE 1.   Different possible routes of deactivation of an electronically excited molecule.

A large number of photoinitiators have been developed during the last decade in order to obtain ever more performing resins which will cure within a fraction of a second of UV exposure. The photochemical behavior of these compounds has been extensively studied, and several comprehensive reviews have been published recently.[26-33] The various photoinitiators mainly used today can be classified into three major categories, depending on the kind of mechanism involved in their photolysis.

a. Radical Formation by Photocleavage

This class includes aromatic carbonyl compounds that undergo a Norrish type I fragmentation when exposed to UV light:

$$\text{ArC–CR}_3 \xrightarrow{h\nu} \text{ArC}^{\cdot} + \, ^{\cdot}\text{CR}_3$$
$$\overset{\|}{\text{O}} \qquad\qquad \overset{\|}{\text{O}}$$

b. Radical Generation by Hydrogen Abstraction or Electron Transfer

Aromatic ketones, when promoted to their excited states by UV irradiation, can undergo a hydrogen abstraction from an H-donor molecule to generate a ketyl radical and the donor radical:

$$\text{Ar}_2\text{C=O} + \text{DH} \xrightarrow{h\nu} \text{Ar}_2\text{C}^{\cdot}\text{–OH} + \text{D}^{\cdot}$$

The initiation of the polymerization usually occurs through the H-donor radical (D·), the ketyl radical disappearing mainly by a radical coupling process. Most often, tertiary amines are used as H-donors. Upon UV irradiation, an electron from the nitrogen lone pair is transferred to the carbonyl oxygen; the exciplex thus formed can further transfer a proton to ultimately generate the ketyl and amino-alkyl radicals:

$$\text{Ar}_2\text{C=O} + \quad \begin{matrix} \text{R}' \\ \diagdown \\ \text{N–CH}_2\text{–R}'' \\ \diagup \\ \text{R} \end{matrix} \quad \xrightarrow{\ h\nu\ } \quad \left[\text{Ar}_2\dot{\text{C}}\text{–}\bar{\text{O}}\dots \overset{+}{\underset{\ }{\text{N}}}\begin{matrix} \text{R}' \\ | \\ \text{–CH}_2\text{–R}'' \\ | \\ \text{R} \end{matrix} \right]$$

$$\rightarrow \text{Ar}_2\dot{\text{C}}\text{–OH} + \quad \begin{matrix} \text{R}' \\ \diagdown \\ \text{N–}\dot{\text{C}}\text{H–R}'' \\ \diagup \\ \text{R} \end{matrix}$$

c. Cationic Photoinitiators

Aryldiazonium salts ($\text{ArN}_2{}^+\text{X}^-$) have been shown to undergo a fast fragmentation under UV irradiation, with formation of free Lewis acids (BF_3, AsF_5, PF_5, etc.) which are known as efficient initiators for the polymerization of epoxy monomers:

$$\text{ArN}_2^+\text{BF}_4^- \xrightarrow{\ h\nu\ } \text{ArF} + \text{N}_2 + \text{BF}_3$$

Because of their poor thermal stability and the evolution of nitrogen that causes bubbles in the UV-cured films, these systems have now been replaced by diaryliodonium ($\text{Ar}_2{}^+\text{X}^-$) or triarylsulfonium ($\text{A}_3\text{S}^+\text{X}^-$) salts that generate strong Brönsted acids upon photolysis in the presence of an H-donor:

$$\text{Ar}_2\text{I}^+\text{BF}_4^- + \text{RH} \xrightarrow{\ h\nu\ } \text{ArI} + \text{Ar}^{\cdot} + \text{R}^{\cdot} + \text{HBF}_4$$

Different types of these powerful protonic acids can thus be formed (HBF_4, HAsF_6, HPF_6, etc.), that are all known to be very active initiators for cationic polymerization.[28,34]

2. The Polymerization Reaction

Once the radicals or ions have been formed by irradiation of the photosensitive resin, the basic chemistry involved will be essentially the same as in conventional polymerization, except that the rate of initiation is considerably larger in UV curing, which usually takes place within less than 1s.

Acrylic monomers are among the most widely used curable systems today because of their high reactivity, moderate cost, and low volatility.[1,2] A large variety of functionalized prepolymers is now available on the market and permits the creation of networks with tailor-made properties, depending on the specific end use. Different types of chemical structures can be used for the prepolymer backbone, most notably, polyurethanes, epoxy derivatives, polyesters, and polyethers with a molecular weight ranging typically between 500 and 3000. Because of the high viscosity of these oligomers, reactive diluents have to be introduced in the formulation; these are usually mono- or multifunctional acrylates that provide good application properties and at the same time increase the cure speed. The mechanical properties of the cured polymer will be highly dependent on the chemical structure of the crosslinked segments, as well as on the functionality of the monomer and oligomer used.

Light-induced cationic polymerization is the most appropriate method to cure rapidly cyclic saturated monomers like epoxides, acetals, or lactones.[34] In the presence of a protonic acid or Lewis acid species, ring-opening polymerization proceeds through the oxonium ion. Efficient crosslinking can be obtained by employing as starting material either difunctional epoxides or epoxide-substituted polymers. Besides its specificity, cationic polymerization

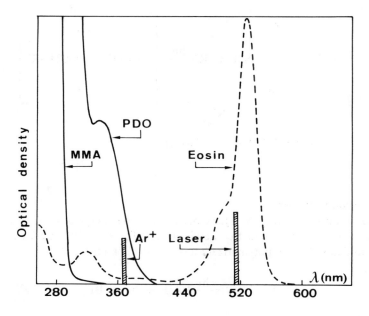

FIGURE 2.   Typical absorption spectra of a photoinitiator (PDO) and a photosensitizer (Eosin) and usual operating wavelengths of an Ar^+laser.

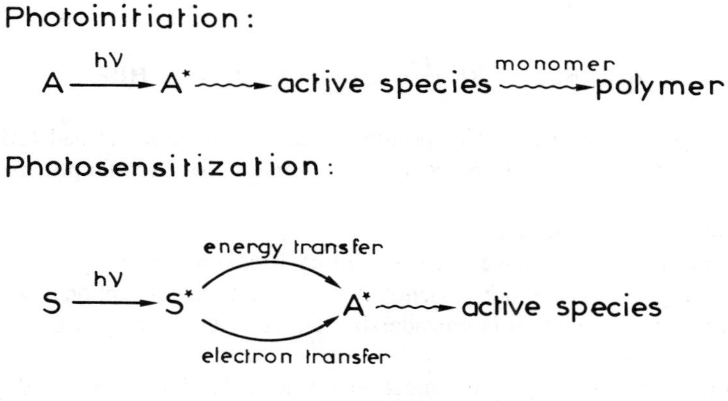

Photoinitiation :

$$A \xrightarrow{h\nu} A^* \rightsquigarrow \text{active species} \xrightarrow{monomer} \text{polymer}$$

Photosensitization :

$$S \xrightarrow{h\nu} S^* \underset{\text{electron transfer}}{\overset{\text{energy transfer}}{\rightleftarrows}} A^* \rightsquigarrow \text{active species}$$

SCHEME I

has the advantage of being little sensitive to atmospheric oxygen; in the absence of nucleophilic agents, the chain reaction will thus continue to develop after the illumination and
provide a beneficial postcure effect that can be enhanced by thermal treatment.

B. SENSITIZATION OF INITIATOR DECOMPOSITION

It has been mentioned that the absorption of photoinitiators usually takes place in the
UV part of the spectrum. Technological requirements gave rise to research endeavors directed
towards the use of visible light- (or near IR) sensitive systems and the best matching between
the photosensitizer (or the photoinitiator) absorption and the laser emission wavelength
(Figure 2). Photopolymerization under visible light can be induced through two different
processes, as shown by Scheme I.

The overall efficiency is strongly dependent on the processes involved in the excited
states of the sensitive systems; this important point has been fully discussed elsewhere[18,26,29,35-37] in the case of UV light photoinitiation.

Table 1
Typical Laser Sources

Source	Type	Operating mode	λ (A°)	P (W)	Δt (s)	E/pulse (J)
Gas lasers	HeNe	CW	6328	0.05	—	—
	Kr^+	CW	6471 (5309, 5682, 7525 6764, 5208, 7993...)	0.1	—	—
	Ar^+	CW	4880—5145 (4765, 4965, 4579, 5017...)	1		
	HeCd	CW	4416	0.05		
	N_2	Pulsed	3371	10^6	10^{-8}	10^{-3}
	Excimer laser	Pulsed	193 249 308 353	10^7	10^{-8}	10^{-2}
Semiconductor lasers	Ga Al As	CW	8000—8500			
Dye lasers		CW Pulsed Mode-locked Modulated	λ_i	Dependent on the pumping laser		
Solid lasers	NdYAG	Pulsed Mode-locked Modulated	10600 5300 2650 3530	10^7 10^9 10^3	10^{-9} 10^{-12}	10^{-2} —
	Ruby	Pulsed Mode-locked Modulated	6943 3472	10^3	10^{-12}	

C. LASER SOURCES AND BASIC INTEREST OF LASER CURING

The characteristics of some typical laser sources are shown in Table 1. Attractive characteristics are available (e.g., in energy, power density, excitation time, emission wavelength) and will probably promote or suggest new developments in the field of photopolymerization or/and photocrosslinking.

In consideration of the intrinsic characteristics of the laser emission, these powerful light sources present many advantages which make them very attractive for curing applications. First, the generally encountered spatial coherence of the emission provides a great directivity to the laser beam that can be focused down to a tiny spot of extremely high intensity. With such a sharp "light pencil", it becomes then possible to directly write complex relief patterns, at micronic resolution, by scanning the photosensitive plate at very large speeds.

Another consequence of this high directivity is that the light-intensity remains essentially constant along the laser beam; by contrast to conventional light sources, the light intensity is thus not decreasing as the source-object distance increases, so that a uniform illumination of large three-dimensional samples can be carried out by an even scanning with the laser beam.

Owing to the temporal coherence of the laser emission which occurs at a well-defined wavelength, the cure penetration can be better controlled than with polychromatic light sources, thus allowing a homogeneous through-cure of thick samples to be worked out conveniently. Besides, the narrow bandwidth of the laser emission line will reduce the extent of the undesirable secondary photochemical reactions, while at the same time eliminating the energy wastage in the nonabsorbing parts of the spectrum and the related heating of the sample that are common to all the industrial UV sources.

Finally, one of the most important characteristics of the laser emission is the large power output that is concentrated in a very narrow beam. It permits to reach extremely high fluence rates which were not accessible so far, even by using the best performing UV-curing equipment, and thus to operate the production line at much greater speed.

The basic mechanism of laser-induced polymerization will now be discussed in more detail, concentrating especially on two important aspects of these processes, namely, the different types of photosensitive systems which have to be used and the kinetics of these ultrafast reactions. In the case of the UV-laser initiation, more emphasis will be laid on the kinetic approach, since the resins used here are essentially the same as in conventional UV curing. By contrast, for the polymerization induced by visible laser radiation, the kinetics proved to be much like the one encountered in UV-laser initiation, so that special attention will be directed toward the various types of photosensitive systems which need to be employed to make the polymerization work under those more stringent conditions.

III. UV-LASER-INDUCED RADICAL POLYMERIZATION

A. LASER-INITIATED RADICAL PRODUCTION AND PHOTOCROSSLINKING POLYMERIZATION

The laser emission is used to produce electronically excited molecules which split into reactive radicals with the highest possible quantum yield. Since the substrate usually behaves as a poor photoinitiator, an additional molecule must be introduced in order to enhance the radical production, much in the same way as in conventional photoinitiated reactions. If the monomer contains more than one reactive double bond, the process develops in the three dimensions, leading to a highly crosslinked and insoluble material that can be used as protective coating or negative-working photoresist. Thanks to the large power output available in the laser emission, the exposure time can be considerably shortened, down to the milli- or even microsecond range.[10-15] Figure 3 shows a schematic representation of the polymer network obtained by laser curing of a photoresist based on a bis-phenol A epoxy-diacrylate and tripropyleneglycol diacrylate, assuming equal reactivity of the acrylate end group of both the epoxy-oligomer and the ether monomer. The crosslink density of this material reaches very high values, since for each acrylate group that polymerizes one crosslink unit will be formed. This leads to a complete insolubilization of the thoroughly cured polymer, which shows no swelling in organic solvents and exhibits remarkable optical and mechanical properties.

Even when the coated photoresist is exposed to the unfocused beam of an argon ion laser emitting only 50 mW in a CW mode at 363.8 nm, the crosslinking-polymerization develops rapidly, once the initial induction period due to oxygen inhibition is over. Complete insolubilization usually requires up to 10 ms of irradiation. The laser scanning speed (S) was found[15] to vary between 6 and 50 cm s^{-1}, depending on the formulation and primarily on the type of photoinitiator used, as described in Figure 4.

B. KINETICS OF THE LASER-INDUCED PHOTOPOLYMERIZATION

The extent of the reaction can be followed continuously by IR spectroscopy in order to evaluate the rate and quantum yield of the laser-induced polymerization of these photoresist systems. Two basic types of lasers emitting in the UV range were employed, either a CW argon-ion laser,[10] or a pulsed nitrogen laser.[12]

1. Argon-Ion laser

Both the polyester and the epoxy-based multiacrylate photoresists were found to polymerize, within a few milliseconds when exposed in the presence of air, to the argon-ion laser beam tuned to its CW emission line at 363.8 nm and operated at an incident photon

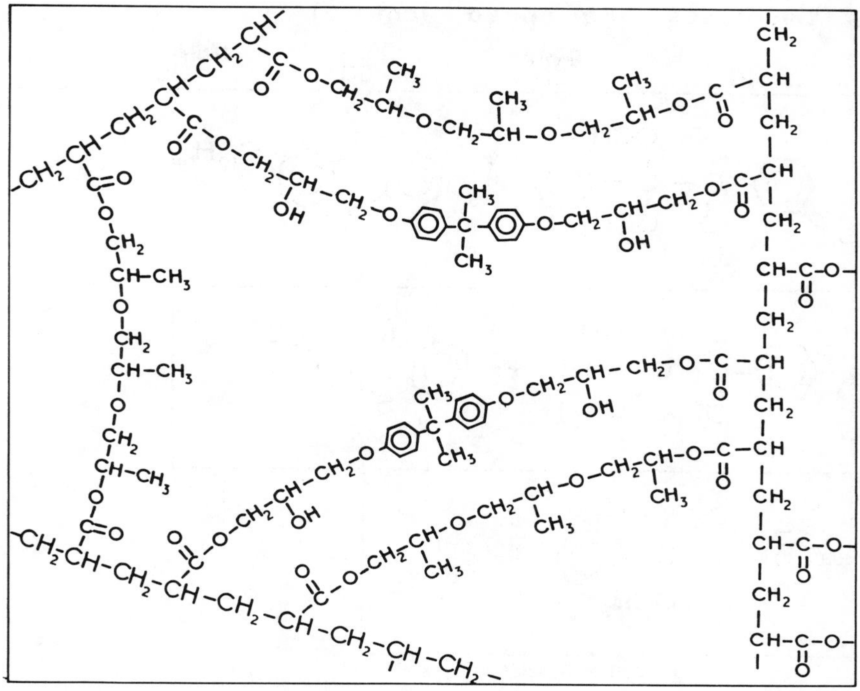

FIGURE 3. Schematic representation of a TPGDA-epoxy diacrylate network. (From *ACS Symp. Ser.* 266, 207, 1984. With permission.)

flux of 7.5×10^{-6} E s^{-1} cm^{-2}. Characteristic S-shape kinetic curves were obtained by plotting the normalized thickness of the insoluble polymer film formed against the duration of the laser exposure, in a semilogarithmic scale (Figure 5).[14] As expected, the induction period observed in the early stages of the irradiation is primarily due to the presence of atmospheric oxygen, which is known to strongly inhibit the photopolymerization of acrylic systems by reacting with the initiating radicals as well as with the growing polymer radicals.[22,38,39]

$$\text{Photoinitiator} \xrightarrow{h\nu} (\text{PI})^* \longrightarrow R^{\cdot} \xrightarrow{M} P^{\cdot} \xrightarrow{M} \text{polymer}$$

$$(\text{PI}) \qquad\qquad\qquad \downarrow O_2 \quad \downarrow O_2$$

$$RO_2^{\cdot} \quad\; PO_2^{\cdot}$$

As shown by Figure 5, the induction period disappeared almost completely when the laser irradiation was carried out under a dry nitrogen purge; total insolubilization of the photoresist was then reached within 2 ms, compared to about 10 ms in the presence of air.

As the polymerization proceeds, the viscosity of the polymer system increases steadily to finally yield a solid, highly crosslinked polymer in which the segmental mobility is quite restricted. Consequently, the encounter probability of the polymer radicals with the reactive acrylate double bonds is sharply reduced, which accounts for the rate slowing down observed in the later stages of the irradiation.

In photoimaging technology, the sensitivity (S) of a negative working photoresist is usually defined as the amount of incident energy required to transform 50% of the initial film into a totally insoluble material. By taking into account the corresponding photon fluxes, S values were calculated from exposure times to be 5 and 1 mJ cm^{-2} for the laser-induced polymerization in air and N$_2$, respectively. The second important parameter of a photoresist

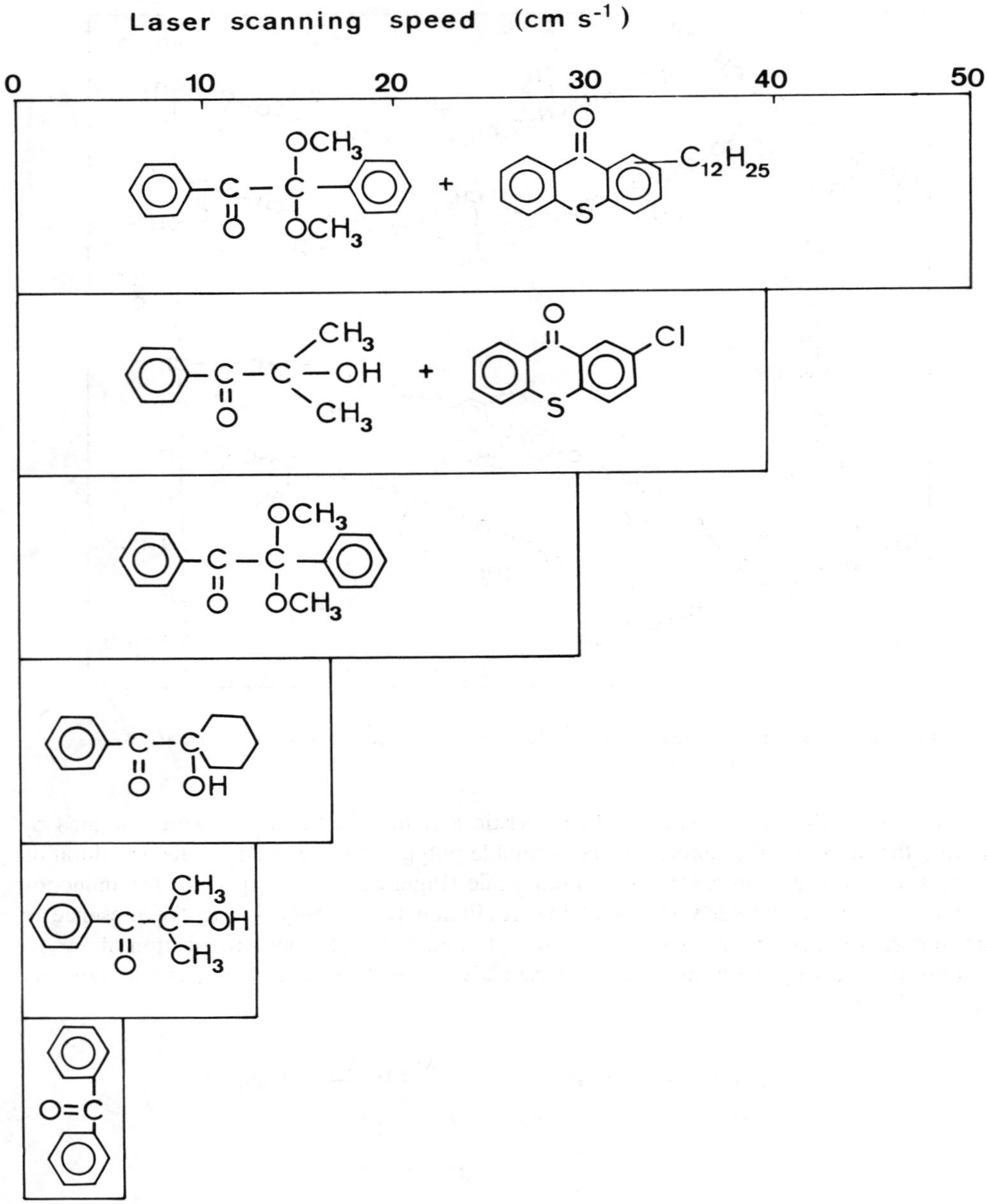

FIGURE 4.   Influence of the photoinitiator on the scanning speed in the UV curing of an epoxy-acrylate photoresist by an argon-ion laser beam ($\lambda = 363.8$ nm).

material is the contrast (γ) that characterizes the sharpness of the high-resolution relief image needed in microlithography. It was evaluated from the slope of the linear portion of the kinetic curves shown in Figure 5 by using the equation:

$$\gamma = \frac{e}{e_0} (\log t - \text{lot } t_i)^{-1}$$

where e_0 and e represent the thickness of the coating before and after laser exposure during

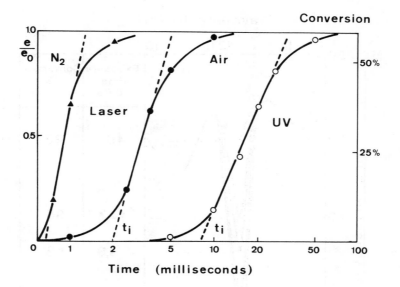

FIGURE 5. Normalized thickness of polymerized epoxy acrylate film vs. duration of the exposure to the argon-ion laser bean (Δ in N_2, ● in air) or to conventional UV radiation (○ in air). (From *ACS Symp. Ser.* 266, 207, 1984. With permission.)

time t, respectively, and t_i is the induction period due to oxygen inhibition. Values of γ range from 1.8 to 2.5, thus revealing a fairly sharp contrast for these negative acrylate photoresists. This remarkable feature may result from the very high crosslinking density of the cured polymer (>6 mol L^{-1}) that prevents it from swelling during solvent development of the patterns.

Laser-induced photopolymerizations can be greatly accelerated either by increasing the power output or by finely focusing the laser beam, down to the micrometer range. For instance, by using a 20μm laser spot and a corresponding light intensity of 10^{-2} E s^{-1} cm^{-2}, the rate of polymerization was increased by 3 orders of magnitude for both the polyester and epoxyacrylate photoresists.[14] Scanning speeds of up to 10 m s^{-1} were then reached,[15] which corresponds to exposure times in the microsecond range. If necessary, faster scanning rates can still be reached either by increasing the laser power output or by focusing more sharply the laser beam down to the micrometer range. Although one would expect ablation to become the limiting step beyond a certain power density, it can hardly occur under the present conditions because the laser radiation is absorbed exclusively by the photoinitiator and not by the substrate. Such a scanning spot technique has already been reported for highly focused beams of lasers emitting in the visible[40] or UV[41] range, but operated at much lower scanning rates (a few millimeters per second) it offers the advantage of successfully eliminating the detrimental speckle effect due to interference patterns in laser lithography.[42] Direct maskless writing of complex patterns on the wafer by a computer-directed laser becomes then feasible, much in the same way as in electron beam lithography.

2. Pulsed Nitrogen Laser

Although the scanning spot technique appears to be well suited for the direct writing with a focused laser beam, one of the major problems that plague CW mode-operated lasers is the lack of power available in the UV range. This is not the case for pulsed lasers, such as nitrogen or excimer lasers, that can deliver in the near- or deep-UV region extremely high power densities, up to 10 W cm^{-2}, during the few nanoseconds duration of the pulse.

By using a pulsed nitrogen laser which emits at 337.1 nm, a few laser shots, 8-ns wide

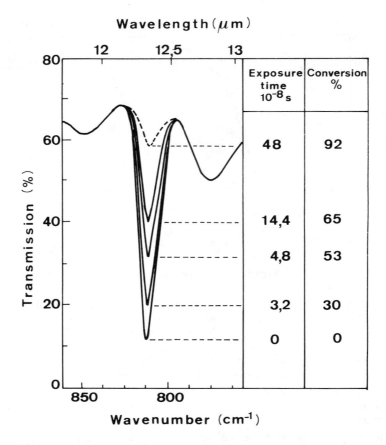

FIGURE 6.   IR spectra of a polyester-acrylate photoresist exposed for various times to nitrogen-laser pulses in the presence of air. (– – –) polyurethane-acrylate. (From *ACS Symp. Ser.* 266, 207, 1984. With permission.)

each, proved to be sufficient for complete insolubilization of the resist in the presence of air, whereas in N_2-saturated systems one single pulse was sufficient to give 100% polymer insolubility.[12]

In a more quantitative approach, the decrease of the IR absorption band at 810 cm^{-1}, characteristic of the acrylate double bond, was followed as a function of the exposure time (Figure 6). This permits to precisely evaluate how many acrylate functions have polymerized as a function of dose and then to deduce the degree of conversion of the polymer formed. Figure 7 shows the reaction profile of the laser-induced polymerization that was obtained by plotting the degree of conversion against the cumulative exposure time. In the presence of air, typical S-shape kinetic curves were again observed for both the epoxy- and the polyester-based photoresists which show comparable reactivities, with a slight advantage for the latter. It is interesting to note that half of the original photoresist layer was recovered as an insoluble polymer when the overall degree of conversion reached 30% and that complete insolubilization required about 60% conversion (Figure 5). Upon extended laser exposure, the degree of conversion leveled off at about 70% for the epoxy and polyester oligomers, while it exceeded 90% using the more flexible polyurethane chain, despite a lower rate of polymerization.[43]

C. LIGHT INTENSITY EFFECT

An important quantity that can be deduced from the reaction profile is the rate of the crosslinking polymerization (R_p), i.e., the number of double bonds polymerized or of cross-

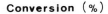

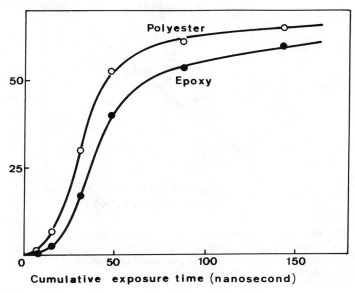

FIGURE 7.   Kinetics of the polymerization of polyester- and epoxy-acrylate photoresists under pulsed laser irradiation at 337.1 nm in the presence of air. (From *ACS Symp. Ser.* 266, 207, 1984. With permission.)

Table 2
Rates and Quantum Yields of Polymerization of an Epoxy-Diacrylate Photoresist Exposed to Conventional or Laser UV Irradiation

UV-light source	Irradiation wavelength (nm)	Light intensity (I_0) (E s^{-1} cm^{-2})	Polymerization rate R_p (mol l^{-1} s^{-1})	Quantum yield (ϕ_p) (mol E^{-1})
Mercury lamp (Med. press)	>250	7×10^{-8}	29	3200
CW Ar ion laser		1.5×10^{-6}	5.8×10^2	2900
		4×10^{-6}	2×10^3	1700
	363.8	8×10^{-4}	1.2×10^5	600
		1.2×10^{-2}	1.6×10^6	540
Pulsed nitrogen laser	337.1	1.1	10^8	420
		12	10^9	400

links formed per second. R_p values were determined from the maximum slope of the kinetic curves (usually reached for conversion degrees between 20 and 40%). Table 2 summarizes the R_p values for the two photoresists tested under various conditions, namely, conventional UV and continuous or pulsed laser irradiation at different light intensities. According to these kinetic data, R_p increases almost as fast as the light intensity; the ratio I_0/R_p, which is directly related to the product of the light intensity and the required exposure time, was found to vary only in the range 10^{-8} to 2.6×10^{-7}E cm^{-2} l mol^{-1} when the light-intensity value was increased by 8 orders of magnitude. Such a small departure from the reciprocity low was quite unexpected, especially for a polymerization, which is basically a chain reaction process and thus strongly dependent on the initiation rate. This light-intensity effect becomes clearly apparent from a plot of R_p against I_0 in a logarithmic scale (Figure 8). The value of

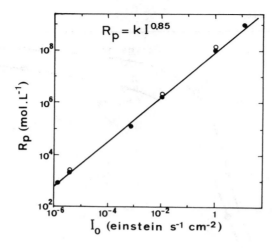

FIGURE 8.   Dependence of the rate of polymerization (R_p) on the light-intensity (I_0) upon UV-laser irradiation of polyester ($\bigcirc$) and epoxy ($\bullet$) multiacrylate photoresists in the presence of air. (From *ACS Symp. Ser.* 266, 207, 1984. With permission.)

the slope of the straight line obtained, 0.85, reveals a close to first-order process for the polymerization of these multifunctional acrylate systems instead of the expected half-order relationship between R_p and I_0.

For an accurate evaluation of the efficiency of UV photons in initiating the polymerization of acrylate photoresists, it is necessary to determine the polymerization quantum yield, ϕ_p, that corresponds to the number of acrylate functions which have polymerized per photon absorbed. ϕ_p can be expressed as the ratio of the rate of polymerization to the absorbed light intensity. The high values of ϕ_p reported in Table 2 indicate that, despite the high rate of initiation provided by the laser irradiation, the propagation chain reaction still develops effectively in these multifunctional systems; each photon induces the polymerization of up to 1700 monomer units in air-saturated systems and up to 10,000 in the absence of oxygen.[43] From a practical point of view, such a large amplification factor is highly recommended when one considers the high cost of laser photons.

D. MECHANISM OF LASER-INDUCED POLYMERIZATIONS

Valuable information about the basic mechanism of the polymerization can be inferred from kinetic studies, in particular from the dependence of the rate of polymerization on the rate of initiation. In conventional light-induced polymerizations, where termination occurs by molecular interaction of polymer radicals, a half-order kinetic law is usually observed. The quasilinear relationship between R_p and I which has been found in the UV curing of multiacrylate systems can be accounted for by considering that only one polymer radical is involved in the termination step, as, for instance, in a radical-occlusion process:

$$\text{Initiation:} \quad \text{In} \xrightarrow{h\nu} R^{\bullet} \qquad r_i = \phi_i I$$

$$\text{Propagation:} \quad R^{\bullet} + nM \xrightarrow{k_p} RM_n^{\bullet} \text{ or } P_n^{\bullet} \qquad R_p = k_p[R^{\bullet}][M]$$

$$P_n^\bullet + P_n^\bullet \xrightarrow{k_t} \text{products} \qquad r_t = k_t[P^\bullet]^2 = r_i$$

$$R_p = \frac{k_p}{k_t^{0.5}} \phi_i^{0.5}[M]I^{0.5}$$

Termination

$$P_n^\bullet \xrightarrow{k_t'} \text{occlusion} \qquad r_t' = 2k_t'[P^\bullet] = r_i$$

$$R_p = \frac{k_p}{2k_t'} \phi_i[M]I$$

r_i = rate of initiation $\qquad\qquad \phi_i$ = quantum yield of initiation

R_p = rate of polymerization $\qquad I$ = absorbed light intensity

r_t and r_t' = rates of termination

Based on the assumption that termination of the growing polymer chains occurs by both mono- and bimolecular pathways, the overall rate of polymerization can then be expressed, at a first approximate, as the sum of two terms which depend on the first power and on the square root of the absorbed light intensity, respectively:

$$R_p = \alpha \frac{k_p}{k_t'} [M]\phi_i I + (1 - \alpha) \frac{k_p}{2k_t^{1/2}} [M]\phi_i^{1/2}I^{1/2}$$

where k_t' is the rate constant of the unimolecular termination process, i.e., the reciprocal of the polymer radical lifetime, and α is a coefficient that reflects the relative contribution of the unimolecular pathway in the termination step, i.e., the probability for a given polymer radical to become occluded.

It should be remembered that, in the polymerization of multifunctional monomers in bulk, the rate constants k_p and k_t are both decreasing when gelification occurs,[44] since the latter process will sharply reduce the encounter probability of the growing polymer radicals with a monomer molecule or with another radical. For an accurate comparison, all the R_p measurements have thus been made at the same degree of conversion, ~20%, when R_p reaches its maximum value.[14] Similarly, the quantum efficiency of the initiation reaction is also likely to be affected by the severe changes in the physical state of the system that occur upon laser irradiation.

Among the various explanations which have been put forward in order to account for a unimolecular termination process, the one that appears to be the most feasible at the present time assumes that termination of the chain growth results from a radical-trapping process.[44,45] As soon as the crosslinking polymerization starts upon UV exposure, the growing polymer radicals become attached to the network under formation and are therefore losing their reptation mobility. Under those conditions, polymer radicals have little chance to come close together to undergo bimolecular termination; they will continue to grow as long as monomer molecules can pass close by and be scavenged by the radical site. When steric hindrance prevents free access to the radical chain end, the macromolecule will stop growing and the living radical site will remain trapped in the polymer network. Reaction Scheme II shows an overall representation of the crosslinking-polymerization process induced by UV laser irradiation of multifunctional monomers.

The presence of trapped radicals in UV-cured multiacrylate networks has recently been demonstrated by ESR spectroscopy;[46,47] such radicals, occluded in the polymer matrix, can

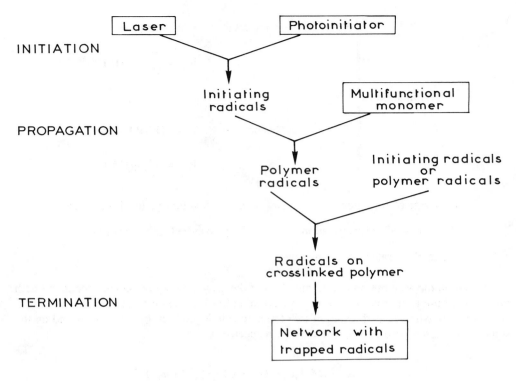

INITIATION

PROPAGATION

TERMINATION

Laser · Photoinitiator → Initiating radicals; Multifunctional monomer → Polymer radicals; Initiating radicals or polymer radicals → Radicals on crosslinked polymer → Network with trapped radicals

SCHEME II

last for days in the absence of oxygen at room temperature. Posteffect experiments have also confirmed that reactive species are still present in the cured polymer well after the end of the UV exposure, as shown by the substantial dark polymerization observed upon moderate heating (50°C) in an inert atmosphere.[48] A remarkable feature revealed by these experiments is that the radical trapping process already occurs at the very beginning of the laser exposure, while the resin appears to be still liquid. Such a result strongly suggests the early formation, in UV curable resin, of microgel particles where mobility restrictions already exist; a significant amount of insoluble polymer has indeed been detected at degrees of conversion as low as 5% in UV-cured multiacrylic coatings.[48]

Kinetic considerations also show that a first-order termination reaction is still feasible under the experimental conditions used, despite the high intensity of the laser beam that, by generating large concentrations of radicals in a short time, was *a priori* expected to favor second-order termination. Actually, it appears that the concentration of initiating radicals still remains relatively low. Even if we assume that all the polymer radicals were to accumulate due to the trapping process, and none of them would disappear by the usual bimolecular termination, their total concentration at mid-reaction (50% conversion) would still be in the order of 10^{-4} mol l^{-1}, as compared to a concentration of 4 mol l^{-1} for the remaining unsaturations. Statistically, each radical site appears thus to be surrounded by over 10,000 acrylic double bonds, and as many crosslinks, a ratio which will highly favor propagation over bimolecular termination.

It is worth mentioning that such a radical trapping process leading to first-order kinetics is actually the major reason for the very high speeds of cure which are now attainable by using intense laser beams. It also explains the poor polymerization efficiency observed upon intense illumination of monofunctional monomers where no network formation and thus no radical trapping can occur.

IV. VISIBLE LASER LIGHT-INDUCED POLYMERIZATION

This section deals with the reactivity of typical sensitizers used to start a polymerization under visible laser light. Evidence is available that many of them are efficient under polychromatic conventional light exposure.[49] On account of their absorption spectra, they can easily be excited by laser beams. Accordingly, we will not give a thorough treatment of these systems; the attention will rather be focused on selected efficient or potentially interesting systems. Moreover, photopolymerization processes will be mainly considered, since systems acting through photoreticulation or photomodification are tackled in another chapter of this book.[25]

A. PHOTOSENSITIVE SYSTEMS

As stated above, different initiation mechanisms can be encountered, according to the type of light-absorbing organic molecule used.

1. Initiation with Photoinitiators

Various examples are reported in the literature.

a. Substituted Usual UV Photoinitiators

Appropriate substitution leads to molecules exhibiting a spectral shift towards the higher wavelengths of the spectrum. Examples are found in thioxanthones (e.g., see **1**)[50-52] and, to a slighter extent, in dialkoxyacetophenones, shown in **2**.[53] These compounds undergo the well-known processes, either an electron transfer process from an amine, followed by a proton transfer reaction (**1**), or a cleavage process (**2**) (see Part II).

1

2

b. Imidazole Derivatives

Upon irradiation, compound **3** (sensitive up to 400 nm)[54] yields the corresponding triarylimidazolyl radical by cleavage of the single covalent bond. Ortho sensitization yields sensitivities up to 500 nm, and addition of an electron or of a hydrogen donor extends the efficiency.

3

c. Unsaturated Aryl Ketones

Unsaturated conjugated ketone sensitizers shown in **4** absorb radiations in the broad spectral range of about 300 to 700 nm[55] (in conjunction with imidazolyl dimers and a free radical-producing hydrogen or electron donor.)

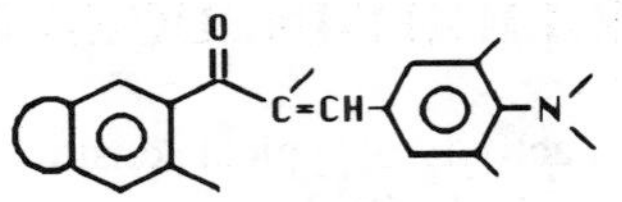

4

d. N-Hydroxyimide or Hydroxyamide Sulfonates

Such compounds (See **5**) induce effectively cationic photopolymerization in the range of 250 to 500 nm (through a cleavage process with production of the corresponding sulfonic acid).[56]

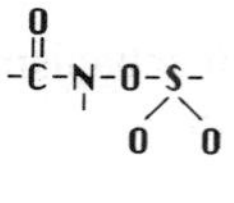

5

e. Monomeric Photoinitiators

A monomeric carbamic ester photoinitiator **6** which is sensitive up to 500 nm (and, in addition, has the advantage of being readily soluble in monomers and oligomers) was recently suggested.[57]

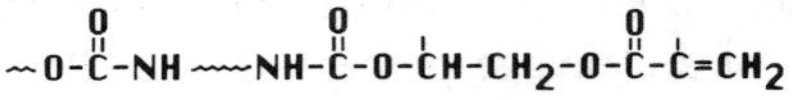

6

f. Ketocoumarines

Substituted ketocoumarins **7** match the imposed irradiation wavelength of a large variety of light sources. The light sensitivity ranges from 350 to 550 nm (Figure 9). In contrast with coumarin derivatives, which exhibit a strong fluorescence and display a good ability as laser dyes, ketocoumarins were shown[58,59] to be efficient triplet sensitizers for photo-crosslinkable polymers.

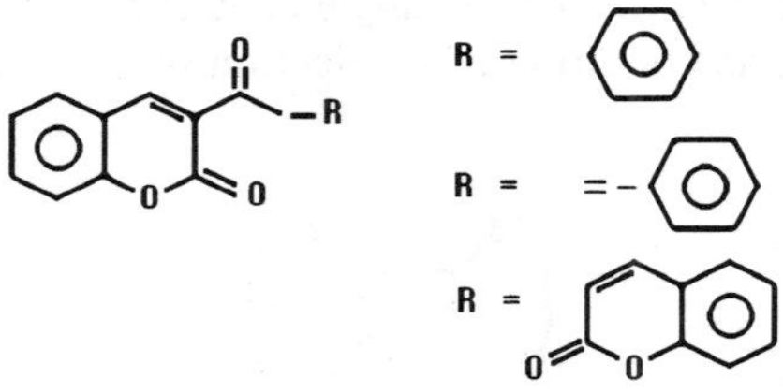

7

In addition, if combined with amines, phenoxyacetic acid, alkoxy-pyridinium salts, they are able to induce a free radical polymerization of acrylic monomers, whose initiation step was explained in terms of an electron transfer between the additive and the ketocoumarin.

g. Metal Salts

The main advantage of metal salts, complexes, or chelates (with nickel, cobalt, chrome, iron, copper ions) when used as photoinitiators must be seen in their special sensitivity in visible light, since they are generally colored. Their limitation lies partly in their water

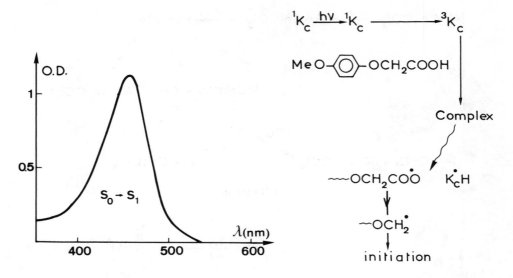

FIGURE 9. Typical absorption of ketocoumarin and mechanism proposed for initiation.

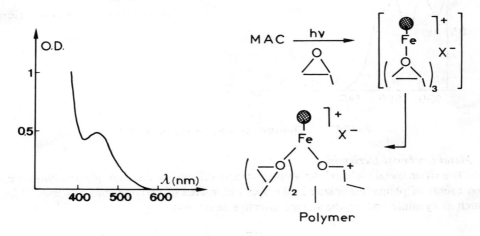

FIGURE 10. Mechanism proposed for decomposition of metal arene complexes (MAC).

solubility, which governs the nature of the polymerization medium. Photopolymerization of acrylamide in the presence of Cr(VI) salt has been recently reconsidered.[60]

h. Organometallic Compounds

Metallocenes are attractive systems. Arene complexes **8**[61] or titanocene derivatives[62] were shown recently to be suitable photoinitiators for cationic polymerization. Their direct photosensitivity extends up to 550 nm (Figure 10).

8a   **8b**

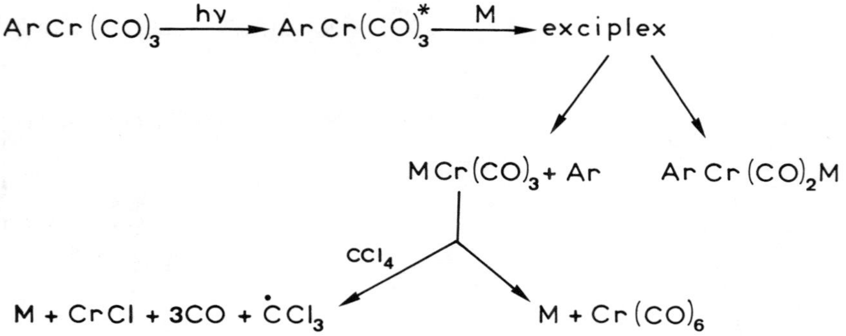

$$\text{Ar Cr (CO)}_3 \xrightarrow{\text{h}\nu} \text{Ar Cr(CO)}_3^* \xrightarrow{\text{M}} \text{exciplex}$$

$$\text{M Cr(CO)}_3 + \text{Ar} \qquad \text{Ar Cr (CO)}_2\text{M}$$

$$\text{M} + \text{CrCl} + 3\text{CO} + \overset{\bullet}{\text{C}}\text{Cl}_3 \xleftarrow{\text{CCl}_4} \qquad \xrightarrow{} \text{M} + \text{Cr (CO)}_6$$

SCHEME III

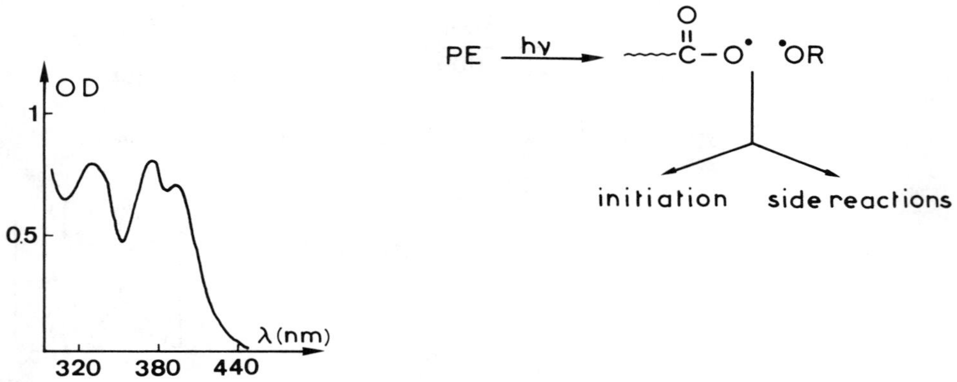

$$\text{PE} \xrightarrow{\text{h}\nu} \overset{\overset{\text{O}}{\|}}{\sim\sim\sim\text{C}}-\text{O}^\bullet \quad {}^\bullet\text{OR}$$

initiation   side reactions

FIGURE 11.   Characteristics of a perester photoinitiator.

i. Metal Carbonyl Derivatives

Transition metal carbonyl derivatives are also efficient initiators of photopolymerization and agents of photocrosslinking in the application area of photographic processes.[63] Dyes (such as cyanines and xanthenes) are effective sensitizers.

$$\text{R} -\!\!\!\!\bigcirc\!\!\!\!- \text{Cr(CO)}_3$$

9

A simplified scheme of the radical initiation process is shown in Scheme III.

j. Nonketonic Perester Structures

Such structures **10a**[64] and **10b**[65] are effective in vinyl polymerization, yielding alkoxy and aryloxy free radical pairs through homolytic decomposition (Figure 11).[64] According to the nature of the chromophore moiety (aryl groups-dyes, etc.), a strong absorption can be achieved over a wide spectral range.

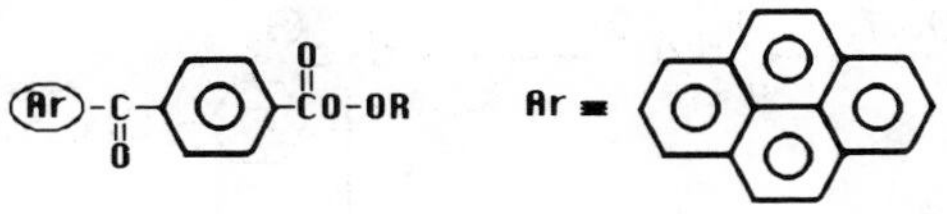

10a

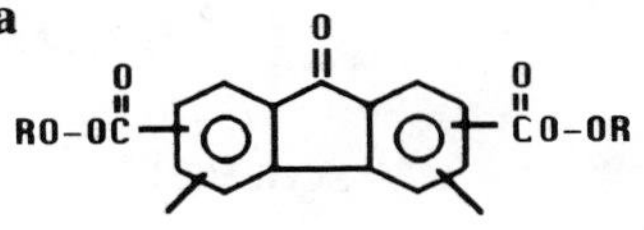

10b

k. Quinones and α-Diketones

Anthraquinonic structures have been recognized as polymerization initiators. Quinone **11a**[66] and camphorquinone **11b**[14,67] can be suitable for irradiation with an argon ion laser or polychromatic visible light. Applications of naphthoquinones are found in the field of microelectronics[68] and in curable sealants.[69] A large variety of α-diketones (e.g., **11c**) are efficient and suitable photosensitive guests in organic medium for recording volume phase holograms.[70] Quinone derivatives are also able to sensitize the decomposition of onium salts.[71]

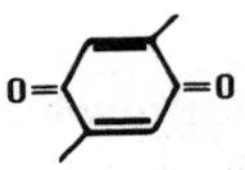

11a 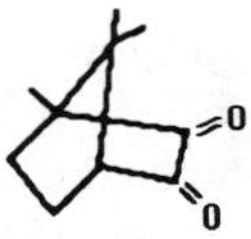   **11b** 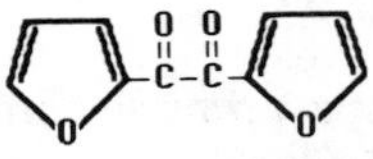   **11c**

l. Dyes

Dyes make up a large class of molecules extensively used to start or sensitize the polymerization of acrylic monomers. They belong to well-known structures, xanthene **12**, thiazine **13**, acridine **14**, anthraquinone **15**, cyanine **16**, and merocyanine **17**.

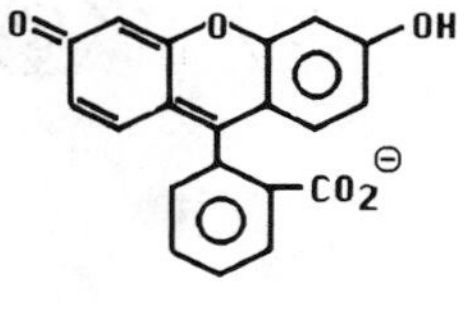

12 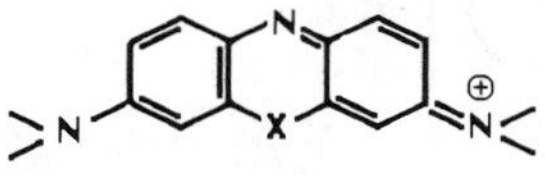   **13**

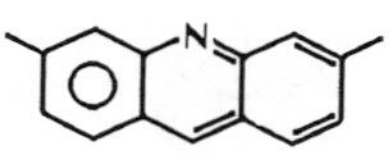

14 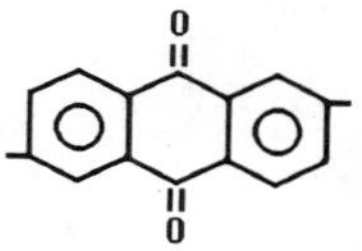   **15**

16 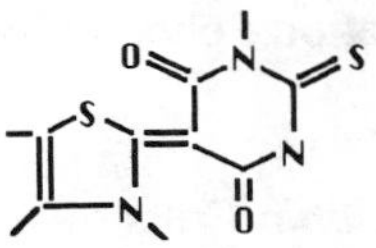   **17**

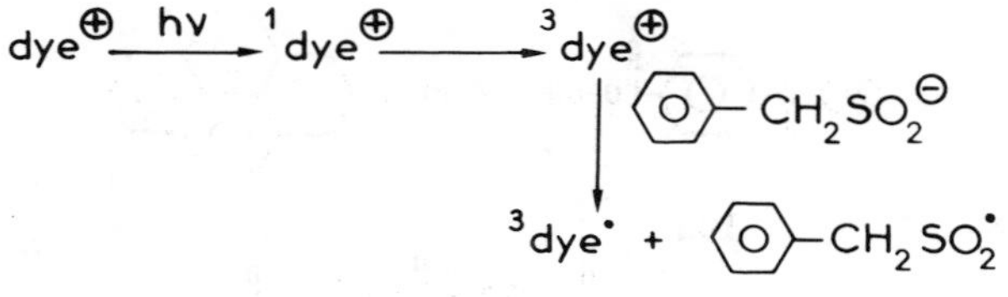

$$dye^{\oplus} \xrightarrow{\;h\nu\;} {}^1dye^{\oplus} \xrightarrow{\quad\quad} {}^3dye^{\oplus}$$

SCHEME IV

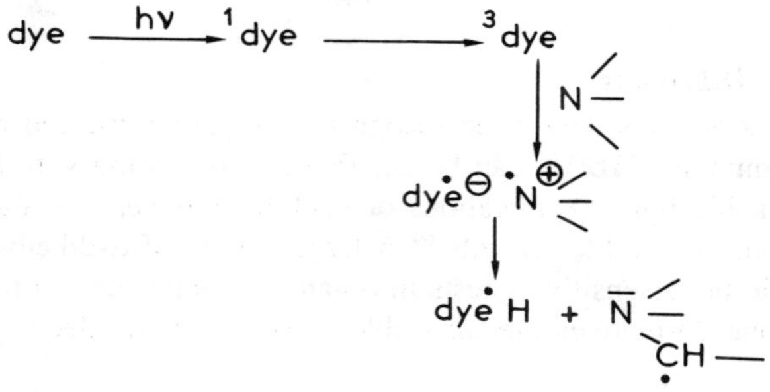

$$dye \xrightarrow{\;h\nu\;} {}^1dye \xrightarrow{\quad\quad} {}^3dye$$

SCHEME V

The reaction occurs primarily through photoreduction of the dye, which may be performed by using toluenesulfonic acid (Scheme IV), amines (Scheme V), ascorbic acid, or diketones. Typical applications have been recently reviewed.[49] Rate constants of the processes involved in the system eosin/methyldiethanol amine in anaerobic solutions have been determined through laser spectroscopy.[72] Improvement of the photoreactivity has been obtained by adding a benzoyl oxime ester **18a**[73] or a phenylacetophenone **18b**[74] derivative to a mixture of eosin and amine and used in the photopolymerization of glass reinforced polyester resins in visible light[72] or of acrylic monomers under laser beams.[75]

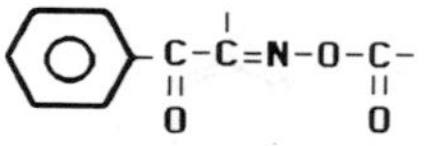

18a 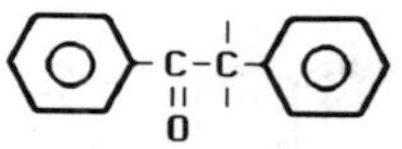   **18b**

Addition of amino ketones **19** to eosin[76,77] leads to printing plates with high resolution, which are useful in coating technology or photographic relief imaging. Various additives (benzotriazole, halogenotriazine, leuko dye) enhance the rate of polymerization.

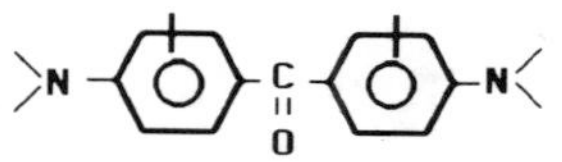

19

2. Sensitization Processes

Radical and cationic photopolymerization can be sensitized by different absorbing organic structures.

a. Sensitized Cationic Processes

Cationic photopolymerization is currently initiated by a large variety of organic salts; when they fail to absorb in the visible wavelength range, photosensitization is generally possible. Among them, the best-known systems[78,79] are based on the following structures: diaryliodonium **20**, triarylsulfonium **21**,

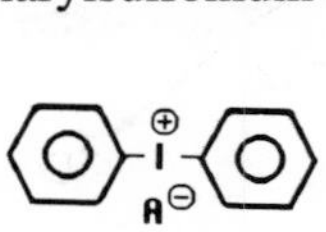

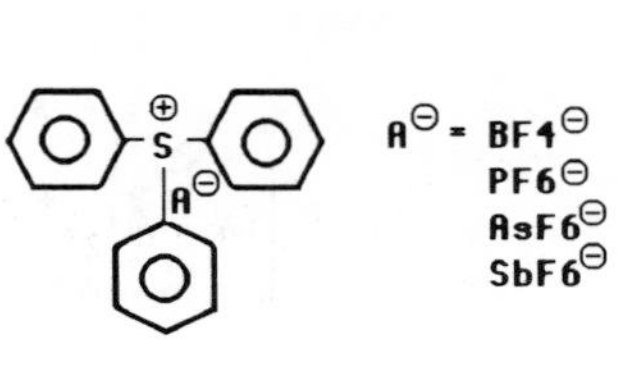

20	**21**	

aryldiazonium **22**

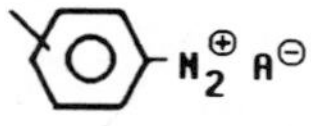

22

triarylpyrylium **23**, benzylpyridinium thiocyanate **24**, dialkylphenacyl-sulfonium **25**, and dialkylhydroxyphenylsulfonium **26** salts.

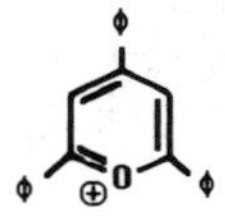

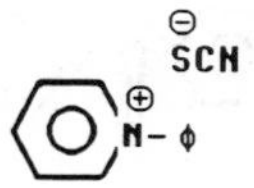

23	**24**

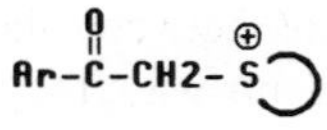

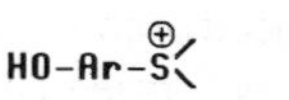

25	**26**

Diazonium salts are photolyzed with N_2 release. Their decomposition can be sensitized[80] by pyrazolone dyes, metal oxalate, tetraphenylporphyrin, ketones, or hydrocarbons. The primary steps of the initiation mechanism in the presence of onium salts still involve an electron transfer and a subsequent hydrogen abstraction on the solvent to generate the Bronsted acid HX; the photosensitivity was extended to higher wavelengths, generally through an electron transfer reaction in the presence of aromatic rings of dyes, or through a radical reaction in the presence of ketones.[72,78-80]

In the presence of onium salts,[72,78] the singlet state of eosin is quenched through an electron transfer process. The cation radical thus generated is responsible for the initiation of cationic polymerization, although this process is in competition with the formation of Bronsted acid by reaction with impurities, as well as by chain transfer processes (Scheme VI).

b. Sensitized Radical Initiation

Recent experiments have been carried out on the system eosin-methyl diethanolamine (MDEA)/α-acyloximino cetone (PDO) **27** (or α,α-dialkoxy-acetophenone DMPA **28**).[72]

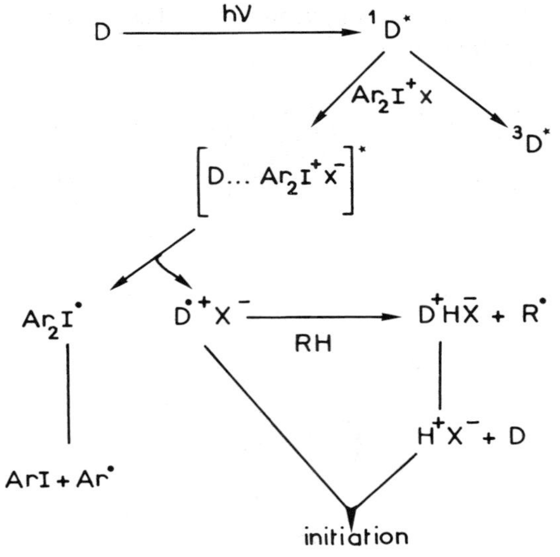

SCHEME VI

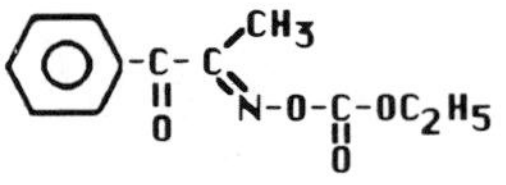

27 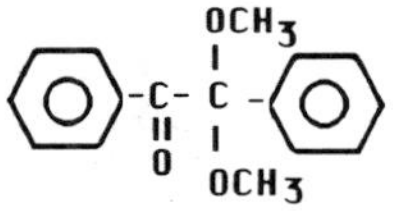   28

Scheme VII shows the primary processes involved in the excited states of eosin, as detected by laser spectroscopy. Subsequent reactions between the excited transient states and the radical species formed cannot be followed in this way; they are suspected of playing a substantial role in the mechanisms of polymerization initiation in the presence of eosin, which acts both as a photoinitiator and a photosensitizer.

Trialkylstannanes are a class of polymerization initiators which may be sensitized by excited photoreducible dyes D* (Scheme VIII).[81]

c. Solid-State Polymerization

Dyes such as cyanine dyes sensitize the photochemical polymerization of diacetylenes in multilayers, according to an electron transfer from the dye to the polymer chain.[82,83]

3. Energy Transfer in Photosensitizers

Benzophenone/Michler's ketone represents a usual system. A more recent and powerful system consists of a thioxanthone derivative **29** and a UV photoinitiator. **30**[84] The spectral sensitization can be extended up to 450 nm (Figure 12), and polymerization occurs by exposure under a short pulse of dye laser at 440 nm;[85] electron and energy transfer processes (the balance of which is dependent on the derivatives used and the polarity of the film matrix) account for the mechanism of initiation.[85]

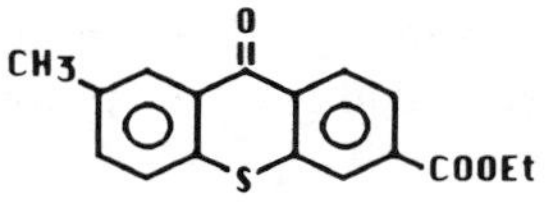

29 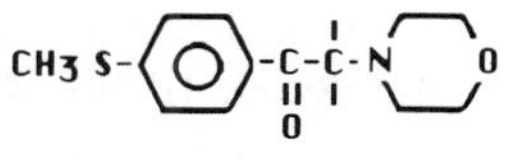   30

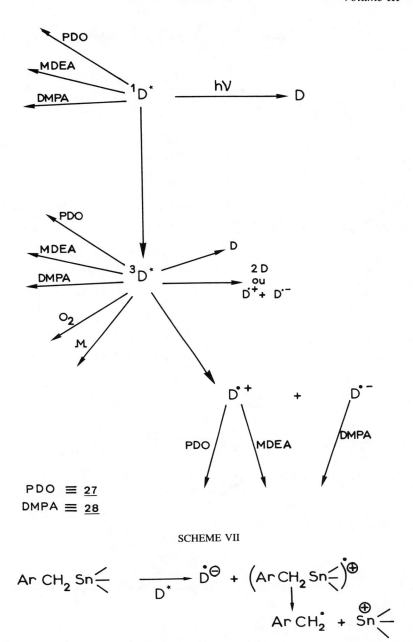

SCHEME VII

SCHEME VIII

4. Initiation through Charge Transfer Complexes

This type of initiation corresponds to the formation of radical ions through an electron transfer process between the polymerizable monomers or/and the photosensitizer combination. Examples of such initiation have been recently found, e.g., in the maleic anhydride-styrene mixture[8] (which is sensitive to an Ar^+ laser exposure), and is discussed in this book.[25]

5. Two-Photon-Sensitive Systems

The different systems which were reviewed above operate through direct or sensitized

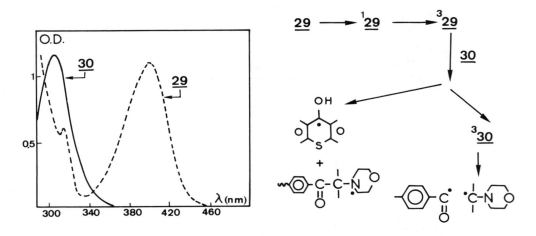

FIGURE 12. Energy and electron transfer in a combination of photoinitiators.

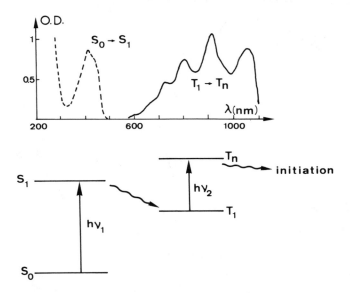

FIGURE 13. Four-level-two-photon plots chemistry for holographic recording in the near IR.

absorption of a photon exhibiting an appropriate energy, corresponding to the energy difference between the ground and the excited singlet state. In specific applications such as holography, it may be worthwhile to find systems working under two simultaneous irradiations of different wavelengths. To meet this requirement, several systems involving two photons have been developed.

One of them is called "two photon-four level" (Figure 13).[86] For example, biacetyl (in a polymer matrix containing monomer) is excited by a noncoherent UV light source ($S_0 \rightarrow S_1$ transition). After intersystem crossing, the $T_1 \rightarrow T_n$ transition is induced by two red laser light beams to produce the interference pattern; the initiation of the polymerization is thought to occur from the upper excited triplet state.

Another alternative consists in a "two photon-two product" system (Figure 14, Scheme IX),[87] in which a ketone is formed *in situ* by specific oxidation of a starting compound with singlet oxygen generated through an energy transfer between a dye (e.g., methylene blue or cyanine) and ground-state oxygen.

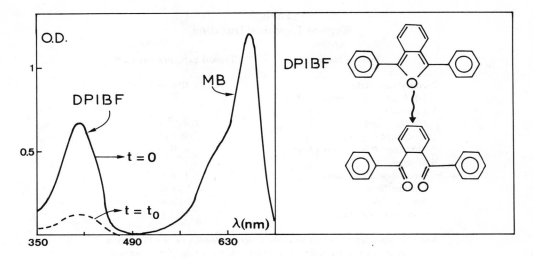

FIGURE 14. Two-product-two-photon system for holographic recording in the near IR.

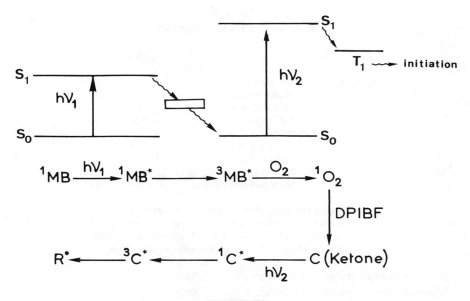

SCHEME IX

B. SPECIFIC FEATURES OF VISIBLE LIGHT-INDUCED REACTIONS

Generally speaking, visible light-induced reactions behave as the corresponding UV polymerizations. Typical exposure duration and relative efficiency (RE) of a given polymerizable mixture, defined by the reciprocal value of relative luminous energy which has to be absorbed $I_0 f_a t$ (where I_0 is the incident light intensity, f_a the absorption function of the film, and t is the exposure time) in order to obtain the same average values of disappearance of the double bonds, are shown in Tables 3 and 4,[72] respectively, for various laser excitation sources:

1. 2-kW continuous lamp (about 10^{18} photons cm^{-2}s^{-1})
2. 15-W CW Ar$^+$ laser, working either in the blue range (50 mW at $\lambda = 514$ nm)
3. Mode-locked ruby laser (6-ps pulses, 300-ps half-width, 10 ns separated, 80 mJ, $\lambda = 347$ nm)

Table 3
Typical Exposure Duration

Type of beam	Typical exposure duration
Polychromatic light	10^{-2} s
2-KW continuous lamp	
Visible lights	
CW Ar⁺ laser	3×10^{-2} s
Modulated CW Nd laser	0.7 s
Modulated CW Cu vapor laser	5×10^{-2} s
Blue lights	
Mode-locked ruby laser	1 shot
Excimer laser	1 pulse
Nitrogen laser	1 pulse
Continuous Ar⁺ laser	10^{-2} s

Note: Incident light intensity absorbed to achieve the drying of a PETA film under various excitation light beams, as a function of the photosensitive formulation:[19,72,75] $>2 \times 10^{16}$ phot cm^{-2} s^{-1}.

Table 4
Role of the Laser Sources[72]

		RE	PS
Visible lasers	Ar⁺ laser	7	(2)
	Modulated Nd laser	3	(2)
	Cu Vapor laser	10	(2)
	R6G dye laser	2	(3)
	HeNe laser	1	(4)
UV lasers	Excimer laser	70	(1)
	Nitrogen laser	50	(1)
	Ar⁺ laser	100	(1)
	Mode-locked ruby laser	40	(1)

Note: Relative efficiency (RE) obtained in presence of different photosensitive systems (PS) (whose concentrations are well selected) (1) without amine PDO **27**; (2) with amine Eosin-PDO; (3) Thionine-PDO; and (4) methylene blue-PDO.

4. Excimer laser (1 pulse 15 ns, 50 mJ, λ = 308nm)
5. Nitrogen laser (1 pulse 8 ns, 3 mJ, λ = 337 nm)
6. Modulated Nd laser (30 mW, pulse half-width: 150 ns, repetition rate: 12500 Hz, λ = 3530 nm)
7. Cu vapor laser (3 W, pulse half-width, 25-ns, repetition rate, 5000 Hz, λ = 510 nm)

The values of RE are obviously a function of the type of sensitizer used (Table 5).[75] It is thus possible to have a sensitive formulation, whatever the wavelength of the laser source used.

As in UV-laser curing (Part III), the exposure times can be greatly reduced by focusing the laser beam and increasing the incident luminous power density. It was recently shown[75] that the rate of polymerization exhibits a linear dependence on the light intensity. Reciprocity effects are presumably weak if one takes into account the recent data reported[42] for photoresists irradiated with an excimer laser. Loss in sensitivity was at worst a factor of about 3 in comparison with conventional UV exposure.

Taking into account the incident light intensity I_0 required to polymerize the monomer

Table 5
Role of the Initiating System in Laser-Induced
Polymerization of PETA at λ = 488 nm

Initiator	t(s)	RE	
PDO 27	0.2	100	
DMPA 28	1.5	12	Eosin/initiator
UVH	5	4	
CTX	10	2	

Dye	t(s)	RE	
Eosin	0.03	100	
Phloxine	0.15	100	Dye/PDO/MDEA
Thionine	0.7	80	
Rose bengale	0.15	120	

Additive	t(s)	RE	
PDO 27	0.15	20	
MDEA	0.05	60	Eosin/additive
PDO/MDEA	0.03	100	

Note: (CTX) 2-chlorothioxanthone and ((UVH) 2-hydroxy-2-methyl-1-phenyl-propan-1-one.

($\sim 10^{20}$ phot + cm^{-2} s^{-1} at λ = 488nm with a beam diameter about 1.6 mm and a power density in the range of 1 W), relief images of typically 16 μ with a scanning speed about 10 m s^{-1} might be obtained by focusing the continuous or quasicontinuous visible laser beam. The increase of the power density by a factor of 10^4 will presumably reduce the exposure time to the microsecond range.

On the contrary, if the wavelength matching is not so crucial, the incident energy available plays an important role and explains, for example, why the low-energy HeNe laser used was not sufficient to induce the polymerization, whereas the more powerful R6G dye laser was adequate. A demonstration of this behavior is easily found by attenuating the Ar$^+$ laser beam: the exposure time t increases as expected, but the energy dose required increases much more rapidly than would be predicted from the expression $I_0 f_a t$ because of O_2 inhibition which results in a decrease of the overall quantum yield of initiation.

As observed previously in UV laser-induced polymerization,[19] the time-intensity profile of the exciting light has little influence on RE. Similar values are recorded, indeed, under continuous low pumping ($\sim$ 1 W cm^{-2}) with an Ar$^+$ laser, as well as undermodulated light beams (150-ns FWHM pulses; instantaneous peak power density: $\sim$kW cm^{-2}) or mode-locked laser emission (train of picoseconds pulses; peak power: MW cm^{-2}). In the same way, a change in the repetition rate of the Nd-modulated laser between 10 and 1250 Hz (thus reducing the peak power by a factor of about 100) does not lead to a significant alteration of RE. Due to the potential high power available with metal vapor laser (such as the Cu vapor laser), shorter exposure times are expected in the near future.

Comparison[72] between UV and visible excitation (such as that delivered by an Ar$^+$ laser working in the blue or visible range of the spectrum) shows that with the most efficient system proposed in Table 4, the relative difference of RE is about 15. This factor is encouraging, and more efficient formulations can presumably be found. For example, in presence of a complex photosensitive system (at λ: 488 nm) consisting in a ketocoumarin, an amine, and an iodonium salt, a quite similar value of RE was obtained very recently (Table 6).[88]

Table 6

**Curing-Time and Absorbed Energy Dose for the
Photopolymerization of a PETA Film (in Presence of Two
Different Photosensitive Systems) under an Ar⁺ Laser
Exposure at λ = 488 nm**

Photosensitive system	t(s)	$I_0 f_a t$ (phot cm^{-2})	RE
Eosin-MDEA-PDO *27*	2	4×10^{17}	0.05
Ketocoumarine-MDEA-$\phi_2 I^+$	0.05	2×10^{16}	1
PDO	0.1*	$2 \times 10^{16*}$	1

Note: Power, 100mW; spot diameter, 6mm; and * under Ar⁺ laser exposure at
λ × 364 nm.

The role of the initiating species has also been stressed. In a recent study,[15] it was demonstrated that the 488-nm laser emission can indeed initiate the polymerization of multiacrylate monomers if benzoquinone is used as photoinitiator; the quantum efficiency was found to be much lower than with UV laser initiation. Recent polymerization experiments[14] carried out with camphoroquinone appear to be more promising, since they show a tenfold increase of the rate of polymerization of epoxy-acrylate photoresists, as compared to benzoquinone-based systems. Such a large improvement is assumed to result primarily from two favorable factors: (1) a higher absorbance of the 488-nm emission band, since camphoroquinone exhibits its maximum absorption in the 460 to 480–nm range, compared to 430 to 450 nm for benzoquinone and (2) a larger initiation efficiency of camphoroquinone, especially when it is complexed with a tertiary amine such as dimethylethanolamine. Scanning rates of a few meters per second were thus obtained for the curing of acrylate photoresists by visible laser radiation.

Table 5 shows the curing times and the relative efficiency of different light-sensitive formulations in a pentaerythritol triacrylate (PETA) film matrix under laser exposure at λ = 488 nm.[75] When the optical density of the dye at the irradiation wavelength increases, the curing time decreases, whereas the absorbed energy increases. Thus, a low value of RE, in terms of the energy absorbed, must not be associated with the concept of a poor sensitizing formulation, since the factor "curing time" could be attractive. It is also apparent that the three concentrations (dye-amine-photoinitiator) are not unrelated. This shows how difficult is the problem of optimizing formulations. Moreover, the chemical nature of the film matrix also affects the overall reactivity of the system and, if more complex industrial formulations are used for the coating film, the cure speed may be enhanced.

For the different systems studied (Tables 4 to 6) and under the experimental conditions used so far,[19,75,88] multiphotonic processes are considered to be negligible to a first approximation, since RE values are not significantly different, although the power densities are in the ratio 1 (CW Ar⁺ laser)-10 (Nd laser)-100 (Cu laser). This does not necessarily hold true in other experimental conditions.

V. PROSPECTS OF LASER CURING

In considering the potential applications of laser processing in the UV curing area, it is important to keep in mind that, owing to the intrinsic characteristics and high cost of the laser photons, this technology is most likely to be used in photochemical processes having high quantum yields, where both small areas have to be treated and high added-value materials are to be produced. Some of the current and potential end uses of laser curing can be divided into two main sectors: (1) surface treatments (coating of optical fibers, modeling in three dimensions, composite materials, pigmented systems) and (2) microlithography (printing plates, microcircuits, video disks, holography).

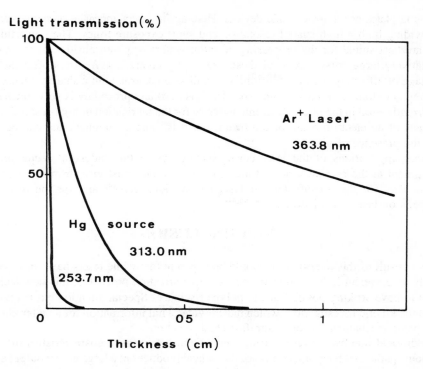

FIGURE 15.   Light-penetration profile in a photoresist containing 0.04% of Irgacure 651 for various emission lines of a mercury lamp and argon ion laser. (From *Radiation Curing in Polymers*, Randell, D., Ed., Royal Society of Chemistry, London, 1987. With permission.)

In the surface treatment applications, it is mainly taken advantage of the extremely high speeds of cure afforded by the laser irradiation, like, for instance, for the ultrafast coating of optical fibers or of sheathed cables. The laser technology permits important savings of space because, owing to the great directivity of the narrow beam, the instant photochemical reaction usually occurs on a few square millimeters area.

Another distinct advantage of laser curing consists in the great penetration of the laser beam, which should help overcome one of the most severe limitations of conventional UV curing. Figure 15 shows, for instance, the penetration profile in an acrylic resin containing 400 ppm of photoinitiator for the 363.8-nm emission line of the argon ion laser and for the 254- and 313-nm emission lines of the mercury lamp. Because of the large variations in the resin intrinsic absorbance with the wavelength, near-UV radiations are penetrating much more into the sample than deep-UV radiations. Half of the incident light is absorbed within 1 cm for the 363.8-nm laser emission, as compared to about 1 mm for the 313-nm Hg emission, and only 10 μm for the Hg line at 253.7 nm. By contrast to conventional poly-chromatic UV sources that create a sharply decreasing cure gradient in thick specimens, laser-induced polymerization permits to obtain a fairly uniform through-cure, even for sam-ples of a few centimeters thickness. The lower cure rate resulting from the necessary use of very low initiator concentrations is partly offset by the high intensity of the laser beam, which allows to still operate at short exposure times. Current users of the UV curing technology should thus be strongly tempted in the near future to take profit of the striking possibilities of lasers to perform deep through-cure, in particular for the modeling of three-dimensional objects by intersecting laser beams, as well as for the curing of opaque samples and of composite resins with enhanced adhesion on both glass and polymer substrates.

The second major area of potential applications of laser curing concerns the microlith-ography industry where photopolymer relief images are needed for the fabrication of microprinting plates or microelectronic devices. Here again, lasers appear as attractive candidates by providing both a high spatial resolution and short exposure times. The powerful pulsed lasers are best suited for the processing of entire wafers or photosensitive plates, owing to the relatively large cross-section of these beams. By contrast, CW lasers offer the unique advantage of affording the direct writing of small-size patterns with a sharply focused beam operating at extremely high scanning speeds. This maskless procedure is very much like the one currently used in the microelectronic industry for the fabrication of high-resolution masks by means of an electron beam or ion beam, but it is more economical and can be worked out in the presence of air.

Other applications of the laser curing technology in this industrial sector are under development at the present time and are likely to exhibit a fast growth in the near future, in particular, for the manufacture of laser-vision video disks[89] and for the recording of holograms on laser-sensitive media.[86,90,91]

VI. CONCLUSION

As a result of this overview, it should be apparent that while lasers have only been used recently as powerful light sources to induce polymerization processes, this new technology proved to have striking possibilities in polymer science. Special attention has been directed toward the kinetic investigation, which has provided vital information for a better knowledge of the basic mechanism of these ultrafast chain reactions.

Additional interest for laser curing stems from the intrinsic characteristics of the laser emission: spatial and temporal control of the radical production; a large power output allowing instantaneous and deep-section polymerizations to be carried out; the spatial coherence that permits to draw high-resolution images by means of sharply focused laser beams; and the narrow bandwidth that can be matched with the maximum absorption of the photosensitive system and thus eliminates undesirable secondary photochemical reactions. Nonlinear processes may, however, become important at high power densities, depending on the photosensitive system used.

The high intensities afforded by lasers has opened up unexplored research areas, while providing at the same time new impetus to existing applications of the UV curing technology. Some of the most promising end uses are expected to concentrate mainly in the electronic industry for the direct writing of complex relief patterns at micronic resolution and for the instant curing of optical fiber coatings. Many other possibilities still have to be explored in the coming years, with the chances of ending with successful applications depending primarily on the creativity and innovative skill of scientists of today. The economic prospects of laser curing are directly related to future progress in our understanding of the various phenomena involved in these ultrafast processes, as well as to the development of reliable and low-cost lasers that can be safely operated in the chemical industry environment.

REFERENCES

1. **Pappas, S. P.,** *UV Curing: Science and Technology,* Technology Marketing Corp., Stamford, CT, 1978.
2. **Roffey, C. G.,** *Photopolymerization of Surface Coatings,* John Wiley & Sons, New York, 1982.
3. **Buddenhagen, D. A., Haeff, A. V., Smith, C. F., Oster, G., and Oster, G. K.,** Observations of ruby laser beam-intensity patterns with dye-sensitized photopolymers, *Proc. Nat. Acad. Sci. U.S.A.,* 48, 303, 1962.
4. **Frigerio, G. E. and Stefanini, A.,** Experimental results on the photopolymerization of acrylamide induced by the light of He-Ne and ruby lasers, *Lett. Nuovo Cimento,* 2(15), 810, 1971.
5. **Decker, C.,** Laser imaging photoresist systems for microcircuits, *Microcircuit Eng.,* 1, 9, 1986.

6. **Castle, P. M.,** Laser-induced photopolymerizations reactions, Proc. IUPAC 28th Macromol. Symp., Amherst, MA, 1982, 282.

7. **Williamson, M. A., Smith, J. D. B., Castle, P. M., and Kauffman, R. N.,** Laser-initiated polymerization of charge-transfer monomer systems, *J. Polym. Sci., Polym. Chem. Ed.,* 20, 1875, 1982.

8. **Sadhir, R. K., Smith, J. D. B., and Castle, P. M.,** Laser-initiated copolymerization of maleic anhydride with styrene, vinyltoluene and *t*-butylstyrene, *J. Polym. Chem. Ed.,* 21, 1315, 1983.

9. **Tsao, J. Y. and Ehrlich, D. J.,** UV-laser photopolymerization of volatile surface-adsorbed methyl methacrylate, *Appl. Phys. Lett.,* 42, 997, 1983.

10. **Decker, C.,** Laser-induced polymerization of multifunctional acrylate systems, *Polym. Photochem.,* 3, 131, 1983.

11. **Decker, C.,** Laser-curing of photoresist systems for imaging of microcircuits, *SME Techn. Paper,* FC83, 265, 1983.

12. **Decker, C.,** Ultra-fast polymerization of epoxy-acrylate resins by pulsed laser irradiation, *J. Polym. Sci., Polym. Chem. Ed.,* 21, 2451, 1983.

13. **Decker, C.,** Kinetic approach of the UV-curing of acrylate coatings by high intensity photon beam, *Polym. Mater. Sci. Eng.,* 49, 32, 1983.

14. **Decker, C.,** Laser-induced polymerization, in *Materials for Microlithography — Radiation Sensitive Polymers,* (ACS Symp. Ser., Vol. 266), Thompson, L. F., Willson, C. G. and Fréchet, J. M. J., Eds., American Chemical Society, Washington, D.C., 1984, 207.

15. **Decker, C.,** UV-curing of acrylate coatings by laser beams, *J. Coat. Technol.,* 56, 29, 1984.

16. **Hoyle, C. E., Hensel, R. D., and Grubb, M. B.,** Laser-initiated polymerization of a thiol-ene system, *Polym. Photochem.,* 4, 69, 1984.

17. **Hoyle, C. E., Hensel, R. D., and Grubb, M. B.,** Temperature dependence of the laser-initiated polymerization of a thiol-ene system, *J. Polym. Sci., Polym. Chem. Ed.,* 22, 1865, 1984.

18. **Fouassier, J. P., Jacques, P., Lougnot, D. J., and Pilot, T.,** Lasers, photoinitiators and monomers: a fashionable formulation, *Polym. Photochem.,* 5, 57, 1984.

19. **Fouassier, J. P., Lougnot, D. J., and Pilot, T.,** Visible laser light in photoinduced polymerization. I. A quantitative comparison with UV laser irradiation, *J. Polym. Sci., Polym. Chem. Ed.,* 23, 569, 1985.

20. **Sadhir, R. K., Smith, J. D. B., and Castle, P. M.,** Laser-initiated polymerization of epoxies in the presence of maleic anhydride, *J. Polym. Sci. Polym. Chem. Ed.,* 23, 411, 1985.

21. **Decker, C.,** Laser-induced crosslinking polymerization for microcircuit applications, in *Strahlenhartbäre Bindemittel und deren Applicationstechnologien,* Hinterwaldner Verlag, Munich, 1985, 161.

22. **Decker, C. and Jenkins, A.,** Kinetic approach of O_2 inhibition in UV and laser-induced polymerizations, *Macromolecules,* 18, 1241, 1985.

23. **Decker, C.,** Laser-curing of acrylic coatings, in *Radiation Curing of Polymers,* Randell, D. R., Ed., Royal Society of Chemistry, London, 1986, 16.

24. **Hoyle, C. E., Trapp, M., and Chang, C. H.,** Laser-initiated polymerization. The effect of pulse repetition rate, *Polym. Mater. Sci. Eng.,* 57, 579, 1987.

25. **Castle, P. M.,** Potential applications of lasers in photopolymers, in *Applications of Lasers in Polymer Science and Technology,* Vol. 3, Fouassier, J. P., and Rabek, J. F., Eds., CRC Press, Boca Raton, FL, 1989, chap. 2.

26. **Schnabel, W.,** Flash photolysis of benzoin and benzophenone containing systems, *Photogr. Sci. Eng.,* 23(3), 154, 1979.

27. **Berner, G., Puglisi, J., Kirchmayr, R., and Rist, G.,** Photoinitiators. An overview, *J. Rad. Curing,* 6,2,1979.

28. **Crivello, J. V.,** Photoinitiated cationic polymerization by sulphonium salts, in *Developments in Polymer Photochemistry,* Vol. 2, Allen, N., Ed., Applied Science Publishers, Englewood, NJ, 1981, 1.

29. **Fouassier, J. P.,** Excited state properties of photoinitiators: lasers and their applications, in *Photopolymerization Science and Technology,* Allen, N. S., Ed., Elsevier, London, 1989.

30. **Hageman, H. J.,** Photoinitiators for free radical polymerization, *Prog. Org. Coat.,* 13, 123, 1985.

31. **Li Bassi, G.,** Photoinitiators in the curing of coating materials, *Double Liaison,* 361, 17, 1985.

32. **Fouassier, J. P.,** Les systèmes photosensibles dans les procédés de réticulation sous UV: réactivité photochimique, *Double Liaison,* 356, 173, 1985.

33. **Rabek, J. F.,** Mechanisms of photophysical and photochemical reactions in *Polymer: Theory and Practical Applications,* John Wiley & Sons, New York, 1987.

34. **Crivello, J. V.,** Photoinitiated cationic polymerization, in *UV-curing: Science and Technology,* Pappas, S. P., Ed., Technology Marketing Publication, Stamford, CT, 1978, 23.

35. **Fouassier, J. P.,** Improvement of photopolymer reactivity: a fundamental approach, *J. Chim. Phys.,* 80, 339, 1983.

36. **Lougnot, D. J. and Fouassier, J. P.,** Laser spectroscopy of the excited states of water soluble photoinitiators of polymerization, in *Applications of Lasers in Polymer Science and Technology,* Vol. 2, Fouassier, J. P. and Rabek, J. F., Eds., CRC Press, Boca Raton, FL, 1989, chap. 6.

37. **Schnabel, W.,** Application of laser spectroscopy to the study of photopolymerization reactions in nonaqueous solutions, in *Applications of Lasers in Polymer Science and Technology,* Vol. 2, Fouassier, J. P. and Rabek, J. F., Eds., CRC Press, Boca Raton, FL, 1989, chap. 5.
38. **Decker, C., Bendaikha, T., Fizet, M., and Faure, J.,** Oxygen inhibition in UV-curing, *SME Techn. Pap.,* FC85, 432, 1985.
39. **Decker, C., Faure, J., and Fizet, M.,** Oxygen effect on UV-curing and photodegradation of organic coatings, *Org. Coat. Plast. Chem.,* 42, 710, 1980.
40. **Becker, R. A., Sopori, B. C., and Chang, W.,** Focused laser lithographic system, *Appl. Opt.* 17(7), 1069, 1978.
41. **Loh, I., Martin, G., Kowel, S., and Kornreich, P.,** Laser lithography of silicone polymers, *Polym. Prepr.,* 23(2), 195, 1982.
42. **Jain, K., Wilson, C. G., and Lin, S.,** Ultrafast deep UV-lithography with excimer lasers, *IEEE Electron. Device Lett.,* 3(3), 53, 1982.
43. **Decker, C. and Bendaikha, T.,** Photopolymérisation de macromères multifonctionnels. I. Etude cinétique, *Eur. Polym. J.,* 20, 753, 1984.
44. **Tryson, G. R. and Shultz, A. R.,** A calorimetric study of acrylate photopolymerization, *J. Polym. Sci., Polym. Phys. Ed.,* 17, 2059, 1979.
45. **Decker, C. and Moussa, K.,** Mechanistic approach to the photopolymerization of multifunctional macro-monomers, *Polym. Mat. Sci. Eng.,* 55, 552, 1986.
46. **Kloosterboer, J., Van de Hei, G., Gossink, R., and Dortant, G.,** The effects of volume relaxation and thermal mobilization of trapped radicals on the final conversion of photopolymerized diacrylates, *Polym. Comm.,* 25, 322, 1984.
47. **Decker, C. and Moussa, K.,** Radical trapping in photopolymerized acrylic networks, *J. Polym. Sci., Polym. Chem. Ed.,* 25, 739, 1987.
48. **Decker, C. and Moussa, K.,** Photopolymerization of multifunctional monomers in condensed phase, *J. Appl. Polym. Sci.,* 34, 1603, 1987.
49. **Ketley, A. D.,** Dye sensitized photopolymerization — a review, *J. Radiat. Curing,* 35, 1982.
50. **Fisher, W., Kuita, V., Zweifel, H., and Felder, L.,** French Patent 8108981, 1981.
51. **Meier, K. and Zweifel, H.,** Thioxanthone ester derivatives: efficient triplet sensitizers for photopolymer applications, *J. Photochem.,* 35, 353, 1986.
52. **Davis, M. J., Gawne, G., Green, P. N., and Green, W. A.,** The synthesis and properties of a novel series of water soluble thioxanthone photoinitiators, *Chem. Spec.,* 1986.
53. **Eichler, J., and Neisius, D. E.,** German Patent 3126433, 1983.
54. **Tanaka, T.,** German Patent 32113129, 1982.
55. **Anderson, A. G.,** Du Pont, European Patent 811029792, 1981.
56. **Renner, C. A.,** Du Pont, U.S. Patent 4371605, 1983.
57. **Skoultchi, M. M.,** U.K. Patent 8208897, 1983.
58. **Williams, J. L. R., Specht, D., and Farid, S.,** Ketocoumarines as photosensitizers and photoinitiators, *Polym. Eng. Sci.,* 23, 1022, 1983.
59. **Herkstroeter, W. G. and Farid, S.,** Photopolymerization relevant triplet state parameters of methyl metacrylate, diethyl 1-4' phenylenediacrylate and methyl-1 naphtyl acrylate, *J. Photochem.,* 35, 71, 1986.
60. **Robert, B., Bolte, M., and Lemaire, J.,** Comportements photochimiques des systèmes chrome VI et III — acrylamide en solution aqueuse, *J. Chim. Phys.,* 82(4), 361, 1985.
61. **Meier, K. and Zweifel, H.,** Photopolymerization of epoxides — A new class of photoinitiators based on cationic iron arene complexes, Tech. Paper FC 85 417, in *Proc. Radcure Europe,* SME Ed., Dearborn, MI, U.S.A., 1985.
62. **Riediker, M., Meier, K., and Zweifel, H.,** European Patent 186626, 1986.
63. **Wagner, M. H., and Purbrick, M. D.,** Transition metal carbonyl derivatives as photopolymerization initiators and photocrosslinking agents in photoresist, *J. Photogr. Sci.,* 29, 230, 1981.
64. **Abu Abdoun, I. I., Thys, L., and Neckers, D. C.,** Nonketonic perester photoinitiators, *Macromolecules* 17(3), 283, 1984.
65. **Humphreys, R. W. R.,** European Patent 147226, 1985.
66. **Castle, P. M.,** Quinones as photoinitiators of laser induced polymerization, Proc. IUPAC, Amherst, MA, 1982.
67. **Ratcliffe, M. J., Shaw, D. L., and Robinson, P. A.,** European Patent EP 90493, 1983.
68. **Stinson, S. C.,** Electronic industry opens frontiers for photoresist chemistry, *Chem. Eng.,* 7, 1983.
69. **Wang, W. L.,** European Patent 823033717, 1983.
70. **Bloom, A., Bartolini, R. A., and Ross, D. L.,** Organic recording medium for volume phase holography, *Appl. Phys. Lett.,* 24(12), 612, 1974.
71. **Baumann, H., Oertel, U., and Timpe, H. J.,** Photopolymerisation mit der Initiatorkombination 9-10 phenanthrenchinon-acrylonium Verbindungen, *Eur. Polym. J.,* 22, 313, 1986.

72. **Fouassier, J. P. and Chesneau, E.**, Unpublished data, 1986.
73. **Bader, S.**, European Patent 793030784, 1980.
74. **Patel, R.**, European patent 97012, 1983.
75. **Chesneau, E. and Fouassier, J. P.**, Polymérisation induite sous irradiation laser visible, *Angew. Makromol. Chem.*, 135, 41, 1985.
76. **Baumann, R., Singer, W., and Fritzsche, K.**, German Patent 160084, 1983.
77. **Bauer, S.**, European Patent 99856, 1984.
78. **Crivello, J. V.**, Cationic polymerization iodonium and sulfonium salt photoinitiators, in *Advances in Polymer Science 62*, Saegusa, T., Ed., Springer-Verlag, Berlin, 1984, 3.
79. **Pappas, S. P.**, Photoinitiation of cationic and concurrent radical cationic polymerization, *Prog. Org. Coat.*, 13, 35, 1985.
80. **Baumann, H. and Timpe, H. J.**, Lichinitierte polymer and polymerizations reaktionen, *Acta Polym.*, 37, 309, 1986.
81. **Eaton, D. F.**, Dye sensitized photopolymerization: activation by Trisalkyllbenzylstannanes, *Photog. Sci. Eng.*, 23, 150, 1979.
82. **Tieke, B. and Wegner, G.**, Photosensitation of the solid-state polymerization of a diacetylene by formation of a mixed crystal with phenazine, *Makromol. Chem.*, 179, 2573, 1978.
83. **Fouassier, J. P., Tieke, B., and Wegner, G.**, The photochemistry of the polymerization of diacetylenes in multilayers, *Isr. J. Chem.*, 18, 227, 1979.
84. **Rutsch, W., Berner, G., Kirchmayer, R., Husler, R., Rist, G., and Buhler, N.**, New photoinitiators for pigmented systems, Tech. Paper FC 84 252, in *Proc. Radcure*, Atlanta 1984, SME Ed., Dearborn, MI.
85. **Fouassier, J. P., Lougnot, D. J., Payerne, A., and Wieder, F.**, Time resolved laser pumped dye laser spectroscopy: energy and electron transfer in ketones, *Chem. Phys. Lett.*, 135, 30, 1987.
86. **Bräuchle, C., Wild, U. P., Burland, D. M., Bjorklund, G. C., and Alvarez, D. C.**, *IBM Res. Dev.*, 26, 217, 1982.
87. **Moisan, J. Y., Gravey, F., Fouassier, J. P., and Lougnot, D. J.**, Milieux d'enregistrement pour holographie et procédé d'enregistrement pour hologramme, Brevet n°84 14 403 déposé le 18/09/84 par le CNET.
88. **Fouassier, J. P., Chesneau, E., and Le Baccon, M.**, Polymérisation induite sous irradiation laser visible. III. Un nouveau systèmes photosensible performant, *Makromol. Chem., Short Commun.*, in press.
89. **Van den Broek, A., Haverkorn von Rijsewilk, H., Legierse, P., Lippits, G., and Thomas, G.**, Manufacture of laser vision discs by a photopolymerization video process, *J. Radiat. Curing*, 11, 2, 1984.
90. **Bräuchle, C.**, Photochemical laser studies in polymer films using holography, *Polym. Photochem.*, 5, 121, 1984.
91. **Carré, C., Ritzenthaler, D., Lougnot, D. J., and Fouassier, J. P.**, A new biphotonic process for recording holograms with cw lasers in the near infrared. *Opt. Lett.*, 12, 646, 1987.

Chapter 2

POTENTIAL APPLICATIONS OF LASERS IN PHOTOCURING, PHOTOMODIFICATION, AND PHOTOCROSSLINKING OF POLYMERS

Peter M. Castle and Rajender K. Sadhir

TABLE OF CONTENTS

I. INTRODUCTION

Incoherent sources of UV light are used extensively in photocuring, photomodification, and photocrosslinking in a wide variety of applications ranging from paints and coatings to printing and microlithography. In most of the applications the coatings are applied as a liquid or paste, but must change to a solid and nontacky state before the painted or coated article can be used. The change is known as curing or drying. Sometimes it occurs by physical means, for example, the evaporation of a solvent or dispersion medium, and sometimes by chemical changes such as polymerization and crosslinking. When a UV source is used to carry out the polymerization or crosslinking, it is referred to as photocuring or photo-crosslinking. These chemical processes connect the many relatively small molecules of the original liquid or paste into a large network of insoluble solid which may be either rigid or rubbery in consistency, depending upon the requirements of a particular application. There are several good reviews and books on this subject.[1-5]

Use of light as the energy source for crosslinking functional monomers, oligomers, and polymers constitutes the basis of important commercial processes with broad applicability. UV curing offers unique advantages such as rapid polymer network formation on heat-sensitive substrates, as well as reduced energy consumption, low emissions, and minimal space requirements. The use of UV lamps such as mercury lamps has been made in many applications. Curing thick sections of coatings by incoherent sources has been a problem. One of the reasons is the light absorption by photoinitiator in the upper layer of a coating, which limits the availability of light to the photoinitiator in the lower layer. There are several advantages in using the laser beam for photocuring of polymers.

The use of UV light in polymerization processes takes advantage of the fact that exposure of organic molecules to irradiation results in the generation of reactive species. The role of UV light has generally been restricted to the initiation of polymerization. However, propagation may also involve species generated by photoexcitations. The photoinitiator molecule is raised from its ground state to an excited singlet state by absorption of light energy. The photoinitiator molecule can undergo photodeactivation from its excited singlet state in a variety of ways. For example, it may undergo chemical reaction or may return to the ground state by either a radiative process (fluorescence) or a nonradiative process (internal conversion and vibrational relaxation). Additionally, it may undergo a transition (intersystem crossing to a triplet state). A typical energy level diagram for aromatic ketones is presented in Figure 1. Photochemical reactions are known to occur from both singlet and triplet excited states, and generally occur from the lowest excited singlet (S_1) and triplet (T_1) states. However, due to the short lifetime of S_1 states ($<10^{-8}$ s) many photochemical reactions, particularly intermolecular processes, occur only via the T_1 states ($>10^{-6}$ s) which are longer lived. After absorption of light energy, a photoinitiator can be converted in a number of ways into reactive intermediate species capable of initiating polymerization, as follows:

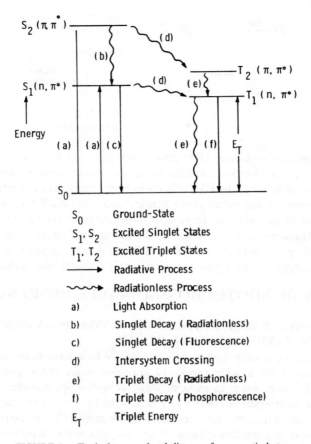

FIGURE 1. Typical energy level diagram for aromatic ketones.

1. Photofragmentation:

$$I^0 \xrightarrow{h\nu} I^1 \longrightarrow I^3 \xrightarrow{\text{cleavage}} \text{reactive intermediate}$$

2. Hydrogen abstraction from a hydrogen-donor molecule, RH:

$$I^0 \xrightarrow{h\nu} I^1 \longrightarrow I^3 \xrightarrow{RH} IH + R^{\cdot} \text{ (free radical)}$$

3. Energy transfer from photosensitizer (donor molecule) to acceptor molecule:

$$D^0 \xrightarrow{h\nu} D^1 \longrightarrow D^3 \xrightarrow{A^{\cdot}} D^0 + A^3$$

reactive
intermediate

4. Electron transfer from donor(triplet state) to acceptor (ground state); formation and subsequent fragmentation of an exciplex:

$$D^0 \xrightarrow{\ h\nu\ } D^1 \longrightarrow D^3 \xrightarrow{\ A^0\ } [D^+ \ldots\ ^- A]^*$$

exciplex

↓ fragmentation

reactive
intermediate

where I = photoinitiator molecule; D = donor molecule; and A = acceptor molecule.

Superscripts 0, 1, and 3 refer, respectively, to the ground, singlet, and triplet states.

The degradation of polymers by UV radiation is most commonly associated with chemical changes and is sometimes denoted as photochemical degradation (PCD). On the other hand, if the PCD pertains to an oxidation process, it is described as photooxidative degradation (POD). Controlled radiative degradation and modification of polymers can be used for various emerging industrial applications and analytical techniques, such as photo- and electron beam lithography, and polymer characterization using highly focused laser-induced degradation.

II. TYPES OF PHOTOCROSSLINKABLE RESIN SYSTEMS

A. PHOTOCROSSLINKABLE SYSTEMS INVOLVING POLYMERIZATION OR COPOLYMERIZATION OF VINYL RESINS

Vinyl resins that are curable by irradiation with UV light have been used commercially during the past decade for a wide variety of purposes in the paint, printing, electronic, dental, and adhesive industries. In most of these industrial applications, the solvent-free photocrosslinkable resin systems consist of mixtures of (1) a reactive viscous oligomer or prepolymer containing at least two polymerizable double bonds; (2) a mono- or polyfunctional monomer, which acts as a reactive diluent of low viscosity and is copolymerizable with the viscous prepolymer; (3) a photoinitiator; and (4) an inhibitor to prevent polymerization during storage at ambient temperatures. In addition fillers, pigments, and other special components — to enhance adhesion or to improve wetting of the substrate — may also be incorporated. Photopolymerizable prepolymers can be divided into three types, depending on the position of the reactive double bonds in the prepolymer molecules, where

1. The reactive double bonds form part of the chain's backbone as, for example, in the case of the conventional unsaturated polyesters based on maleic or fumaric acid.
2. The polymerizable double bonds are present in the end groups of a linear or branched prepolymer as, for example, in the so-called epoxy acrylates and urethane acrylates and in polyesters having acrylate end groups.
3. A polymer possesses pendant polymerizable double bonds. Such prepolymers can, for instance, be made by reaction of a polyol or a polycarboxylic acid with a compound (e.g., vinyl isocyanate or glycidyl acrylate) which contains both a photopolymerizable double bond and a second reactive group of a different kind.

Most of the photoinitiators which are currently of technological value for free radical polymerization or crosslinking fall into one of the four general classes: (1) benzoin ethers and esters, benzyl ketals, and (more recently) benzoylphosphine oxide derivatives containing the unit aryl $-CO-P(O){<}$; (2) acetophenone derivatives; (3) α-acyl oxime esters; and (4) combination of aromatic ketones with amines.

Table 1 lists a number of important current examples of photoinitiators.

TABLE 1
Typical Commercial Photoinitiator Suppliers and Absorption Maxima

Generic class/product structure	Trade name	Supplier	Optimum absorption (nm)
Benzoin isobutyl ether	Vicure 10	Stauffer Chemical	240—270
1-Phenyl-1,2-propanedione-2-(*o*-ethoxycarbonyl) oxime	Quantacure PDO	Aceto chemical	275—400
2,2-Dimethoxy-2-phenyl-acetophenone	Irgacure 651	Ciba-Geigy	250—350
Diethoxyacetophenone	DEAP	Upjohn	240—350
Benzophenone	None	Upjohn	240—350
Halogen-substituted alkyl aryl ketone	Sandoray 1000	Sandoz	240—360

B. THIOL-ENE SYSTEMS

A very large number of photocrosslinkable systems based on a polythiol plus a polyene have been described in literature. The basic reaction most often involved in the cure of these systems is the addition of thiol group across the C=C double bond:

$$R^1 - SH + R^2-CH{=}CH-R^3 \rightarrow R^1{-}S{-}\underset{\underset{R^2}{|}}{CH}{-}CH_2{-}R^3$$

Crosslinking occurs if the polythiol and/or the polyene is of a functionality greater than 2. In practice, the polythiols most used are the thioglycolate and 3-mercaptopropionate esters of polyols. A variety of polyenes has been used, e.g., allyl ester of polycarboxylic acids, adducts of allyl alcohol with isocyanate-tipped prepolymers, and adducts of allylisocyanate with polyols. Aromatic ketones are frequently used as photoinitiators for thiol-ene systems.

C. SYSTEMS WHICH CROSSLINK BY CYCLOADDITION REACTIONS

Polyvinyl cinnamate is one of the best-known examples of a material which becomes insoluble through photocycloaddition. Crosslinking involves activation of the cinnamate ester chromophore by UV light, followed by reaction of the activated chromophore with a second cinnamate unit in the ground state. As in the case of cinnamate, chalcone type groups also dimerize to form cyclobutane rings on UV irradiation. Coumarin-type, maleic-type derivatives, and anthracene derivatives also undergo cycloaddition on UV irradiation.

D. AZIDO COMPOUNDS

A primary step involved in the crosslinking of materials containing azide groups is photodissociation of the azide groups to form nitrene-type intermediates which may undergo a variety of reactions. Bisazides containing two aromatic azide units linked by an unsaturated residue are often used with cyclized polyisoprene, particularly in photoresist systems.

E. PHOTOINDUCED CROSSLINKING OF EPOXIDES AND OTHER CATIONICALLY CURABLE RESINS

The most significant photoinitiators for the polymerization of epoxy resins, from a commercial point of view, are aryldiazonium, diaryliodonium, and triarylsulfonium salts, which possess anions of poor nucleophilicity such as hexafluoroborate, hexafluorophosphate, hexafluoroarsenate, and hexafluoroantimonate. On photolysis, these salts liberate either a Lewis acid or a Brönsted acid, which then rapidly effects crosslinking of the epoxy resin. A list of various photoinitiators used for curing epoxies is given in Table 2.

III. LASERS AS PHOTOCHEMICAL ENERGY SOURCES

Lasers are unique sources of photochemical energy. Since the photochemical process is governed by the statistics of photons interacting with molecules within the radiation field, the properties of the radiation field are extremely important. The important properties of the field are intensity (photon flux), individual photon energy (the wavelength or frequency of the light), irradiated volume (geometry), and temporal behavior (continuous wave, CW, or pulsed). If these properties are considered in detail, lasers are quite different when compared to the usual incoherent light sources which have been employed in photochemical processing.

One of the outstanding features of laser radiation is its spectral purity. The worst wavelength spread one may experience is of the order of several nanometers, when working with free-running dye lasers. At the other end of the scale, linewidths as narrow as 0.01 Å can be routinely achieved. In most instances in organic photochemistry a very narrow line is not necessary, since the absorption bands for organic molecules in solution are generally broad compared to laser linewidths. However, these absorption bands are considerably narrower than the broad bandwidth output available from most incoherent light sources. If a photopolymerization reaction is driven by a laser, the overall quantum efficiency can be much higher, since the laser wavelength can be chosen to match the photochemically active absorption band. When a broadband source is used, much of the radiation falls outside the absorption band of the photochemically active molecule and is wasted. It is not uncommon for less than 10% of the available radiation from an incoherent source to be absorbed by the photochemically active molecule. In the case of a laser, conditions can be adjusted in a manner which allows all of the radiation to be absorbed by the target molecule.

TABLE 2
Structure and Properties of Some Typical Cationic
Photoinitiators

Cation	Anion	Absolute maxima (nm)
2,5-Dichlorobenzene	SbF_6^-	238,358
CH_3O–⟨O⟩–I^+–⟨O⟩	BF_4^-	246
CH_3–⟨O⟩–I^+–⟨O⟩–CH_3	PF_6^-	237
[⟨O⟩]$_3$–S^+	BF_4^-	230
[CH_3O–⟨O⟩]$_3$–S^+	AsF_6^-	225,280
[⟨O⟩]$_3$–Se^+	BF_4^-	258,266,275

In comparing intensities of lasers and incoherent sources, the question of geometry plays an important role. While it is reasonably easy to obtain multikilowatt CW lamps compared to multikilowatt CW lasers (which are uncommon), the unidirectional nature of the laser radiation can make a lower power laser easier to employ than the higher power lamp, which has an isotropic distribution of its radiation. Extremely high intensities can be easily obtained by focusing a laser beam with very simple optics. This is possible as a result of the low angular divergence of a laser beam. The task is much more difficult to accomplish with an incoherent isotropic light source. As a result of these properties, a greater intensity per unit area may be achievable with a laser with an output of only 50 W when compared to an incoherent source rated at kilowatts. Spot intensities of megawatts per square centimeter can be achieved with small lasers.

The topic of spot intensities leads to a consideration of the geometrical differences between laser radiation and incoherent radiation. Most incoherent sources (lamps) emit their light in a more or less spherically symmetric (isotropic) distribution. Lasers do not. The light from a laser is highly directional with beam divergences of several milliradians or less. A typical argon ion laser operating at a wavelength of 5.14.5 nm and 10 W has a beam diameter of about 2 mm. The power density in this beam is more than 300 W/cm^2. If the beam diameter is reduced by a simple focusing element, the power density will be inversely dependent on the square of the radius of the beam and can achieve very high values at small spot sizes. The highly directional nature of the laser beam makes it easy to construct optical systems which shape the beam and direct it with very little energy loss. The optical geometry of an incoherent source is much less convenient and efficient. It is virtually impossible to collect all of the light from an isotropic source and direct it to a target.

Lasers can also be very different from incoherent sources in their temporal behavior. The fastest pulsed lamp systems operate on a microsecond timescale. Pulsed lasers have pulse widths as broad as milliseconds and as narrow as tens of femtoseconds (10^{-14} s). The repetition rates of the laser pulses can vary from 1 to 10^{10} Hz.

The variety of available lasers makes the process of matching the laser wavelength with

TABLE 3
Common Lasers with Wavelengths Useful for Photopolymerization

Laser	Wavelength (or range in nm)	Mode (CW/pulsed)	Power (J/pulsed, W/CW)	Price (or range $k)
Argon ion	457—530	CW	0.015—30	6—60
Dye	400—1000	CW[a]	0.5—1.5	8—50
Dye	350—1000	Pulsed[b]	10^{-4}—20	5—60
Excimer lasers				
F$_2$	157	Pulsed	0.01—0.3	
ArF	193	Pulsed	0.0001—0.6	
KrCl	222	Pulsed	0.1—0.25	
KrF	248	Pulsed	0.02—15	
XeCl	308	Pulsed	0.002—1.5	
XeF	351	Pulsed	0.0002—1.0	
Copper vapor	511,578	Pulsed	40 W @ 5 kHz	

[a]   CW Dye lasers require a CW pump laser, e.g., an argon ion laser.
[b]   Some pulsed dye lasers require a pump laser, e.g., a nitrogen, Nd YAG, or excimer laser.

the absorption band of a photochemically active molecule a fairly easy (but possibly expensive) task. There is a wide variety of lasers that offers a selection of wavelengths which cover the region for the IR at 16μm to the UV at 194 nm in a continuum. This covers the wavelengths most commonly used for photochemistry and analytical spectroscopy. There is, however, a caveat that must be considered. While there may be laser radiation available at these wavelengths, this does not imply that the laser will be economically efficient at a specific wavelength.

The wavelength range of greatest interest in photopolymerization extends from the blue-green into the UV (500 to 190 nm). A number of lasers have usable output in this wavelength range. Table 3 lists the laser types which cover this region of the spectrum, as well as some of their important characteristics.[6] The highest average and peak powers are those associated with the excimer lasers. These lasers are not tunable and have narrow linewidths. As a result, they cover only a very small portion of the spectral region of interest. If it is desirable to use an excimer laser, the target molecule must be chosen with care and tailored to absorb the radiation of the laser. The excimer lasers, as a class, are the highest efficiency lasers in this wavelength region. As a result, they will be the lasers of choice in the immediate future for carrying out process photochemistry in the near UV and UV.

If the photopolymerization process under consideration does not require high average power and/or efficiency, there are several other options, depending on the desired wavelength. In the visible and near UV (400 to 250 nm) there are tunable dye lasers which can be tuned over the entire range by combining fundamental mode with second harmonic (frequency doubled) operation. The widest continuous tuning range found in dye laser systems is supplied by the pulsed dye lasers, which can be driven by flash lamps, excimer lasers, or a variety of YAG laser harmonics. These systems offer a wide tuning range, excellent beam quality, and high peak powers. However, the wall plug efficiencies (conversion of electricity to usable laser light) for dye lasers are often very poor. Flash lamp-pumped dye lasers can be an exception.

There are instances in which a thermal initiation, as opposed to a photolytic initiation, would be the preferred polymerization mechanism. The obvious laser choice in such a situation would be an IR laser. The best developed technology in IR laser systems can be found in CO_2 and Nd:YAG lasers. A large number of industrial machining and shaping systems have been built using these lasers. From a processing systems point of view these systems tend to be the most advanced and reliable. The CO_2 laser has its most efficient output in the vicinity of 10.6 μm, but can produce significant output over most of the region from 9 to 11 μm. Many liquid monomer systems have enough absorption in this region to be heated effectively by a CO_2 laser. It is often more difficult to employ a Nd:YAG laser for this purpose. The wavelength produced by Nd:YAG lasers falls at 1.06 μm, a wavelength which overlaps harmonics of the fundamental vibrational frequencies of only some molecules. These absorption features are also much less intense than the fundamental absorptions and are less efficient at coupling laser energy into the system.

In spite of the obvious optical advantages of the laser, when compared to incoherent energy sources, there are drawbacks which must be given serious consideration in developing a polymerization process based on a laser system. Factors such as capital investment and operating efficiency are areas in which lasers compare unfavorably with other energy sources. Another factor which is important is the technical expertise required to keep a laser-based system operational. In addition, laser systems are more complex than most other energy sources and, therefore, tend to have more "down" time and require more maintenance. These types of factors will influence the application of lasers to polymer processing. The subsequent sections will describe a variety of types of photopolymerization reactions and processes to which lasers have been or might be applied. On that basis some conjectures about further developments in the application of lasers to polymerization processes will be made.

IV. LASER PHOTOCHEMICAL REACTIONS AND THEIR APPLICATIONS

Photopolymerization of monomers is conventionally classified into four types: (1) polymerization by direct irradiation of monomers; (2) polymerization in the presence of photoinitiators; (3) photoinduced charge-transfer polymerization; and (4) four-center type polymerization in the solid.

Direct irradiation of aryl vinyl monomers under deaerated conditions generally leads to cyclodimerization or free radical polymerization. For example, styrene or vinylnaphthalene produces head-to-head cyclodimers as main products. With regard to photopolymerization of type 2, photoinitiators absorb photoenergy, giving rise to either free radical or cationic species via homolytic bond cleavages, hydrogen abstraction, or electron transfer. In photoinduced charge transfer polymerization, the monomers are involved in charge transfer interaction. Photopolymerization of monomers using initiator systems may also involve electron transfer in the process of the generation of active species for polymerization.

A. LASER-INDUCED CROSSLINKING POLYMERIZATION FOR PHOTORESIST APPLICATIONS

Light-induced crosslinking polymerization of photosensitive resins is important to the surface protection of materials because of the rapid cure and excellent quality of the coating obtained.[1,3] Crosslinking polymerization is particularly appropriate for microlithographic applications in the production of high-resolution relief images that are needed in the manufacture of integrated circuits or printing plates. The source of UV radiation most commonly used is the medium pressure mercury lamp with which power densities of 30 W/cm^2 can easily be achieved in UV curing equipment by means of semielliptical focusing reflectors. Higher light intensities and therefore higher cure rates can be obtained by replacing the mercury lamps by powerful lasers that emit in the UV range and that are now commercially available.

The photocrosslinkable systems that involve the polymerization of vinyl monomers usually consist of three main components: (1) a multifunctional oligomer that constitutes the backbone of the network, (2) a reactive monomer that acts as a diluent, and (3) a photoinitiator that, by action of light, decomposes into reactive species.

Decker[7-10] has evaluated several formulations comprising these oligomers: epoxy-di-acrylate compound derived from the glycidyl ether of bisphenol A (Ebecryl 605A), an aliphatic polyurethane acrylate (Setalux UV 2260), a polyester hexa-acrylate (Ebecryl 830), and three monomer diluents, tripropyleneglycol diacrylate (TPGDA), trimethylolpropane triacrylate (TMPTA), and pentaerythrito triacrylate (PETA). Five photoinitiators studied were 2, 2-dimethoxy-2-phenylacetophenone (Irgacure 651 from Ciba Geigy), 1-benzoyl-cyclohexanol (Irgacure 184 from Ciba Geigy), 2, 2, -dimethyl 2-hydroxyacetophenone (Dar-ocure 1173 from Merck), benzophenone, and 2-chlorothioxanthane (both from Adlrich). The last two photoinitiators were associated with a hydrogen donor compound, *p*-dimethyl amino-ethyl-benzoate (DMABE) at a concentration level of 2 to 5%. These formulations were cured with an Argon ion CW laser (363.8 nm) and an N_2 pulsed laser (337.1 nm). Photocrosslinking polymerization proceeded as follows:

Cross-linked polymer

The multiacrylate resins were found to polymerize within a few milliseconds upon

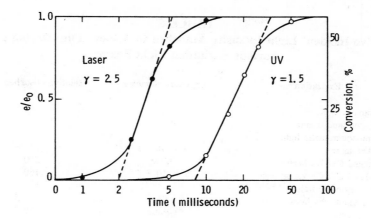

FIGURE 2. Normalized thickness of the polymerized film vs. duration of the exposure to the Ar⁺ laser or UV beam.

TABLE 4
Rates (R_p) and Quantum Yields (ϕ_p) of Polymerization of Multiacrylate Photoresists

Type of UV source	Light intensity ($Es^{-1} cm^{-2}$)	R_p(mol $l^{-1} s^{-1}$)		ϕ_p(mol E^{-1})	
		Epoxy acrylate	Polyester acrylate	Epoxy acrylate	Polyester acrylate
Hg lamp $\lambda > 250$ nm	1.5×10^{-6}	5.8×10^{2}	—	2900	—
Ar⁺ laser $\lambda = 363.8$ nm	4×10^{-6}	2.0×10^{3}	2.5×10^{3}	1700	2100
	8×10^{-4}	1.2×10^{5}	—	600	—
	1.2×10^{-2}	1.6×10^{6}	2×10^{6}	540	700
N₂ laser $\lambda = 337.1$ nm	1.1	1×10^{8}	1.7×10^{8}	420	550
	12	$\sim 10^{9}$	—	400	—

exposure to 363.8-nm emission of an argon-ion laser; the crosslink density of the totally insoluble polymer formed was estimated to be 5 mol l^{-1}. The kinetic curves obtained from the normalized thickness of the insoluble polymer film plotted against the irradiation time showed the induction period in the early stages due to the presence of atmospheric oxygen which inhibits these radical-induced polymerizations (Figure 2). The induction period disappeared if the irradiation was performed under a dry nitrogen purge. The kinetic curve obtained by argon ion laser has been compared with the curve obtained by using a powerful and highly focused medium pressure mercury lamp. The reaction proceeds four times more slowly with a mercury lamp, as expected from the decrease in light intensity.

When a pulsed nitrogen laser was used, polymerization developed even more rapidly. In air, a few shots, 8 ns wide, proved to be sufficient for a complete insolubilization of the crosslinked polymer network in acetone solvent. In a nitrogen atmosphere one single shot from a nitrogen pulsed laser at 337.1 nm was sufficient for complete insolubilization. Table 4 summarizes the rates of polymerization (R_p) and quantum yield (ϕ_p), which is the efficiency of UV photons to initiate the polymerization, for two formulations using TMPTA and Irgacure 651. The rate of polymerization increased almost as fast as the light intensity. Even at very high intensities the multiacrylate polymerization proceeds at close to a first-order kinetic law. This work was conducted to determine the utility of lasers in microlithographic applications. It was concluded that it is possible to cure multiacrylate photoresists at extremely high light intensities, as with pulsed lasers, without losing much of the quantum efficiency

TABLE 5
Relative Incident Light Intensity Absorbed to Achieve Film Drying under Various Excitation Light Beams

Excitation	Exposure duration	Relative absorbed dose
Polychromatic light		
2-KW continuous lamp	10^{-2} s	1
Visible monochromatic light		
CW Ar$^+$ laser	3×10^{-2} s	210
Modulated CW Nd laser	0.7 s[a]	550
Modulated CW Cu vapor laser	5×10^{-2} s[b]	160
UV monochromatic light		
Mode-locked ruby laser	1 Shot	2.5
Excimer laser	1 Pulse	0.7
Nitrogen laser	1 Pulse	0.5
Continuous Ar$^+$ laser	10^{-2} s	1

[a]   During a 0.7-s exposure, the system has received 8750 pulses of light which yields an irradiation time of 1.3ms.

[b]   True irradiation time: 6.2 μs.

or photosensitivity. High-resolution relief images can be obtained by a single laser pulse, thus making feasible ultrafast photoimaging for microcircuit applications.

A photopolymerizable composition containing unsaturated monomers and a tertiary amine or a disubstituted aminoaryl carbonyl compound have been shown to have high sensitivity to argon laser.[11] A mixture of 2-(benzoylmethylene)-3-ethyl-5-(3, 5-dimethylbenzothiazol-2-ylidene) thiazolidin-9-one, triethanolamine, and chlorinated polyethylene showed high argon laser sensitivity, curing at an irradiation intensity of 1.81×10^5 erg/cm^2.

There are some reports[12] about polymers which are sensitive to visible light emitted by lasers. These polymer systems have been explored as argon laser photoresists. Preparative methods to introduce the photosensitive phenylenediacrylate (PDA) in the polymer chain have been explored. Among these, the substitution reaction of poly(chloromethylstyrene) (CMS) copolymer with the potassium salt of the monoester of phenylenediacrylic acid (PDAA) was found to be better due to the availability of the potassium salt and the physical properties and storage stability of the resultant polymers. The polymers, thus prepared, demonstrated high sensitivity to 488-nm Ar laser light with a quantum efficiency for crosslinking of 0.44 when sensitized with ketocoumarin. Preliminary applications to holographic recording have also been developed.

Fouassier et al.[13-15] have done an extensive study of laser photoinduced polymerization by using a variety of photoinitiators and UV/visible laser excitation sources. The experiments were conducted on 30-μm films of monomers containing photoinitiators. The two formulations evaluated were (1) PETA + 5% DMPA and (2) PETA + EOSIN + methyldiethanolamine + PDO. The reaction was followed by IR spectroscopy. Relative incident light intensity absorbed to achieve the drying of the film under various excitation light beams is shown in Table 5. Visible laser irradiation resulted in a sharp decrease of the photosensitivity (by a factor of about 300), which is related to the low efficiency of the energy transfer processes taking place between the dye and the photoinitiator and to the poor efficiency of the overall processes between the dye and the amine. The average quantum yield corresponding to the disappearance of the double bonds of the monomer shows that much has still to be done to improve the efficiency of visible photosensitive systems. However, the system developed for visible light has been shown to be quite competitive as far as the exposure duration is concerned.

The time-resolved laser spectroscopy led to the conclusion that the higher reactivity of

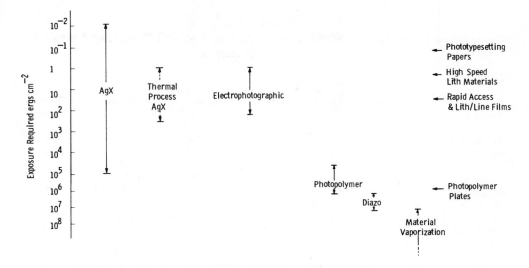

FIGURE 3. Chart showing the range of sensitivities (in ergs/cm²) of various photosensitive systems.

multifunctional resins upon UV and visible light irradiation may be accounted for by chemical and physical effects, e.g., macromolecular structures, viscosity, matrix effect, diffusion processes, and polymer chain motions, rather than by an enhancement of the efficiency of the photoinitiation processes. Moreover, only weak interactions are suspected to take place between the excited state of the photosensitizer and the initiator, hence, an inefficient energy "channeling" and the poor reactivity of such systems when used under visible laser irradiation.

B. LASER USE IN THE PRINTING INDUSTRY

The important features[16] of laser exposure from the point of view of photosensitive imaging systems are

- the duration of a spot exposure is typically 10^{-6}s or less
- the exposure of broad areas of the light-sensitive material is accomplished by scanning the spot across the material.

A complete image may take up to 1 min or more for complete exposure. Although individual areas are only exposed for a microsecond or less, the time to expose an image will depend on a number of factors, including spot size and scan rate. The spot size can vary from 0.001 to 0.010 in. and the number of scan lines from 200 to 1200 lines per inch or more.

Many systems have been used for laser imaging. Figure 3 charts the range of sensitivities (in ergs /cm²) of various systems which have been used in graphic arts or printing. The nature of laser light allows some of the less-sensitive materials to be used.

Some of the lasers used to expose these materials are

NdYAG and CO_2	IR wavelengths
Helium neon	Red wavelength
Helium cadmium	Blue wavelength
Argon ion	Blue and UV wavelengths

Several workers have used acrylamide and other crosslinking polymers to form negative images for printing industry. The above-mentioned systems are UV sensitive, but the spectral sensitization into the visible region is being pursued. Iwata et al.[17] have developed polyesters

TABLE 6
Comparison of the Efficiency of Polymerization by Laser and Mercury Lamp UV-Induced Polymerization

Monomers[a]	Monomer feed ratio	Argon laser irradiation		UV initiation		Efficiency laser/UV
		Energy (J)	Copolymer yield/kJ (%)	Input (J)	Copolymer yield/kJ (%)	
Sty/MAH	4.5:1	900	29.7	7560	0.03	990
Sty/MAH	9:1	1560	7.8	7560	0.003	236
		—	—	18,900	0.047	—
VT/MAH	4:1	858	28.8	7560	0.05	576
VT/MAH	8:1	1408	15.5	7560	0.047	330
t-BS/MAH	3:1	1560	17.5	7560	0.07	250
t-BS/MAH	6:1	1560	9.5	7560	0.013	1192
CHO/MAH	4:1	240	308	408	2.7	114
CHO/MAH	2.6:1	180	473	1,021	2.55	185

[a]   Sty — styrene, VT — vinyltoluene, t-BS — t-butylstyrene, CHO — cyclohexenoxide, and MAH — maleic anhydride.

of p-phenylene bis (α-cyanobutadiene carboxylic acid). The sensitivity defined as the exposed energy sufficient to get a spot diameter equal to that of the laser beam was found to be 1.5 mJ/cm². In regard to laser printing, its sensitivity is high enough to permit its use as a material sensitive to a relatively low power argon ion laser (5-W, 488-nm, 1 .25-mm beam diameter).

C. LASER-INITIATED POLYMERIZATION VIA CHARGE TRANSFER COMPLEXES

At Westinghouse we have explored the applicability of the laser-initiated polymerization of charge-transfer monomer systems. The photochemical initiation in these cases involves the formation of charge transfer complexes. These charge transfer complexes are formed by the removal of one electron from the donor organic compound (D) in the presence of an electron acceptor molecule (A):

$$D + A \rightarrow (D....A) \rightarrow (D^{+\delta}....A^{-\delta}) \rightarrow D^{+} \cdot \overline{\cdot A}$$

When a polymerizable monomer is directly involved in the electron transfer step, either as the acceptor or the donor, radical ions can be produced which can lead to polymer formation. In the instance of two different monomers being utilized, one an acceptor and the other a donor, an alternating 1:1 copolymer can result.

Using a pulsed nitrogen laser and an argon ion laser, 24 donor-acceptor monomer charge transfer systems were screened for polymerization in different solvents.[18] Polymerization by laser initiation was achieved in only two systems, i.e., 2-vinylnapthalene/fumaronitrile and 9-vinylanthracene/fumaronitrile. The best polymer yields were obtained with the 2-vinyl-naphthalene/fumaronitrile system in sulfolane solvent. The polymer contained a high percentage of sulfolane, presumably arising from solvent transfer to the growing polymer chains during the propagation phase of the polymerization. A detailed study[19] of the argon ion laser-initiated polymerization of styrene/maleic anhydride, vinyltoluene/maleic anhydride, and t-butylstyrene/maleic anhydride showed that the copolymer yield is directly proportional to the laser irradiation time and the molar concentration of maleic anhydride in the feed. Enhanced rates of polymerization were obtained by increasing the electron-donating capabilities of the donor monomer. The argon ion laser-initiated polymerization in these systems was found to be more energy efficient than UV/visible-induced polymerization, as measured by the energy absorbed by the polymer system (Table 6).

An argon ion laser was also shown to cure epoxies in the presence of maleic anhydride.[20] Short irradiation times of - 1 to 3 -min duration gave very high yields of epoxy polymer. IR spectral studies and quantitative chemical analyses of the polymer products indicated the formation of polyether linkage at lower levels of conversion and an adduct of polyether and maleic anhydride at higher polymer conversions. The UV-visible spectroscopic results indicated that the polymerization was initiated by the excited charge transfer complex between the electron donor, cyclohexene oxide, and the electron acceptor, maleic anhydride. Laser-initiated polymerization of cyclohexenoxide/maleic anhydride was shown to be several hundred times more efficient than UV-initiated polymerization, as measured by the energy absorbed by the polymer system (Table 6). The mechanism of polymerization has been suggested on the basis of the following observations:

1. Development of yellow color appears after irradiation with argon ion laser.
2. UV spectra show the presence of a charge transfer complex.
3. Inhibition of polymerization is observed in the presence of a small amount of triethylamine in the monomer solution before irradiation.
4. Rate of polymerization is enhanced by the addition of a Lewis acid.
5. There is a higher rate of polymerization in polar solvent as compared to nonpolar solvent.
6. Polymer obtained in the beginning (irradiation time, about 90 s) shows the presence of polyether linkage only, and higher irradiation time produces an adduct of polyether with maleic anhydride.

Photodissociation of some of the polymers has been explained by a simple charge transfer mechanism by making use of nanosecond laser photolysis, ion photodissociation of poly(N-vinylcarbazole) quenched by N,N-dimethyltoluidine, 1, 4 dicyanobenzene, and dimethyl terephthalate studied in dimethylformamide by Masuhara et al.[21] Nanosecond laser photolysis showed that the dissociation yield decreased as the degree of polymerization increased. The results have been discussed in terms of simple electron donor-acceptor systems.

IV. THERMAL EFFECTS OF LASERS IN CURING AND CROSSLINKING OF POLYMERS AND THEIR APPLICATIONS

Thermal energy of lasers can be effectively used for activating a free radical polymerization or for crosslinking a polymer. The driving force for the potential use of lasers is the need for reducing the hardening time of adhesive materials. Lasers offer advantages in providing localized heating, and they are capable of curing materials in a complex configuration and hard to reach places. Some applications of the thermal effects of lasers in curing and crosslinking of polymers are summarized in the following section.

A. HIGH-SPEED HARDENING/CURING OF RESINS BY LASER IRRADIATION

It has been 25 years since the creation of the laser, touted as the greatest invention of this century. During this time various kinds of lasers and their application techniques have been developed, and the areas of laser applications are expanding every year. For example, the CO_2 laser and the Nd:YAG laser have found practical applications as energy sources in laser processing; semiconductor lasers have been used as light sources for optical fiber communication, while He-Ne lasers and Ar ion lasers have been used as light sources in data processing equipment. Applications for medical and therapeutic instruments have arisen in recent years, and various therapeutic systems, such as a laser retina-freezing apparatus, laser scalpels, and laser surgical microscopes have been developed. Diagnostic techniques utilizing lasers have also made progress. In dentistry, painless removal of carious portions by laser irradiation has been studied,[22,23] and some clinical applications have been realized.

In many industries which utilize adhesive materials, the need for reducing the hardening time of adhesive materials is increasing every year due to high-speed conveyor lines. Although quick hardening is possible in cyanoacrylate or hot melt-type adhesives, the adhesives may not be suitable for certain temperature-sensitive substrates. At present, UV light or electron beams are used for the high-speed hardening of polymeric materials. These polymers have attracted attention because, in the hardening process, they are the sole recipients of the applied energy. They can be utilized for materials which do not withstand heating, such as paper, wood, and plastic, and because of their energy saving features. However, the UV hardening method has problems; for example, the light transmission of these materials is low, the thickness of the hardened layer is restricted, and the method cannot be applied to opaque materials. For this reason, applications are limited to the hardening of thin films. The electron beam hardening method is confined to the hardening of thin films, and it is reported that an inert gas is necessary during hardening. Also, precautions against secondary X-rays and ozone are necessary.

On the other hand, in the laser hardening method, light penetration into the material is excellent with some lasers, and the hardened layers can attain substantial thickness. Applications are not confined to the hardening of films and may be expanded to areas where a substantial thickness is required. Since lasers have high directional specificity, they can be used for spot irradiation of a material to be hardened. Moreover, since hardening can be achieved within a short time, thermal effects on other materials are minimized. Thus, applications will be possible for materials having poor heat resistance. In addition, since remote-controlled or scanning type lasers are possible and laser applications are easily automated, there may be considerable merit in certain applications.

In recent years, many acrylic oligomers having epoxy, urethane, polyester, polyol, silicone, polybutadiene, polyacetal structures, etc. within their molecular skeleton have been developed; they have acryloyloxy or methacryloyloxy groups in the terminal or side chain of the molecule and lend themselves to high-speed hardening by radical reaction, as do common acrylic or methacrylic monomers. These materials may be given various properties depending upon the molecular structure which constitutes the skeleton, and thus are attractive materials.

Materials to which laser hardening is applied generally have the following composition:

- Acrylic or methacrylic monomers or oligomers
- Acrylic or methacrylic monomers for dilution purposes
- Polymerization initiators
- Additives (reaction promoters, stabilizers, etc.)
- Coloring agents, fillers

Hardening behavior under laser irradiation varies to a large extent with the kind of laser used. Applications as energy sources for hardening can be expected from Nd:YAG lasers (solid-state lasers) and CO_2 lasers (gas lasers). These lasers are widely utilized in industry for material processing, etc., and further developments are expected. The preferred laser is Nd:YAG (wavelength 1:06 μm) in which the laser light can be guided through a flexible glass fiber, thus affording the potential for easily irradiating a location with restricted access.

Several resin systems have been evaluated for their hardening characteristics,[24] and their applications in dentistry and microelectronics using ND:YAG lasers. For these studies, the experiments were conducted in Teflon containers (4 × 5 mm). The samples, prepared by adding benzoyl peroxide and fine silica powder to various kinds of methacrylate-type monomers or oligomers, were irradiated by a perpendicularly incident laser. The percentage of hardening was determined by the following formula:

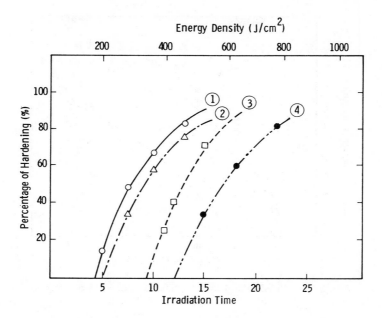

FIGURE 4.   Relationship between irradiation time and percentage of hardening (1) Bisphenol-A diglycidyl methacrylate type, (2) triethylene glycol dimethacrylate type, (3) polyester diacrylate type, and (4) polyurethane diacrylate type.

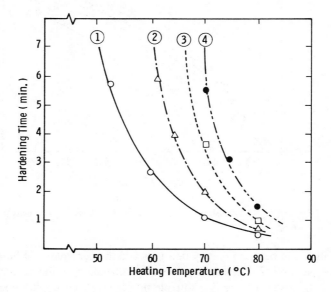

FIGURE 5.   Reactivity of each material (the numerals in the figure are the same as those in Figure 4).

$$\text{Hardening (\%)} = \frac{\text{weight of hardened material}}{\text{weight of packed material}} \times 100$$

The relationship of percentage hardening to irradiation conditions, reactivity of each material, irradiation time, and the energy density are shown in Figures 4 to 7. It is clear from these figures that materials having higher reactivity have a better laser hardening property. Figure 6 suggests that the hardening time can be shortened by increasing the laser

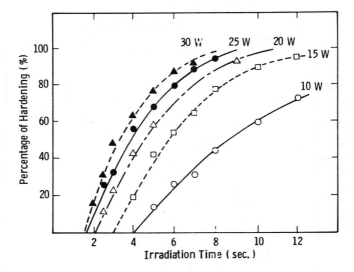

FIGURE 6.   Relationship between irradiation time and percentage of hardening.

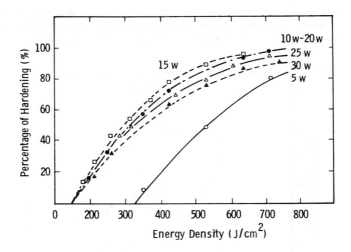

FIGURE 7.   Relationship between energy density and percentage of hardening.

power. The relationship between energy densities and the percentage of hardening is shown in Figure 7. The percentage of hardening can vary depending on the power, even when the energy density is the same. There is a suitable power level, depending upon each material to be hardened; in the case of material 1 (Figure 4) the best power is 15 W.

From the standpoint of energy savings and safety, it is preferable to conduct laser hardening with the lowest possible energy. Increased concentration of initiator, such as benzoyl peroxide in dimethacrylate-type polymers, improves the laser hardening property. However, since the increase in the amount of initiator naturally entails certain disadvantages, such as a decrease in the degree of polymerization of the hardened material, there is a limit to the amount of initiator added. One can achieve the energy savings by adding a polymerizable accelerator, a coloring agent, and an absorptive material, which can enhance the laser absorption efficiency and substantially improve the laser hardening property.

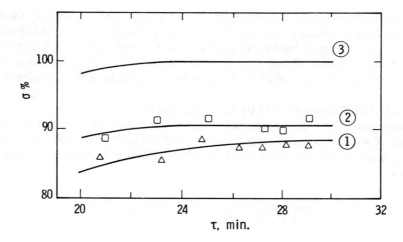

FIGURE 8.   Dependence of the degree of polymerization σ of ML12 enamel on the drying time under identical thermal conditions. (1) Contact heating; (2) contact heating and illumination at λ = 3.066 (average power = 25 MW/cm²); and (3) contact heating and illumination at λ = 3.066 (average power = 0.25 to 0.5 W/cm²).

Laser hardening has excellent features not exhibited by conventional hardening methods, and the use of the method is expected to increase. Future applications of thermal laser hardening will be in fields in which only laser hardening is effective or in which secondary benefits can be expected from the laser hardening method.

B. RESONANT ACTION OF LASER RADIATION ON THE POLYMERIZATION PROCESS

Kal'vina et al.[25] suggested that the most promising method of achieving selected chemical reactions in polymers is the resonant interaction between IR laser radiation, and the vibrations of those end groups whose dissociation results in polymerization. However, in spite of the wide range of wavelengths which can be generated by the currently available lasers, it is difficult to select a laser line which would satisfy two requirements simultaneously. One is the coincidence with the vibration frequency of the selected degree of freedom of a molecule, and the other is a sufficiently high power level to cause selective heating. Following are the two examples in which the CO_2 laser (λ = 3.006μm) has been used.

1. Hardening of Enamel

The formulation of a three-dimensional structure in enamels based on alkyd resin involves functional hydroxyl (OH–) and carboxyl (COOH–) groups for which the wavelength of the valence vibration is 3 μ. Radiation from a pulsed gas laser (average power 25 mW) was focused by a lens on a metal substrate carrying a film of the material. This film was heated by the contact method. The temperature of the sample was measured with a thermocouple. A film 55 to 60 μm thick of alkyd resin was deposited on a metal substrate. Then the substrate temperature was raised to 120°C, and the gel fraction was determined in the illuminated parts of the film. This temperature was maintained the same in all of the experiments, and the gel fraction was extracted from each film by immersion for preset time in a solvent.

Figure 8 shows typical results obtained in the hardening of the ML12 enamel. The degree of polymerization was plotted against the drying time. Curve 1 was obtained by contact drying of the film, curve 2 by the combined effect of contact heating and laser radiation with an average power density 25 mW/cm², and curve 3 was obtained by the combined action of contact healing and laser radiation with an average power density of the order of 0.25 to 0.5 W/cm^2 (focused radiation). The laser energy on its own, without simultaneous heating, was insufficient to harden the polymer. Also, illumination of the film with laser radiation of wavelength 2.02 and 1.5 μm of the same power density as that of the $\lambda = 3.066$ μm radiation failed to change the range of hardening or the degree of polymerization.

2. Imidization of Polyamide Acid by Laser Radiation

Monochromatic ($\lambda = 3.066$ μm) laser radiation was used for imidization[26] (dehydrocyclization) of a polyamide acid prepared from dianhydride of pyromellitic acid and 4, 4'-diamino diphenyl ester in dimethylformamide. The imidization occurred in accordance with the equation

The existing heat treatment methods can ensure only 85% completion of this reaction, which reduces the performance of the final product with respect to its heat stability, strength, dielectric, and other properties. Laser radiation ($\lambda = 3.066$ μm) matched the valence vibration frequency of CONH at 3.048 μm. The changes due to imidization of the polyamide acid were determined by IR spectroscopy. It has been concluded that the resonant action of laser radiation on polyamide acid films increased the number of imide rings considerably and consequently altered the physicochemical properties of these films during subsequent heating.

While it is possible to choose a polymerization initiator with an absorption that matches an IR laser wavelength, chemistry induced by selective bond dissociation is almost certainly not the mechanism by which a reaction will proceed. More recent studies of the IR laser-induced chemistry in large molecules[27] indicates that there is little chance of bond selective chemistry in a large molecule in the gas phase. The liquid phase is an even less likely environment.

In a large organic molecule (ten atoms or more or a density of rotational states of 10^3 to 10^4 cm^{-1} at room temperature), it is not possible to accumulate a large amount of vibrational energy in a specific bond. In bond-selective laser photochemistry, unimolecular dissociation of the target molecule at the laser-excited bond is the goal. As a result of the large state density, excess vibrational energy is rapidly delocalized over the entire molecule. If enough energy is absorbed by the molecule, dissociation will occur through the bond with the lowest thermal stability (the lowest available thermal activation energy channel). There are other complications brought about by the rapid loss of energy (equilibration) of the excited molecule to surrounding molecules.

As a result of these observations, IR lasers appear to provide only a heat source for the initiation of polymerization reactions. Once the laser has supplied enough energy to raise the molecule and its surroundings to a threshold temperature, the initiation step occurs, and polymerization proceeds.

VI. MISCELLANEOUS

A. ABLATION OF POLYMERS BY LASERS

This topic has been described in detail in Chapter 5. We intend to summarize some of the potential applications of the ablation of polymers by lasers. When pulsed UV laser radiation falls on the surface of a polymer, the material at the surface is spontaneously etched away to a depth of 0.1 to several micrometers.[28] In the process the depth of etching is controlled by the width of the pulse and the fluence of the laser, and there is no detectable thermal damage to the substrate. The material that is removed by etching consists of products ranging from atoms to small fragments of the polymer. They are ejected at supersonic velocities.

The obvious application of UV laser etching is in photolithography.[29,30] This process is used widely in semiconductor processing to pattern silicon, oxide, or metal surfaces. A light-sensitive polymer film called "photoresist" acts as the medium for patterning. In present practice, after exposure to a light source that registers a pattern, the photoresist is developed by wet chemical methods. It has been demonstrated that photoresists can be patterned by means of proximity printing[31] or projection printing with UV laser radiation. Resolution of 0.3 µm has been claimed. Other attractive applications at present are those in which UV laser ablation offers unique possibilities. For example, polyimides are industrially important because of their toughness and electrical properties. However, their resistance to chemicals and heat make them difficult to pattern by conventional methods. UV laser etching offers a convenient solution. An important advantage of UV lasers (in contrast to an IR laser) is the lack of carbonization at the new surface that is created, since a surface film of carbon alters the electrical properties of the film in an undesirable way.

IR laser ablation of polymers has been studied by several workers[32] with an intent to achieve pyrolysis with controlled amounts of energy supplied to a specific area of the sample. This has its merits in characterizing polymers.

B. IRREVERSIBLE SOFTENING OF THERMOSET POLYMERS

Allyl diglycol carbonate is a very useful thermoset plastic, but is very difficult to process due to its hardness and brittleness. Conventional methods of machining the thermoset plastic produce cracks, chips, and vents, or lack control, precision, reproducibility, and cleanliness. A low-powered (5- to 15-W) CO_2 laser coupled with astigmatism-free beam-focusing mirrors for processing thermoset plastic such as CR 39 (PPG) eliminates all the above-mentioned problems and offers additional advantages like high speed, adaptability to automation, etc.[33,34] The high efficiency of the CW CO_2 laser (>15%), simple fabrication and operation, and very strong absorption of its 10.6-µm radiation by CR-39 make it the most suitable laser for this application.

For the purposes of this experimental study the power of the CO_2 laser was reduced by changing the gas composition and current to produce a net 5 to 15 W in the focused laser beam. The reason for using low laser powers was to keep the power density below a value at which other effects like uncontrolled combustion of the plastic, a laser-supported absorption wave, and nonlinear phenomenon, etc. may not initiate, even at the minimum possible interaction time. The arrangement for directing and focusing the laser beam is shown in Figure 9. The focused spot diameter of the laser beam was 225 µm and depth of focus about 5 mm. The CR-39 sheet was held horizontally on a turntable mounted directly on a variable speed stepper motor. The processing was carried out under ambient conditions, but to remove the gaseous products released by the sheet a suction arrangement was used.

The decomposition of CR-39 on interaction with the laser beam was evidenced by the deposition of a very small droplet of a yellow oily liquid around the processed region and white smoke spewed out during the processing. The products of laser-induced chemical and/or structural decomposition are gaseous, condensable gases and/or liquids. The exact nature of the decomposed product was not analyzed. The dimensions of the processed zone are observed to extend linearly with increasing power density and beam residence time, in the range of these parameters. The heat-affected zone (HAZ) around the processed region does not depend on the power density but increases with an increasing beam residence time. It

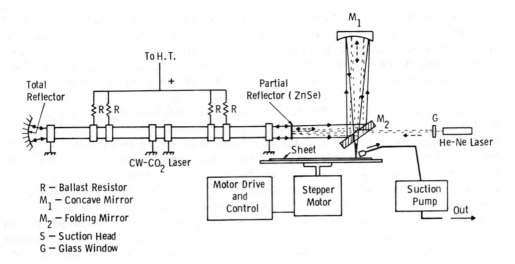

FIGURE 9.   Schematic diagram of the experimental setup.

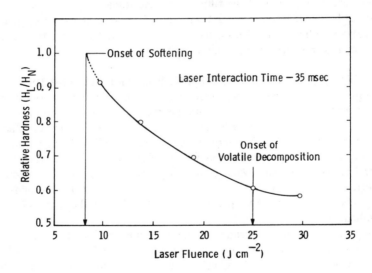

FIGURE 10.   Hardness of the laser treated CR-39 surface (H_L) relative to that of the untreated one (H_N) at various fluences of the CO_2 laser beam used for the treatment.

is therefore possible to get very small HAZ (~10 μm) at very high speeds of processing (~20 m/min).

Another major change which occurs on the surface of the thermoset material such as allyl diglycol carbonate (CR-39) after plasma treatment is the softening of the surface. This process may have practical significance in such applications as improving the printability on the polymer, stronger adhesion with resins, and suitable eye implants like contact lenses and artificial cornea. The relative hardness (H_L/H_N) of the polymer surface in separate pieces treated at different laser fluences is shown in Figure 10.

Under specific conditions of the treatment with respect to interaction time, power density, sample thickness, and wavelength of the laser, surface hardness of the allyl diglycol carbonate can be reduced to about 60% of its original value (Rockwell M95-100) before the onset of

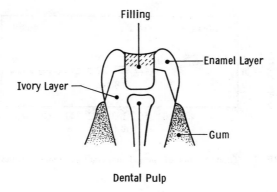

FIGURE 11.   Laser hardening utilized for a deep cavity.

optical degradation. From the experimental data and the theoretical results, it is concluded that the laser softening is basically due to the depolymerization, not to structural decomposition in the polymer. Rapid cooling immediately after the laser heating plays an important role in the formation of the porous heterogeneous microstructure in the treated polymer.

VII. POTENTIAL APPLICATIONS OF LASERS IN CURING RESINS

A. APPLICATIONS OF LASER HARDENING IN DENTISTRY

Metal amalgams have been used in the past as the primary filling material for repairing dental caries, but because the color does not match that of teeth or because of concern about mercury pollution, methacrylate type materials have chiefly been used in recent years. Two methods are being used at present for hardening this kind of material: room temperature hardening and photohardening. These methods are not exempt from problems. Room temperature hardening has the serious shortcomings of a short pot life and a limited working time. Due to the need for hardening the material within a short period of time (2 to 3 min) inside the mouth, a tertiary amine such as dimethyl-*p*-toluidine is added, although it has been contended that such materials damage the dental pulp. On the other hand, photohardening has no limitation with regard to working time because no hardening occurs until irradiation. However, the disadvantage is that the hardening is insufficient and a large amount of unreacted monomer remains in the internal portions inaccessible to light.[35] Such unreacted monomers reportedly irritate the dental pulp, causing serious clinical problems.

It is possible that all of these problems may be resolved if laser hardening methods are introduced. Indeed, there is no limitation with regard to working time because rapid hardening does not occur until laser irradiation is accomplished; furthermore, there is no need to use a noxious substance such as dimethyl-*p*-toluidine.

In dental applications, safety is important. For example, when laser hardening is utilized for a deep cavity, as shown in Figure 11, the concern arises that heat during hardening may be transferred to the neighboring dental pulp, causing it to be damaged. This problem can be resolved by adopting a method whereby only the upper layer portion of the filled material is rapidly hardened by the laser, while the inner portion is allowed to harden gradually inside the mouth. In this manner, the temperature increase inside the material is minimized, as is the effect on the dental pulp.

Figure 12 shows the internal temperature (3 mm below the surface) when a 10-W laser is used. In this case, the material used was prepared by adding a polymerization initiator and silica powder to a monomer comprised of Bisphenol-A diglycidyl methacrylate as a primary component (see Scheme I), which has a good reputation in dental applications.

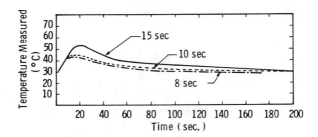

FIGURE 12.   Time variation of internal temperature (3 mm below the surface).

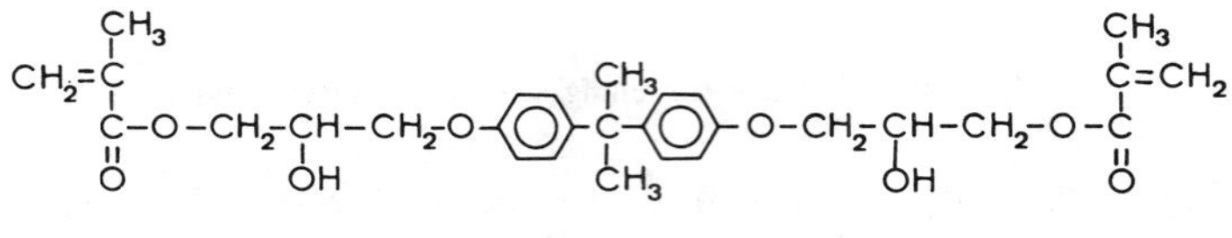

SCHEME I

In the experiment, the upper layer hardens completely after about 8 s of laser irradiation, and the maximum internal temperature during the experiment is about 42 to 43°C. This temperature increase is not seen to exert any appreciable effect on the dental pulp. However, since prolonging the irradiation time leads to higher internal temperatures, excessive irradiation should be avoided whenever possible. Excessive irradiation can be prevented by making an adjustment such that material is decolorized simultaneously with hardening so that hardening can be determined visually.

Safety precautions against mistaken irradiation of other tissues are also important. For this reason, it is necessary to minimize the irradiation power and energy. Thus, the addition of an absorptive substance is very useful. Since such a substance need not color the material, it can be used without any problem for dental materials for which natural tooth color is required. When such an absorptive substance is added, up to 50% hardening (about 2-mm hardening thickness) can be accomplished with a laser power of 2 W and an energy distribution of 100 J/cm^2 even without using a tertiary amine. If the material is colored with a suitable coloring agent, the energy can be lowered further. It has been confirmed that these powers and energy distributions have no effect, even when the laser impinges on teeth or skin.[36]

Another benefit to clinical dentistry can be expected from the application of the laser hardening method. There are reports that the acid resistance of teeth can be improved and dental caries prevented by irradiating teeth with Nd:YAG lasers. According to these reports, such effects can be obtained when laser light is directed at the rate of 100 J/cm^2 onto the surface of teeth which have been coated with a black ink.[36] In materials to which an absorptive substance has been added, this energy level is sufficient to accomplish hardening. At present the irradiation conditions for prevention of dental caries are not the same as those used for thermal hardening. Depending on future studies, the dream of simultaneous hardening and caries prevention may become a reality.

Another potential application of laser hardening can be in electroconductive adhesives used increasingly in microelectronic industry. Electroconductive adhesive materials consist of an electroconductive filler such as metal, metal-coated hollow beads, or carbon, and a binder which is a synthetic resin. Epoxy and acrylic materials are used as the binder. The time usually required to harden the binder is several hours for room temperature-hardened materials and less than an hour for thermally hardened materials; a cumbersome reaction step is required for heating. Thermally hardened materials cannot be used in conjunction

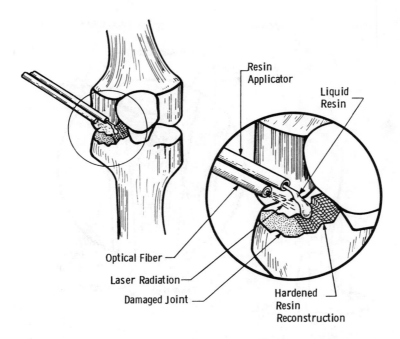

FIGURE 13. Repair of damaged joint surfaces by a laser-induced polymerization process.

with materials having low heat resistance. Laser hardening may be an ideal method whereby hardening is accomplished rapidly without thermal effects on material.

B. LASER-INDUCED POLYMERIZATION IN MEDICINE

During the last decade lasers have been tried as a tool in the treatment of many medical problems with varying degrees of success. Lasers have been used in the treatment of several common eye disorders, e.g., detached retinas and cataracts. They have been used, as mentioned earlier, in dental repair. Research continues in the evaluation of the utility of lasers in the destruction of arterial blockages. Laser-activated drugs are being tested for their efficacy in cancerous tumor reduction. Lasers are also being tested in the treatment of various skin disorders such as port wine marks. There is, as a result, a degree of acceptance of lasers as medical tools. There has been little activity, with the exception of the dental repair example, in the use of laser-initiated polymerization as a medical technique.

There are a number of applications envisioned for photochemically polymerized biocompatible polymers. Many of the applications would be surgical in nature. The ability to transmit laser radiation through fiber optics to very localized areas presents the possibility of coupling laser polymerization with microsurgical techniques.

There are some obvious candidates, such as coupling laser-assisted bone repair with arthroscopic surgical techniques. Repair of a damaged joint (such as a knee) can be visualized with aid of Figure 13. In this figure a probe is shown applying a biocompatible polymer to a damaged surface of the joint with a fiber optic probe directing the laser radiation onto the surface of the unpolymerized material. A repaired surface could be generated by polymerizing as many layers of the material as needed over the affected area. One might envision replacing or repairing damaged cartilage by the *in situ* grafting of an appropriate polymer to the bone surface in the same manner. Techniques such as these could have some distinct advantages when compared to current practices. The repair of a knee joint instead of the replacement of it is a much less traumatic procedure, both in the process and in recovery.

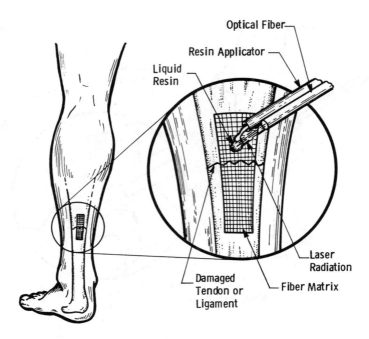

FIGURE 14.   Tendon repair using a laser-induced polymerization technique.

Another type of repair that could be made with laser-polymerized material is the repair or replacement of tendons and ligaments. Figure 14 shows a possible means by which such a repair could be carried out. In the figure, a web of an appropriate reinforcing material (e.g., carbon fiber web) is used as a matrix for additional strength. An appropriate resin is applied in layers and photochemically hardened by laser radiation from a fiber optic probe. An appropriate polymer which would bond strongly to existing tendon, ligament, and bone tissue would allow the repair to take place without sutures, staples, or other attachment devices.

C. CREATION OF SOLID BODIES BY LASER-INDUCED POLYMERIZATION

There is a growing interest in the direct production of solid bodies from computer-generated model data.[37] At present the only system by which a three-dimensional solid model can be created directly from a computer-generated data base accomplishes this in an awkward fashion. In order to create the model a reservoir of monomer is translated under a focused light beam or the light source is translated over the surface of the monomer solution. The light causes a polymerization reaction to occur wherever it touches the monomer solution. A polymerized pattern is thus developed in a thin layer of the monomer solution. It is then necessary to increase the level of the monomer solution after each pass of the light source over the solution surface. The solid object is completed by the layering of many computer-generated two-dimensional images one on top of the other. The thickness of each layer is controlled by adjusting the monomer solution depth over the most recently completed layer (see Figure 15). Since this system depends on a number of motions involving mechanical translation, it is inherently slower than a system of optically translated light beams would be. This has been seen in the comparison of the writing speeds of mechanically vs. optically scanned pattern generators. A mechanically scanned pattern generator may achieve linear writing speeds of meters per second, while an optically scanned system may attain speeds of hundreds of meters per second.

A less cumbersome system could be designed using a laser-based approach. There are

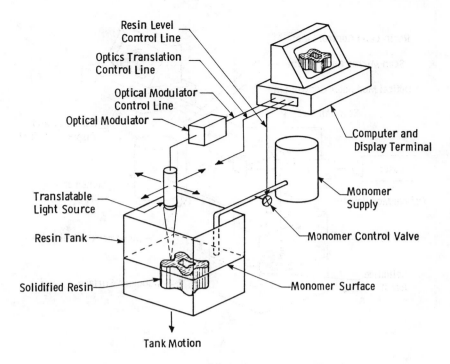

FIGURE 15.   Mechanically scanned, computer-driven 3-D solid pattern generator.

two areas in which a laser-based system would offer advantages, the speed of exposure of the individual layers and the layer height control. The hardware and software concepts already exist for laser pattern generation coupled with polymerization processes in photoresist exposure.[38] Westinghouse Electric Corporation and Eocom 2 have developed a computer-driven system for the direct exposure of photoresists on copper-clad circuit board substrates. The system is based on the use of an argon ion laser and a number of accousto- and electrooptic elements for beam conditioning and translation. By using this type of system instead of the mechanical translation system described above, the speed for the creation of each layer would be dramatically increased. Figure 16A and B indicates the manner in which a combination of optical and mechanical scanning could be used to accomplish the layer exposure.

Figure 16A is a straightforward exchange of a coherent light source for the original incoherent source. In addition to the change in light source, there is a change in the beam transport and optical modulating system. Scanning mirrors are necessary to move the beam over the surface of the monomer solution. An electro- or acoustooptic device is necessary to turn the beam on and off in order to select which areas to irradiate. It is still necessary in this system to change the height of the monomer solution surrounding the model as each layer is formed.

Figure 16B is representative of the elements needed for a system based on a two-photon photoinitiated polymerization reaction. This system would remove all mechanical motion of the model and require only the movements of scanning mirrors which would dramatically reduce the time to produce a model. The photochemistry involved in this type of system is more complicated than that required for the polymerization reactions previously used in these processes. In this two-beam process a polymerization initiator is used which is activated only after the absorption of two photons of different but known wavelengths. The envisioned process could be as follows:

$$In + h\nu_1 \rightarrow h\nu_2 \rightarrow In^{**} \rightarrow In^{\cdot} + X^{\cdot}$$

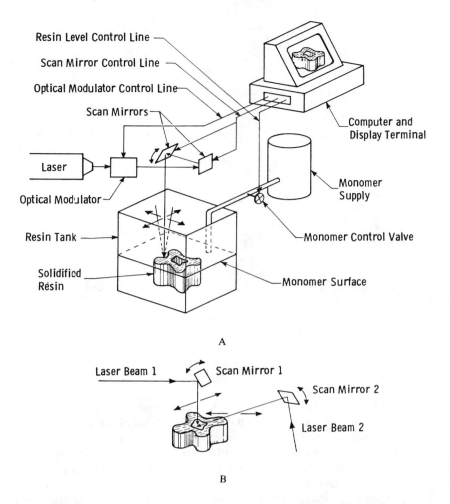

A

B

FIGURE 16.   (A) Single-laser system for computer generation of 3-D solids. (B) Two-laser subsystem.

(In is the ground state initiator and $X^{\cdot}$ is some fragment of In which is eliminated to form $In^{\cdot}$.)

Polymerization will be initiated only in the volume in which two light beams of the proper wavelength intersect.

The concept of a two-photon photoinitiator is simple, but the photochemistry is complex. There are some difficult requirements placed on In and its excited states. In* must be in a long-lived, photochemically unreactive, excited state. In addition, In* must be coupled to In** by an accessible laser wavelength (UV or visible). In** can react to provide a cation or anion to initiate an ionic polymerization. Each of these excitation steps should also have a high quantum efficiency.

D. LASER-INDUCED POLYMERIZATION USING VISIBLE LIGHT

Experience with visible laser-induced polymerization reactions has shown that there are advantages to be found in the use of visible radiation in photopolymerization processes. The molecular electronic excitation processes that occur in the visible region of the spectrum are usually sufficiently different from those excitation processes that occur in the UV to cause large differences in the oscillator strengths. In general, these differences lead to small molar

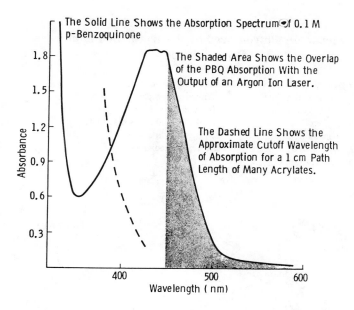

FIGURE 17.   UV-visible absorption spectrum of *p*-benzoquinone (PBQ).

absorbencies for molecular absorptions occurring in the visible spectrum. The advantage to be gained from this is the greater depth of penetration of useful radiation into mixtures of monomers and initiators triggered by visible light.

Another advantage in the use of visible lasers is seen in Figure 17. In the case of *p*-benzoquinone (PBQ) the photochemically active n-π^* transition (which is responsible for the absorption feature in the region from 350 to 520 nm) overlaps the output of an argon ion laser from about 450 to 515 nm. The dashed line in Figure 17 shows the approximate absorption onset for a number of acrylate monomers. The information shown in this figure indicates that argon ion laser will deposit all of its photochemical energy in the PBQ and very little in any acrylate monomer matrix with which PBQ could be mixed. This leads to two distinct advantages over a UV-initiated system. The first is that the quantum efficiencies for the visible process are much higher, since the energy is deposited only in photochemically active molecules. The second is the depth of penetration of the radiation. Since the molar absorptivity for the absorption feature in PBQ is relatively low and there is no absorption by the monomer matrix, the optical density of a solution containing fractions of a percent of PBQ is very low at argon ion laser wavelengths. Polymerization depths of more than 2 in. have been demonstrated. Table 7 summarizes the monomers with which this reaction scheme has been tried successfully.

Figure 18 indicates the experimental arrangement used during the studies of the argon ion laser/PBQ-initiated polymerization reactions. The solutions to be irradiated were purged and then kept oxygen-free. The course of the reactions was followed in two ways, by the temperature rise of the solution during irradiation and by changes in the scattering of focused helium neon laser light. The actual temperature rises were no greater than 15°C. Figure 19 shows a reaction exotherm typical of the system PBQ/hexanedioldiacrylate. The molecular weights given at the various stages of the exotherm were determined using gel permeation chromatography.

The data from Figure 19 indicate that the polymerization reaction occurs very quickly. Calculations based on data generated from hexanedioldiacrylate (HDDA) indicate that a 10-W laser has the capability of polymerizing (to a gel) approximately $^1/_2$ kg of material per minute. The ability of the visible radiation to penetrate inches or more into a solution of an

TABLE 7
Monomers Polymerized by the Irradiation of *p*-Benzoquinone

n-Hexylacrylate	Hexanedioldiacrylate
2-Ethylhexylacrylate	Trimethylolpropanetriacrylate
2-Ethylbutylacrylate	Tetraethyleneglycoldiacrylate
t-Butylacrylate	1,3-Propanedioldimethacrylate
Propanedioldiacrylate	2-Ethylhexylmethacrylate
Acrylic acid	Dicyclopentadiene
N-Vinylcarbazole	

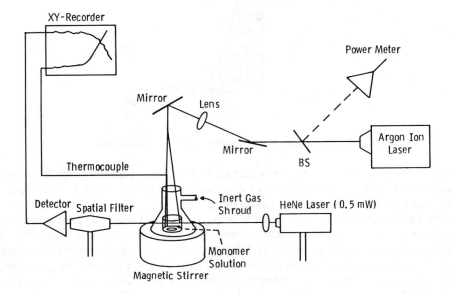

FIGURE 18.   Laser-induced polymerization setup.

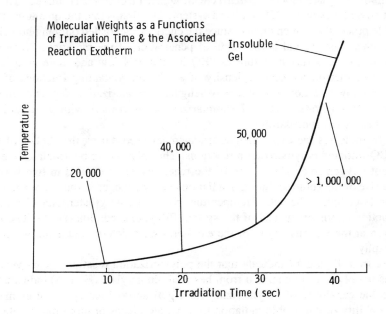

FIGURE 19.   Typical reaction exotherm plot.

appropriate initiator/monomer system offers the opportunity of curing thick section material at an even rate throughout the body of the material. This is in contrast to thermal processes in which the reaction proceeds from the outside inward as a result of thermal conduction from the exterior surfaces.

E. HIGH-SPEED CURING OF POWDERS BY LASERS

Powder coating systems are generally comprised of particles of thermoplastic or thermosettable resins together with curing agents, fillers, pigments, and flow control agents, as appropriate. Such coatings are usually applied by fluidized bed or by electrostatic spray techniques. The resulting films are generally thick (>10 mils) when applied in a fluid bed. Electrostatic techniques allow coatings as thin as 1 mil to be applied to conductive substrates. However, the technique cannot usually be applied when dielectric substrates are involved. A coating applied by either technique must subsequently be heated to effect a cure thereof if the coating is thermosettable, or to allow the powder to flow and fuse it in thermoplastic. In the past, plasma arcs have been used for applying finely divided thermosettable and thermoplastic resin particles to all classes of substrates to form coatings, for example, on metal, paper, plastics, etc. Also, flame spraying of powders has been used on a very limited basis.

Lasers provide a process for applying finely divided epoxy resins or any other thermosettable or thermoplastic resin particles with the other ingredients mentioned above to a substrate to form coatings which are cured or fused as applied and do not require a subsequent baking. The process utilizes a carbon dioxide laser to melt the resin particles and activate the initiator as they are propelled towards the substrate in the laser beam. Such a process has the advantages of producing very rapid cures without having to heat the whole substrate, which may be too large to be heated, may have heat-sensitive portions, or may be permanently fixed to another assembly in a remote area (e.g., power plant). Another advantage is that the amount of energy required for coating and curing on the substrate is very low, since the whole substrate does not have to be heated. The laser curing system described here is portable and would, therefore, be of great utility to effect field repairs on damaged or refurbished equipment.

Finely divided particles, i.e., particles of about 1 to 200 μ equivalent spherical diameter, of thermosettable resins or thermoplastic resins, are introduced into a defocussed carbon dioxide or other suitable laser beam which heats the particles to a temperature greater than the softening point of the resin (Figure 20). The residence time of the resinous particles in the laser beam is maintained for a sufficient period to substantially liquify the resinous particles, activate catalyst, initiate polymerization of the thermosettable resinous particles, or merely to melt thermoplastic particles. The resin particles are introduced into the laser beam with the help of nitrogen or argon gas near the outlet of a laser beam from the tube. The point of introduction will determine the residence time of the resin in the laser beam. To prevent overheating of the resin particles beyond their degradation point, the temperature of the laser beam can be controlled by the amount of the power input to the laser.

F. HOLOGRAPHY

There has been a variety of investigations into polymers and polymer blends which can be used to store holograms. The holographic images have, in many instances, been created in thin photoresist-like films. The interference pattern of two intersecting laser beams causes polymerization to occur more rapidly at the intensity peaks than at the nodes in the interference pattern. The development of the film by dissolution of the unpolymerized material leaves a polymeric image of the laser interference pattern. These types of images have been utilized in a variety of ways, e.g., the manufacture of holographic gratings for dispersive spectrometers and the deposition of capacitor strips on memory chips.

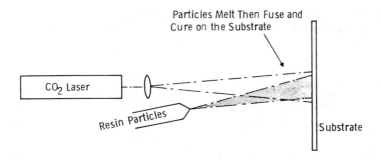

FIGURE 20.   Equipment arrangement for laser coating.

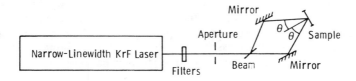

FIGURE 21.   Schematic diagram of the experimental geometry for the production of gratings.

A direct method for the production of high-resolution gratings has been demonstrated[39] using KrF laser irradiation at 248 nm. Gratings with efficiencies of up to 40% have been produced using this technique with spatial frequencies in the 630 to 2900 line pairs per millimeter range. The experimental geometry used for the production of the gratings is shown in Figure 21. The two beams were directed at approximately equal angles of incidence onto the surface of the plastic film. By varying the angle of incidence, spatial frequencies of 630 to 4000 line per millimeter were produced corresponding to grating periods of S = 0.25 to 1.59 μm. The plastic target materials with different UV absorption depths evaluated were polyetylene terephthalate (PET), polyimide, polyethylene (PE), polytetrafluoroethylene (PTFE), and polymethylmethacrylate (PMMA). Of these materials only the PET and polyimide produced grating structures, while the other plastic materials either showed no change for low-intensity illumination or showed signs of damage with no grating pattern visible of high irradiance levels. Further refinements of these techniques will allow the production of more complex structures at submicrometer resolution.

VIII. REQUIREMENTS FOR LASERS TO HAVE A SIGNIFICANT IMPACT ON POLYMERIZATION PROCESSING

There is a variety of reasons why lasers have not had a larger impact on the processing areas of polymer technology. The four major areas of concern have been expense, complexity, reliability, and basic polymer photochemistry. This section will discuss these areas of concern and indicate what will have to occur to lessen their impact on the more widespread use of lasers in polymer processing.

One of the biggest detractors in the consideration of the use of lasers as processing tools in any area of technology has been their initial expense and high cost of operation. The data in Table 3 show the wide range of initial capital investment that might be necessary when considering a laser for a given process. The cost is not as great if the wavelength necessary to drive the photochemistry happens to be coincident with the wavelength of an excimer laser or a CO_2 laser, which are two of the least expensive lasers. Even these lasers can be very expensive for a processing system. Capital investments in these systems can easily exceed several hundred thousand dollars. The wavelength demands of photopolymerization reaction often do not allow an "inexpensive choice" for a laser system.

There is a second component of the cost consideration which relates to the operating expense (including maintenance). Maintenance costs can be considerable for certain types of lasers. An example is the replacement of a plasma tube in an argon ion laser, which may be as much as $14,000 to $16,000 every 2 years. Optics, high-power pulse networks, and high current-high voltage electronics can prove to be expensive to maintain, when compared to standard UV lamp technology. Maintenance is not necessarily the highest operating cost. Very often the low energy conversion efficiency of lasers can lead to an unacceptably high operating expense. In this regard, the efficiency with which electricity is converted to *useful* photochemical energy is a concern. With the exception of the previously mentioned excimer and CO_2 lasers, lasers are inefficient in this respect. Overall electrical to optical conversion efficiency can fall well below 1%. However, the useful conversion efficiencies for UV lamps may not be much larger, if only the energy contained in the wavelengths that interact with the target molecules is considered. There are other costs associated with consumables, such as laser gases and dyes, which can be large depending on the type of laser or process.

When the total of all the costs is computed, it is found that photons are expensive chemical reagents, the expense being wavelength dependent. In the IR (CO_2 or CO) lasers, photons cost a few cents per Einstein (6×10^{23} photons), while in the visible or UV, the costs can be on the order of dollars per Einstein. However, before these numbers are compared directly, it must be remembered that each IR photon carries one tenth (approximately) or less of the energy of a visible photon. Thus, equivalent energy depositions may not be so different in cost. As a result, the quantum efficiency for the process being considered is very important. Free radical and ionic polymerization processes are high quantum efficiency processes with quantum efficiencies ranging from tens to thousands. These high quantum efficiencies in some instances may overcome the high energy costs associated with lasers. Condensation reactions will probably not be of importance, since they lack radical or ionic propagation characteristics.

The issue of system complexity and ruggedness is one of concern for many lasers that might be chosen for a process. While CO_2, Nd:YAG, and excimer lasers have been shown to operate effectively in a production environment, many other laser systems such as dye lasers have not yet been packaged adequately for long-term stable performance. Many lasers, which have not been packaged for the production environment, require too much attention from a skilled technician or professional to maintain peak performance. The need for constant attention to the internal optical alignment of the laser is one of the major drawbacks for a system that has not been packaged for industrial application.

In spite of the above-mentioned concerns, lasers will play an increasing role in polymer processing technology. The infrequent use of lasers at present is due to the current level of development of laser systems. There are indications that over the next decade there will be significant changes in laser technology. Developments in semiconductor lasers can be expected to increase the availability of lower cost, more reliable laser systems with moderate powers. Continued interaction between polymer and laser technologists will broaden the scope of applications and accelerate the development of lower cost, useful laser systems.

IX. SUMMARY

Lasers are unique sources of photochemical energy. They are quite different when compared to the usual incoherent light sources which have been employed in photochemical processing. One of the outstanding features of laser radiation is its spectral purity. The wavelength range of greatest interest in photopolymerization extends from the blue-green into UV (500 to 190 nm). A number of lasers with usable output are available in this wavelength range.

High-resolution relief images can be obtained by a single laser pulse, thus making feasible ultrafast photoimaging for microcircuit applications. It is possible to cure multiacrylate photoresists at extremely high intensities without losing much of the quantum efficiency or photosensitivity. Thermal effects of lasers have been successfully employed in curing and crosslinking of polymers. Lasers have been employed in reducing the hardening time of adhesive materials. Resonant action of laser has been employed in hardening alkyd resin enamels.

Potential applications of lasers in curing resins are highlighted in this chapter. There are a number of applications envisioned for photochemically polymerized biocompatible polymers. Many of the applications would be surgical in nature. The ability to transmit laser radiation through fiber optics to very localized areas presents the possibility of coupling laser polymerization with microsurgical techniques. There are some obvious candidates, such as coupling laser-assisted bone repair with arthroscopic surgical techniques to repair a damaged joint (such as a knee) or to repair or replace tendons and ligaments. An appropriate polymer which would bond strongly to existing tendon, ligament, and bone tissue would allow the repair to take place without sutures, staples, or other attachment devices. Laser hardening can be used in filling a deep cavity in dental repairs. The temperature increase inside the material and its effect on dental pulp is minimized by using a method whereby only the upper layer portion of the filled material is rapidly hardened by the laser, while the inner portion is allowed to harden gradually inside the mouth.

There are two areas in which a laser-based system would offer advantages in creating solid bodies: the speed of exposure of the individual layer and the layer height control. The hardware and software concepts already exist for laser pattern generation coupled with polymerization process in photoresist exposure. Lasers also provide a process for applying finely divided epoxy resins or any other thermosettable or thermoplastic resin particles with the other ingredients, such as fillers, to a substrate to form coatings at a high speed which are cured or fused as applied and do not require a subsequent baking. The laser curing system described here is portable and would, therefore, be of great utility to effect field repairs on damaged or refurbished equipment.

The laser has unlimited potentials for applications in photocuring, photomodification, and photocrosslinking of polymers which have not yet been tapped. It is hoped that applications in many different areas will be developed in the future.

ACKNOWLEDGMENTS

The authors would like to acknowledge Debbie Smoody and Barb Cichon for the preparation of the manuscript.

REFERENCES

1. **Green, G. E., Stark, B. P., and Zahir, S. A.,** Photocrosslinkable resin systems, *J. Macromol. Sci. Rev. Macromol. Chem.,* C21(2), 187, 1981.
2. **Senich, G. A. and Florin, R. E.,** Radiation curing of coatings, *J. Macromol. Sci. Rev. Macromol. Chem. Phys.,* C24(2), 239, 1984.
3. **Pappas, S. P.,** *UV Curing: Science and Technology,* Vol. 1, Technology Marketing Corp., Norwalk, CT, 1978, chaps. 1 and 2.
4. **Shalaby, S. W.,** Radiative degradation of synthetic polymers: chemical, physical, environmental, and technological considerations, *J. Polym. Sci: Macromol. Rev.,* 14, 419, 1979.
5. **Labana, S. S.,** *Ultraviolet Light Induced Reactions In Polymers,* American Chemical Society, Washington, D.C., 1976.

6. Lasers and applications, in *1987 Buying Guide,* High Tech Publications, Torrance, CA, 1986, 112.
7. **Decker, C.,** Laser-induced polymerization of multifunctional acrylate systems, *Polym. Photochem.,* 3, 131, 1983.
8. **Decker, C.,** UV curing of acrylate coatings by laser beams, *J. Coating Technol.,* 56, 713, 1984.
9. **Decker, C.,** Ultrafast polymerization of epoxy-acrylate resins by pulsed laser irradiation, *J. Polym. Sci. Polym. Chem.,* 21, 2451, 1983.
10. **Decker, C.,** Light-intensity effect in laser-induced photopolymerization, *ACS Polym. Prepr.,* 25(1), 303, 1984.
11. **Shunichi, K., Akihiro, M., Akira, U., Akira, S., and Akira, O.,** Photopolymerizable mass, *Ger. Offen. 2753889,* June 8, 1978.
12. **Ichimura, K. and Nishio, Y.,** Photocrosslinkable polymers having p-phenylenediacrylate group in the side chain: argon laser photoresist, *J. Polym. Sci. Part A: Polym. Chem.,* 25, 1579, 1987.
13. **Fouassier, J. P., Lougnot, D. J., and Pilot, T.,** Visible laser light in photoinduced polymerization. I. A quantitative comparison with UV laser irradiation, *J. Polym. Sci. Polym. Chem.,* 23, 569, 1985.
14. **Fouassier, J. P. and Lougnot, D. J.,** Ionic photoinitiator for radical polymerization in direct micelles: the role of the excited species, *J. Appl. Polym. Sci.,* 32, 6209, 1986.
15. **Fouassier, J. P., Jacques, P., Lougnot, D. J., and Pilot, T.,** Laser photoinitiators and monomers: a fashionable formulation, *Polym. Photochem.,* 5, 57, 1984.
16. **Shahbazian, S. M.,** Photosensitive systems for laser exposure in the printing industry, *J. Photo. Sci.,* 32, 111, 1984.
17. **Iwata, K., Hagiwara, T., and Matsuzawa, H.,** A novel photocrosslinkable polymer with argon ion laser, *J. Polym. Sci. Polym. Lett.,* 22, 215, 1984.
18. **Williamson, M. A., Smith, J. D. B., Castle, P. M., and Kauffman, R. N.,** Laser-initiated polymerization of charge transfer monomer systems, *J. Polym. Sci. Polym. Chem.,* 20, 1875, 1982.
19. **Sadhir, R. K., Smith, J. D. B., and Castle, P. M.,** Laser-initiated copolymerization of maleic anhydride with styrene, vinyltoluene, and *t*-butylstyrene, *J. Polym. Sci. Polym. Chem.,* 21, 1315, 1983.
20. **Sadhir, R. K., Smith, J. D. B., and Castle, P. M.,** Laser-initiated polymerization of epoxies in the presence of maleic anhydride, *J. Polym. Sci. Polym. Chem.,* 23, 411, 1985.
21. **Hiroshi, M., Satoshi, O., Noboru, M., Akira, I., Kenichi, O., and Shigekazu, K.,** Ionic photodissociation of poly(*n*-vinylcarbazole) systems in solution, *Kobunshi Robunshu,* 37(4), 275, 1980.
22. **Stern, R. H. and Sognnaes, R. S.,** Laser beam effects on hard dental tissues, *J. Dent. Res.,* 43, 873, 1964.
23. **Yamamoto, H. and Sato, K.,** Prevention of dental caries by Nd:YAG laser irradiation, *J. Dent. Res.,* 59, 2171, 1980.
24. **Sassano, S. and Kamemizu, T.,** High speed hardening of resin by laser irradiation, *Kagaku Kogyo,* 37(3), 223, 1986.
25. **Kal'vina, I. N., Moskalenko, V. F., Ostapchenko, E. P., Pavlovskii, Protsenko, T. V., and Rychkov, V. I.,** Resonant action of laser radiation on polymerization processes, *Sov. J. Quant. Electron.,* 4(10), 1285, 1975.
26. **Il'yasov, S. G., Kal'vina, G. A., Kyulyan, G. A., Moskalenko, V. F., and Ostapchenko, E. P.,** Imidization of polyimide acid by laser irradiation, *Sov. J. Quantum Electron.,* 4(10), 1287, 1975.
27. **Danen, W. C. and Jang, J. C.,** Multiphoton infrared excitation and reaction, in *Laser Induced Chemical Processes,* Steinfeld, J. I., Ed., Plenum Press, New York, 1981, chap. 2.
28. **Srinivasan, R.,** Ablation of polymers and biological tissue by ultraviolet lasers, *Science,* 234, 55, 1986.
29. **Blum, S. E., Brown, K. H., and Srinivasan, R.,** U.S. Patent 4414059, 1983.
30. **Yeh, J. T. C.,** Laser ablation of polymers, *J. Vac. Sci. Technol.,* A4(3), 653, 1986.
31. **Rice, S. and Jain, K.,** Direct high resolution excimer laser photoetching, *Appl. Phys.,* A33, 195, 1984.
32. **Cozzens, R. F. and Fox, R. B.,** Infrared laser ablation of polymers, *Polym. Eng. Sci.,* 18(11), 900, 1978.
33. **Kukreja, L. M., Bhawalkar, D. D., Chatterjee, U. K., and Gupta, B. L.,** Experimental study of laser induced decomposition based processing of a brittle plastic, CR-39 (allyl diglycol carbonate), *Appl. Phys.,* A36, 19, 1985.
34. **Kukreja, L. M.,** Irreversible softening in a thermoset polymer — allyl diglycol carbonate® due to a laser treatment, *J. Appl. Polym. Sci.,* 33, 985, 1987.
35. **Hirabayashi, S.,** Nonuniformity of polymerization in visible light polymerizable composite resins, *Shika Zairyo to Kikai,* 3(5), 665, 1984.
36. **Yamanoto, H.,** Applications of laser light in dentistry, *J. Jpn. Dent. Assoc.,* 37(3), 259, 1984.
37. **Wittenberg, R. C.,** CAD database creates solid models, *Electron. Eng. Times,* 424, 1, 1987.
38. **Tuck, J.,** Direct imaging, *Circuits Manuf.,* 36, February 1984.
39. **Ilcisin, K. J. and Fedosejevs, R.,** Direct production of gratings on plastic substrates using 248-nm KrF laser radiation, *Appl. Opt.,* 26 (2), 396, 1987.

Chapter 3

THREE-DIMENSIONAL MACHINING BY LASER PHOTOPOLYMERIZATION

M. Cabrera, J. Y. Jezequel, and J. C. Andre

TABLE OF CONTENTS

ABSTRACT

There are two different manufacturing processes to make three-dimensional material objects by laser-aided photopolymerization: multiphotonic or monophotonic processes. If the former are still the matter of fundamental research and have not yet led to industrial developments, the latter have already led to experimental systems that enable the study of the practical problems these processes imply. There are two different types of problems which can be either related to the polymerization reaction itself or to the total time the processes require. On the one hand, the nature of the reactants (monomers or oligomers) induce reactions which are usually exothermal and result in highly shrunk polymers. On the other hand, it is necessary to make the objects within "reasonable short times", i.e., times that are similar to those required by other already existing techniques. This leads to technological difficulties which can be overcome by pluridisciplinary research.

I. INTRODUCTION

In the industrial field, prior to the mass production of an object, it is often useful to have several samples of this object at one's disposal to carry out the first tests, the marketing, and the manufacture matrices. To do this, it is necessary first to draw the object plans in a design department and, secondly, to make these samples in a mechanical workshop in a substance which is usually different from the manufactured product one (e.g., wood).

It is known that the study and perfecting of new industrial parts are now made easier and faster by using the technique called "computer-aided design" (CAD). The traditional way of making samples has now become a real handicap because it is often slow and expensive. When CAD techniques are used, the numeric data which define the shape of the object are stored inside a computer memory. Therefore, from the numeric coding of the object, it is possible to machine a sample by using numerically controlled mechanical tools, for instance. Unfortunately, in order to mill, on the one hand, the tools must be given a trajectory tangential to the surface without damaging the surrounding surfaces, and on the other hand, they are wearing fast. Using sharp-pointed mills and soft materials should reduce these difficulties, but they cannot be avoided if hard materials must be used.

Therefore the idea is to use the sole properties of light to make the samples, so as to get rid of all these problems arising from the use of mechanical tools. The recent improvements in the polymer chemistry, as well as in new technologies such as computer science, robotics, or photonics, enable us to think that it is actually possible to set up computer-aided manufacturing (CAM) processes to make industrial samples by using polymers and lasers.

It is usually possible to focus laser beams into a transparent medium with high precision and to deflect them rapidly. If a chemical reaction then occurs due to the light-matter interaction, it must be possible to develop a laser CAM. When compared to the standard mechanical techniques, the laser CAM would be particularly convenient. It would, in particular, be possible to make parts of really any shape without molding, thus eliminating the clearance, shaving, or breaking problems, as well as the sharpening of the tools, and so on. In mechanical processes, mills and drills of different sizes are used. As an example, to make a $10 \times 10 \times 10$-cm plastic object, less than 3 h are required, the precision being better than 0.1 mm. It is thus necessary to set up a largely optimized process so that the laser CAM can compete with the mechanical processes. The low inertia of the light carriers should be a real advantage in that case.

One could imagine that the organic nature of the material would restrict the use of this laser CAM process. In fact, we have already obtained samples made of polymer which have satisfying characteristics, especially from the viewpoint of hardness. Moreover, the polymer

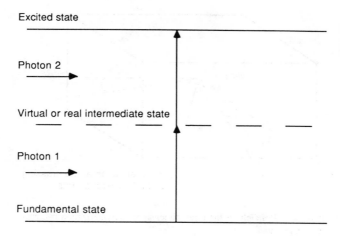

FIGURE 1. Biphotonic absorption.

chemistry is varied enough to enable the feasibility of making objects made of very different materials: colored, charged, luminescent polymers, etc.

II. THE DIFFERENT LASER PROCESSES

To induce the reticulation of ethylenic monomers (such as acrylates), the basic principle consists in the local creation of free radicals which initiate the polymerization chain reaction.

In fact, the idea of using light to make objects is not especially new.[1] More recently, two different types of processes have been propounded and studied. In both cases the molecules are excited by light but the numbers of the required photons differ. If more than one photon is necessary, these are the multiphotonic processes.[2-8] If only one photon is necessary, these are the monophotonic processes.[9-14] The feasibility of both of these methods has been tested by our research group.[15-19]

A. THE THREE-DIMENSIONAL MULTIPHONIC PROCESSES
1. Creation of a Precursor Electronic State for Free Radicals

When some molecules are irradiated by a photon flux that has reached the proper intensity (1 to 100 MW/cm² with lasers), a phenomenon, called two-photon absorption, occurs (Figure 1). This means that even if the energy of each photon is lower than the molecule excited state energy, this energy level can be obtained by adding up the energies of two photons. Several biphotonic absorption paths exist according to the nature of the intermediate state. Either it is a virtual state, and then is a true biphotonic absorption — from the viewpoint of quantum theory — for which the absorption selection rules between two different states are different from the monophotonic absorption rules, or it is a real state, and then can be resonant or metastable.

In practice (Figure 2), two perpendicular pulsed laser beams have to be focused on the same spot so that unstable species can be created, initiating the polymerization reaction. This reaction must not propagate outside the volume delimited by the intersection of the two crossing beams. The three-dimensional object is then obtained by controlling the displacement of the crossing spot.

The absorption probability of n identical photons by a molecule A is proportional to the nth power of the excitation flux I (photon·cm^{-2} s^{-1}). For biphotonic processes, it becomes[20]

$$\frac{dI}{dX} = \delta \cdot [A] \cdot I^2 \qquad (1)$$

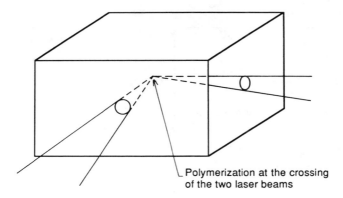

FIGURE 2.   Principle of the three-dimensional polymerization by crossing two laser beams.

in which X (cm) is the distance run by a photon inside the medium containing A, [A] (molecule.cm^{-3}) is the concentration of the absorbing species A, and δ (cm^4.s.photon^{-1}.molecule^{-1}) is the biphotonic absorption coefficient.

This equation can be compared to the usual Beer-Lambert's law for monophotonic processes:

$$\frac{dI}{dX} = \epsilon \cdot [A] \cdot I \tag{2}$$

in which ϵ (L.molecule^{-1}.cm^{-1}) or (cm^2.molecule^{-1}) is the molecular absorption coefficient.

If the two photons have two different frequencies ν_1 and ν_2 (Hz), Equation 1 becomes

$$\frac{dI}{dX} = \delta \cdot [A] \cdot I(\nu_1) \cdot I(\nu_2) \tag{3}$$

where I (ν_1) and I (ν_2) are the intensities of light at ν_1 and ν_2.

In the case of a coherent absorption, the order of magnitude of the molecular absorption coefficient δ is about 10^{-50}cm^4.s.photon^{-1}.molecule^{-1}. This value of δ is much larger if an intermediate state exists.

However, this technique has a major restriction originating in the "shadow" of the part of the piece that has already been polymerized. As a matter of fact, the polymer refractive index always differs from those of the monomer or of the oligomer, resulting in very difficult focusing problems. The angle between the two laser beams can be modified indeed, but then the apparatus setup turns far more complex.

Hence, the laser beams are held perpendicular, and a sequential biphotonic absorption process based on a radical polymerization initiated by two crossing laser beams (Figure 3) has been propounded. In that case the biphotonic absorption coefficient δ can be written as

$$\delta = \frac{4\pi^4}{c^2} \nu_1\nu_2[\lambda_1 \cdot S^{02} \cdot \lambda_2]^2 g(\nu_1 + \nu_2) \tag{4}$$

in which λ_1 and λ_2 are the polarization vectors of the two photons 1 and 2 whose energies are hν_1 and hν_2, respectively, S^{02} is the transition tensor from state 0 to state 2, and g ($\nu_1 + \nu_2$) is the light flux diametrical profile of the laser beam.

Each element of the biphotonic transition tensor can be written as

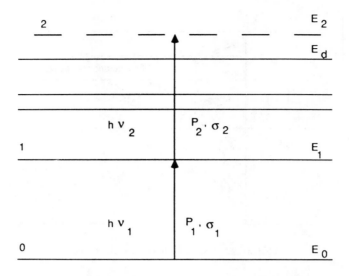

FIGURE 3. Energy diagram of a biphotonic reaction: (0) Initial state; (1) singlet or triplet intermediate state whose energy is E_1; (E_2) excited state; and (E_d) dissociation energy of the molecule.

$$S_{\alpha\beta}^{02} = \sum_i \left[\frac{\langle \Psi_0 | \mu_\alpha | \Psi_i \rangle \langle \Psi_i | \mu_\beta | \Psi_2 \rangle}{E_i - h\nu_1} + \frac{\langle \Psi_2 | \mu_\alpha | \Psi_i \rangle \langle \Psi_i | \mu_\beta | \Psi_0 \rangle}{E_i - h\nu_2} \right] \tag{5}$$

in which μ_α and μ_β are the components of the two-photon dipole moment transition vector μ_{02} and Ψ_0, Ψ_1, and Ψ_2 are the molecular orbital functions whose energies are E_0, E_i (intermediate state), and E_2, respectively.

The energies of the two photons must satisfy the following conditions:

$$h\nu_1 + h\nu_2 > Ed \tag{6}$$

with $h\nu_1 < Ed$ and $h\nu_2 < Ed$.

The excited molecule density of the first energy level reaches its maximum when the $0 \rightarrow 1$ transition is saturated, i.e., if the power P_1 (W/cm²) is

$$P_1 = \frac{h\nu_1}{\sigma_1 T_1} \tag{7}$$

in which σ_1 (cm²) is the $0 \rightarrow 1$ transition cross-section and T_1 (s) is the lifetime of the excited state 1.

The excited molecule photodissociation probability is given by the formula

$$W_2 = \frac{\sigma_2 P_2}{h\nu_2} \tag{8}$$

in which σ_2 (cm²) is the $1 \rightarrow d$ photodissociation cross-section from state 1 and P_2 (W/cm²) is the radiation intensity at the frequency ν_2.

The following condition must be satisfied to have an efficient photodissociation:

$$W_2 \geqslant \frac{1}{T_1} \quad \text{i.e.,} \quad \frac{P_2}{P_1} > \frac{\sigma_1}{\sigma_2} \quad (\text{with } \omega_1 \neq \omega_2) \tag{9}$$

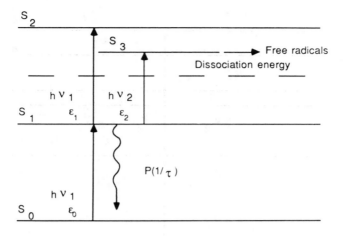

FIGURE 4.   Sequential biphotonic absorption.

Typical values are as follows:

$$\sigma_1: \quad \text{from } 10^{-9} \text{ to } 10^{-11} \text{ cm}^2$$

$$\sigma_2: \quad \text{from } 10^{-18} \text{ to } 10^{-19} \text{ cm}^2$$

which gives P_2/P_1 from 10^7 to 10^{10}.

Last, let us make clear that the following conditions must be satisfied to have a really selective system: (1) the energy $h\nu_1$ of the first photon must be greater than the thermal agitation energy during the $0 \rightarrow 1$ pumping, i.e.,

$$h\nu_1 \gg kT \tag{10}$$

in which h (J.s) is Planck's constant, ν_1 (Hz) is the frequency, k (J.K^{-1}) is the Boltzmann constant, and T (K) is the temperature; (2) if a pulsed laser is used, the pulse width τ_p must be inferior to rotational relaxation time τ_{rot} where

$$\tau_{rot} = \frac{Z_{rot}}{N_0 V_0 \sigma_0} \tag{11}$$

in which Z_{rot} is the number of collisions that are necessary to restore equilibrium, N_0 (molecule cm^{-3}) is the molecular concentration, V_0 (cm.s^{-1}) is the mean speed of the molecules, and σ_0 (cm^2) is the efficient collision cross-section.

The main principle of this type of sequential initiation appears in Figure 4 and was in the beginning studied by the Battelle Institute.[7]

The main difficulty for this type of system lies in the choice of the photoinitiator. It must absorb the *first* radiation to reach its first *excited* state, but it must not be photodissociated. Furthermore, the lifetime of this excited state must be long enough so that the absorption by this excited state of a photon issued from the second laser beam is made probable.

Several products fulfilling these conditions have been proposed.[7,8] An important problem remains, nevertheless: a biphotonic absorption by the monomer or by the initiator may occur somewhere along the paths of each laser beam. This would initiate additional polymerizations in spots different from the planned ones, which are harmful to the precision of this process.

We have performed a computer simulation of this occurrence so that it could be visu-

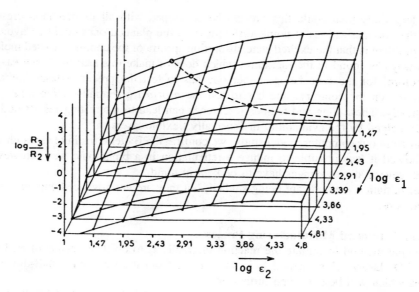

FIGURE 5. Values of the ratio R_3/R_2 as functions of ϵ_1 and ϵ_2. (R_2) Rate of formation of free radicals by absorption of two photons $h\nu_1$; (R_3) rate of formation of free radicals by absorption of one photon $h\nu_1$ and one photon $h\nu_2$; (ϵ_1 and ϵ_2) molecular absorption coefficient, respectively from state S_1 to S_2 and from state S_1 to S_3 (see Figure 4). The energy density F_1 of the first laser beam is equal to 0.02 J. mm^{-2} in 250 ns. The energy density F_2 of the second beam is equal to 0.004 J. mm^{-2} in 8 ns.

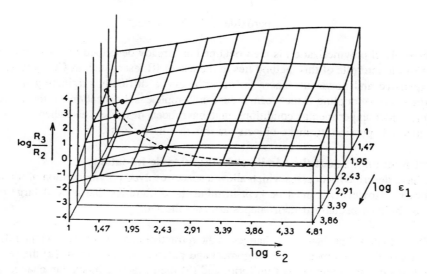

FIGURE 6. Same as Figure 5 with the following conditions: (F_1) 0.005 J. mm^{-2} in 10 ns and (F_2) 0.5 J. mm^{-2} in 100 ns.

alized.[16] Without entering into all the details of this work, it can be seen in Figures 5 and 6 that the ratio of the rate R_2 of formation of free radicals initiated by the absorption of two photons $h\nu_1$ to the rate R_3 of the same phenomenon due to the two photons $h\nu_1$ and $h\nu_2$ is usually larger than one.

This computer simulation points out that the development of this process requires the prior discovery of at least one photoinitiator which could be split only by the biphotonic absorption of the photons $h\nu_1$ and $h\nu_2$, but could be not be split (or very seldom) by two photons $h\nu_1$. Only then could this process be developed with all its precision preserved, because the polymerizations at points different from the planned ones could be avoided.

The problem is that the excited state absorption spectra of the commonly used molecules cannot always be found in the literature, which doesn't make this crucial choice easy.

Therefore, for the time being, either one has to test the precursor-laser interactions randomly or, on the contrary, one has to test the *a priori* potentially interesting systems systematically. These long and tedious researches can lead to interesting results and could be greatly helped by the availability of especially flexible lasers.

Apart from our computer model study, we confirm the Battelle Institute research works, which showed it was possible to induce a reticulation in a biphotonic way. However, as aforesaid, several disadvantages have to be kept in mind.

Other initiation methods are being studied, but they are not advanced enough yet to be described here.

2. Thermal-Initiated Manufacturing Processes

A simple method to initiate the reaction thermally is to do it by means of an IR laser (e.g., a CO_2 laser). This method is in fact drawn from the two and a half-dimensional processes which will be described further on.

The principle of the initiation step is as follows: an absorbing reactant locally transfers the energy of the light to the medium. This one contains thermal initiators of polymerization (e.g., peroxides) which can give free radicals.

$$\text{absorbing reactant} \xrightarrow{\text{IR } h\nu} \text{local heatup}$$

$$\text{peroxide} \xrightarrow{\Delta H} \text{free radicals}$$

A space-resolved polymerization is then obtained as already described by Grisoni et al.[12,16]

One of the interests of this method lies in the use of IR lasers such as CO_2 lasers, which are inexpensive and powerful. There is, nevertheless, a difficulty originating in the heat transfer/reactivity coupling which can induce a complete solidification of the monomer. However, under experimental conditions that have been described elsewhere,[16] we have proved that it is possible to make objects via this thermal process.

3. Short Reminder about the Limits of Multiphotonic Initiations

We have described processes which allow a space-resolved polymerization of monomers or oligomers. However, it must be kept in mind the difficulties that arise if large objects have to be built, especially if their shapes are complicated:

1. The "shadow" problem and changes in the refractive indexes such as (1) the refractive index of the monomer (or the oligomer) and polymer are not equal; (2) the refractive indexes of oligomers are not uniform; and (3) the refractive index can change locally due to temperature rises (a polymerization is often exothermal).
2. Moving and focalizing two laser beams fast onto the same spot inside a large reactor ($10 \times 10 \times 10$ cm) with a precision better than 0.1 mm.
3. High-energy laser beams are required to obtain monophotonic absorptions. This implies the use of optical components having very short focal lengths, which is incompatible with the large size of the part.

For all these reasons we decided to develop apparently less-sophisticated processes first which should be easier to set up. However, we still continue to have a fundamental research activity in this multiphonic processes field.

B. MONOPHOTONIC PROCESSES OF LAYER-BY-LAYER POLYMERIZATION

These processes are based on the surface light absorption which follows an exponential law, the Beer-Lambert's law. To make a three-dimensional object, it is necessary to polymerize the monomer (or the oligomer) layer by layer, hence, the name two and a half-dimensional process. These processes are usually monophotonic, in opposition to the truly three-dimensional processes which are always multiphotonic.

The reactant is always polymerized close to an open surface, which makes the heat transfer easier. Moreover, one single laser is usually enough, which greatly simplifies the system.

In practice, mixtures of monomers or oligomers are used, which do not absorb much of the laser emitted photons, and initiators, which absorb them efficiently, on the contrary. Then, via a photochemical reaction, these initiators produce reactive species, usually free radicals, which initiate the monomer (or oligomer) polymerization. In other words, the liquid monomer solidifies. Two different types of polymerization can occur according to the functional characteristics of the monomer. Either it is monoethylenic, which leads to a linear chain reaction, or it is polyethylenic, which leads to branched chain reactions, i.e., to a reticulation. In that latter case, the more intense the light is, the harder the polymerized product is. Furthermore, the thickness of the reticulated product is usually lower than the light penetration depth (Figure 7), which has been verified both experimentally and by computer simulation.

Therefore, this process consists in polymerizing part of the surface layer of a monomer. Thus, the object to be made is divided into horizontal "slices" of constant thickness. By controlling the displacements of a laser beam irradiating the monomer perpendicular to its surface, one induces the polymerization of the part of the "slice" that has to be solidified. This is done in such a way that before each movement of the laser beam an elementary volume is polymerized, the surface of which is nearly equal to the beam surface. Such an elementary volume is called a "voxel", which stands for "volume element" by analogy with the acronym "pixel" (picture element). When a slice has been made by the side by side solidification of the selected voxels, the next layer of monomer is added, the preselected voxels are polymerized, and so on, as can be seen in Figure 8.

An example diagram of a two and a half-dimensional polymerization apparatus setup can be seen in Figure 9. The polymerization is initiated by a laser beam. Motorized trolleys, micrometrically controlled by a computer, make the beam displacements possible along the x and y axes. The movements along the z axis result from the controlled pumping up of the liquid monomer. The focalizing optics have to be lifted up so that the laser beam is kept focalized on the surface of the monomer.

The superiority of lasers over conventional light sources lies in the great geometrical precision of the focalizing spot. The geometrical precision of the object to be made depends on how deeply the light energy can penetrate into the monomer. This depends on the concentration of the initiator and on its absorption coefficient at the selected wavelength.

The differences must be made between the horizontal and vertical precisions. (1) Vertically, the precision depends on the two following factors: the depth to which the polymerization propagates, which depends on the depth at which the beams penetrate into the medium, and the thickness of the monomer layers. The covering of the polymer by the liquid monomer is limited by its viscosity and the surface tension characteristics of the liquid/solid interface (layers thinner than 50 μm have already been obtained). (2) Horizontally, the precision depends on the propagation of the polymerization reaction which is several 10-μm wide. It is thus possible to expect to reach a precision better than 50 μm, as our experiments showed.

A restriction to this process appears when objects presenting disconnected parts have to

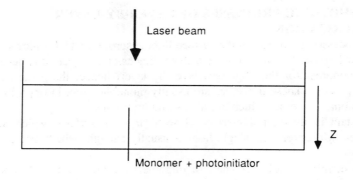

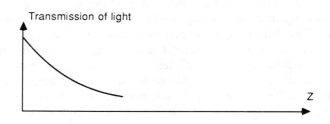

FIGURE 7. The monophotonic process.

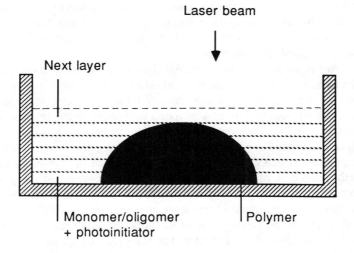

FIGURE 8. Layer-by-layer polymerization.

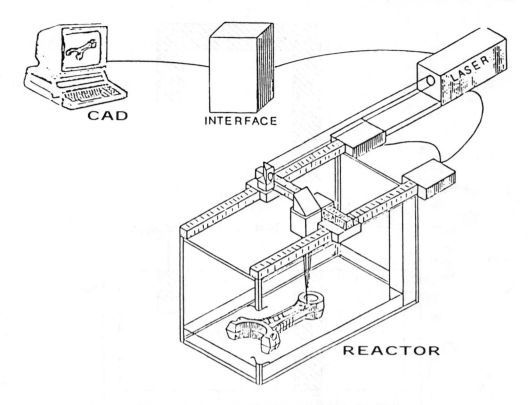

FIGURE 9. Example diagram of a two-and-a-half dimensional polymerization process device.

be made, as shown in Figure 10. Let us imagine this object has to be made the way it stands and not upside down. During the polymerization procedure, two disconnected parts will have to be made and one of them will move since the monomer is liquid.

There are two types of problems posed by the photopolymerization of 3-D objects:

1. The "material" problems. The chemical nature of the monomer and photoinitiator must be carefully adjusted because of the polymer shrinkage in particular.[18] Furthermore, its density, which is usually higher than the monomer one, depends on the reaction progress therefore on the photon flux. As a consequence, the object gets out of shape when being made.
2. The "system" problems. The system has to be very fast. Let us imagine that a filled cube, the sides of which measure 10 cm long, has to be made with a precision better than one tenth millimeter. A slow system, solidifying one voxel every millisecond will need nearly 12 d to create the 10^9 voxels!

These two problems are connected, in fact. Thus, so that a new fresh monomer layer may spread out rapidly on a large polymer surface, its viscosity has to be low. However, on the other hand, the more fluid an acrylic monomer is, the more it usually shrinks, and as a consequence the larger the deformation of the object will be in the end. Therefore, the researcher's behavior will have to be really matter-of-fact, searching for compromises, but always keeping in mind that most of the industrial elementary pieces have simple shapes.

Never forget, either, that if it is easy to achieve flashy demonstrations, it is much more difficult to set up and propose a "perfect" system, especially if large objects have to be made.

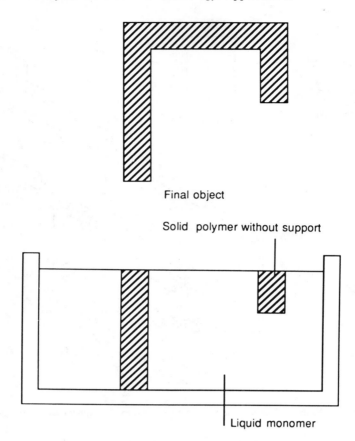

Final object

Solid polymer without support

Liquid monomer

FIGURE 10.   A complication which can arise in the layer-by-layer process.

C. THE PATENTS WAR

The schematic diagram drawn in Figure 9 describes a process setup that has been patented by the French company Cilas Alcatel.[7] For the time being, no apparatus has been marketed, contrary to what happened in the U.S. by 3D Systems (Sylmer, CA). Besides, Japanese[10] and Israeli research teams are working on this subject. Apart from the legal points of view originating in the multiplying of nearly similar ideas, the fact that many research teams aim at the same goal prove the interest that industrials take in the two-and-a-half processes.

III. MATERIALS

A. GENERAL REMARKS

Two-and-a-half photochemical processes usually use a material which is mainly a mixture of an initiator (I) and a monomer or an oligomer (M/O). Polymerization reactions are chain reactions which can be divided into three steps.

1. Initiation Step

In the initiation step, reactive species are produced, e.g., photochemically. The initiator absorbs the light energy and splits into parts to give free radicals (R), which can initiate a chain reaction.

For example,

labile bond

or

2. Propagation Steps

The propagation reaction is a well-known reaction, resulting from the successive additions of an M/O molecule to a macroradical (M_n of length n) in order to produce a longer macroradical (M_{n+1}).

3. Termination Steps

These reactions occur when two radical species are spatially close to each other. The kinetics of these termination steps as well as the kinetics of the propagation steps greatly depend on the surrounding medium, especially on its viscosity insofar as this concept keeps its meaning on the molecular or macromolecular scale.

These three steps can be summed up as follows: (1) initiation:

$$I \xrightarrow{h\nu} 2R'$$

$$R + M \xrightarrow{k_i} M'_1$$

(2) propagation:

$$M'_n + M \xrightarrow{k_p} M'_{n+1}$$

and (3) termination:

$$M'_n + M'_m \xrightarrow{k_t} \begin{cases} M_{n+m} \\ \text{or} \\ M_n + M_m \end{cases}$$

The initiation rate (R_i) depends on the absorbed light flux Φ_{abs}:

$$R_i = \left[\frac{dR}{dt}\right]_{initiation} = 2\Phi_{abs} \tag{12}$$

A classical expression of the polymerization rate (R_p) is

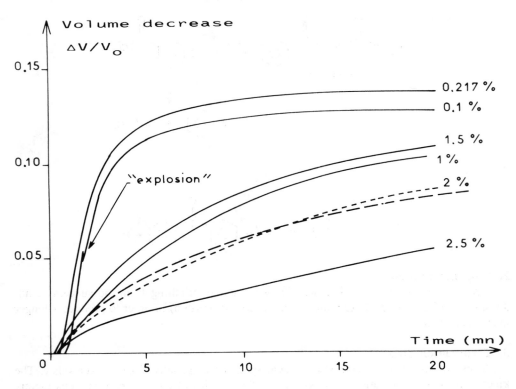

FIGURE 11.   Influence of the weight-to-weight ratio of the photochemical initiator on the reticulation kinetics of ethylene-glycol dimethyl acrylate under continuous excitation at 365 nm. Experimental curves of the volume decreases vs. time.

$$R_p = - \frac{d(M)}{dt} = k_p \left[\frac{R_i}{2k_t} \right]^{1/2} (M) = k_p \left(\frac{\Phi_{abs}}{k_t} \right)^{1/2} (M) \tag{13}$$

Therefore, it is proportional to the square root of the absorbed light flux. Without entering into all the details, let us point out that Equation 13 cannot be applied under the three following circumstances at least; (1) when there are high monomer conversion ratios; (2) if the light flux is intense; and (3) when stabilizing agents have been added to the mixture.

In the case of practical mixtures, one of these conditions at least is, indeed, fulfilled. The general expression of the R_p rate is still valid, but the exponents are different and vary during the reaction. This is due to the modification of the medium physical properties, in particular to the high increase in the viscosity at molecular scale, since the liquid-mixture turns into a solid.

Typical reactions have been experimentally studied.[15] The monomers or oligomers were either monoethylenic or diethylenic acrylates (or their mixture), and the light sources were either continuous or pulsed. Several mathematical models of these systems have been developed using two different techniques. One was a kinetic approach of the problem[18] and the other was a Monte-Carlo modeling based on the percolation theory,[15] but without taking the mobility due to Brownian movements into account. If the progress of the reaction vs. time is characterized by the volume decrease, as well experimentations as computer simulations lead to the same type of curves (see Figures 11, 12, and 13). They point out that the kinetic evolution of the system is complex, but it can be described by the classical laws of chemical kinetics.

Without entering into all the details of both approaches, several conclusions come out:

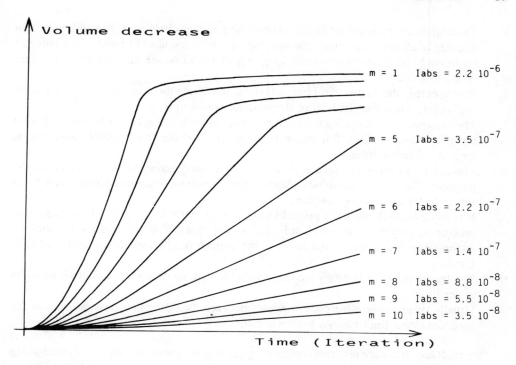

FIGURE 12.   Computed curves of the volume decreases vs. time obtained for different laser powers from several simple kinetic models.

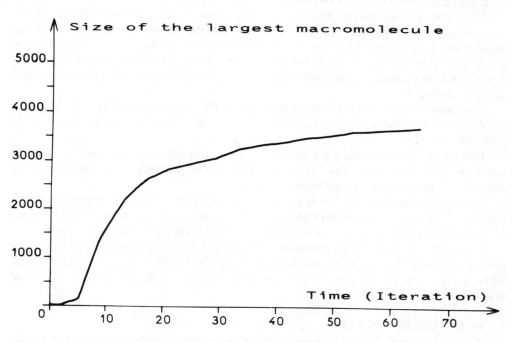

FIGURE 13.   Size of the largest macromolecule vs. time (Monte-Carlo calculations). A laser pulse creates at t = 0 radicals, which give a reticulated polymer by a random process.

1.   The higher the photon flux is, the shorter the polymer chain is. It is well known that one single photon can initiate the opening of a great number of bonds. The polymer chain could be a few micrometers long, which would be the ultimate precision limit of the process.

2.   As expected, the higher the photon flux is, the faster the reaction is, at least at its beginning, when the viscosity of the medium is still low.

3.   The reaction never ends, it goes on, very slowly indeed, a long time after the irradiation has been stopped. Our EPR experiments proved that the free radicals could live as long as several months.

4.   The reaction is never complete; there is always some unreacted monomer inside the polymer. This is a well-known phenomenon in polymerization reactions, and more precisely for reticulation reactions.

5.   It is more sensible to use a polyethylenic M/O to get a very hard three-dimensional reticulated polymer. Moreover, it is well known that during the reaction the solidification point is reached all the faster as the number of ethylenic bonds of the M/O is great.

6.   The hardness of the polymer increases with the number of absorbed photons up to its upper limit.

7.   Usually there are no adherence problems between the different voxels, as long as the hardness upper limit has not been reached.

We shall now study the problems originating either in the exothermicity of polymerization reactions or in the polymer shrinkage in greater detail.

B. EXOTHERMICITY OF POLYMERIZATION REACTIONS

Customary polymerization reactions are exothermal. During acrylates polymerization, for example, the temperature can rise up to 100°C if the reaction is complete, so it can increase the polymerization rate up to a factor of five. During our experiments, it rose up to about 60°C at the layer surface, but it decreases rapidly with depth, because the polymerization is not complete and is not homogeneous, since the light is, in principle, exponentially absorbed.

Heat transfers are slow because polymers are thermal insulators. For acrylate polymers the thermal conductivity is about 5.10 cal.cm^{-1}. s^{-1}. K^{-1}, and the specific heat is about 0.35 cal. g^{-1}. K^{-1}.

One of the most important parameters of the system, the ratio between the depth at which the laser beam penetrates and the thickness of the monomer layer, usually remains unchanged when the temperature rises up locally.

Another phenomenon that must be taken into account is the local expansion of the material. This can induce stresses inside the polymer and lead to an object appreciably out of shape when finished. For example, a flat surface can turn into a parabolic hyperboloid.

Up to now we have had no particular problem dissipating the heat produced during the polymerizations in our conventional apparatus. However, we will be faced with this problem when we use very fast systems. It will then be necessary to search for poorly exothermal reactions.

C. POLYMER SHRINKAGE

Regarding materials, one of the most important current problems is the polymer shrinkage. In our first experiments we used diacrylates, which led to a 15% shrinkage roughly. Moreover, if not carried by an already solidified volume, especially when the photon flux is weak, the hardening material has a definite tendency to bend down towards the bottom of the reactor (see Figure 14). This phenomenon is all the more pronounced, as the viscosity of the carrying fluid is low.

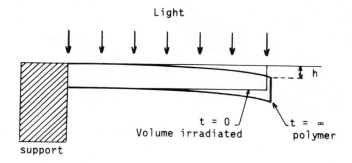

FIGURE 14.   Visualization of the warping when the shrinkage is important.

Now we polymerize commercial products (diacrylic oligomers) leading to shrinkages lower than 6%. The polymer may float at the monomer surface, thanks to surface tension phenomena. In any case, its deformation is strongly decreased by the high viscosity of the reaction mixture which acts as a support.

A model of the effect of shrinkage during the polymerization has been proposed[18] and is summarized hereunder.

During the reaction, the material is considered as viscoelastic. Every voxel is represented as a spring, the stiffness of which is k, in parallel with an absorber, the viscosity of which is η. The shrinkage is a function of $k.\eta$ and of the progress of the reaction. The polymer remembers its past; the deformation of the modelized object depends as well on the sum of the deformation of every voxel as on the way symmetries and limit conditions make it relax. Phenomenological models have been propounded.

The deformation due to the shrinkage clearly appears on objects we have made which are only a few centimeters long. This phenomenon can be reduced by several ways:

1.   One can use new polymers with very low shrinkage.
2.   The internal part of the object should not be polymerized, but only its surface, so the deformation will only concern a small volume of polymer.
3.   The software could take this deformation into account and counteract it by controlling the intensity of the photon flux and irradiation time or the light displacements (dashed polymerization). Computer simulations have already been developed. They lead to the conclusion that the deformation is smaller when pulsed light sources, such as excimer lasers, are used. (However, the system study will show the necessity to use continuous lasers.)

To visualize the inappropriate consequences of the polymer shrinkage, we have made computer simulations of the construction of a horizontal parallelepipedic rail fixed to a vertical support. This rail is supposed to be one voxel thick.

The principle of its construction method appears in Figure 15. The voxel next to the left side of the vat is first polymerized, and we admit that it sticks to it. After a Δt time interval, the next voxel is irradiated, and so on, Δt being kept constant. Therefore, if the laser is continuous, Δt is the irradiation time, and if it is pulsed, Δt is the pulse period.

The results of the computer simulations, in which the force of gravity was not taken into account, appear in Figure 16. The shrinkage effects have been largely exaggerated for a better understanding.

D. FEASIBILITY OF POLYMERIZING AN OBJECT WITH ACCURATE DIMENSIONS

We have shown above that the polymerization shrinkage can put an object out of shape,

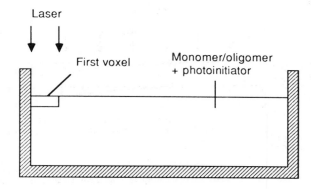

At t=0 : polymerization of the first voxel

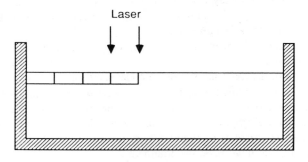

t>0 : polymerization of successive voxels

FIGURE 15.   Principle of construction of a parallelepipedic beam.

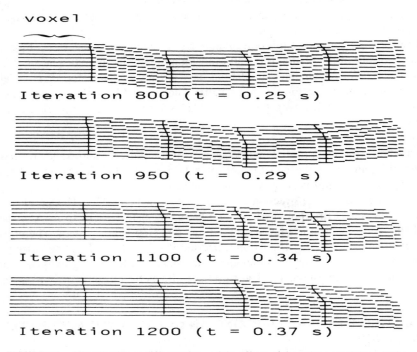

FIGURE 16.   Visualization at different times of the effects of the polymer shrinkage in the case of polymerization of a parallelepipedic beam.

FIGURE 17.   Example of an object made by layer polymerization (largest dimension is 80 mm).

resulting in dimensions shorter than expected, especially if common reactants, such as acrylates, are used.

Now, from an industrial point of view, a process is interesting only if the reactants are easy to find and cheap. Acrylates fulfill both conditions; they cost about 10$/kg. However, because of the shrinkage, a real problem appears when large objects have to be made.

Nevertheless, as it can be seen in the photograph, Figure 17, it is possible to make reasonably large objects (about 100 mm) with an acceptable warp (about 1 mm). This required a sensible choice of the reaction mixtures and of the experimental conditions, although these latter are less critical.

IV. SYSTEM STUDY

As a systems engineer, let us divide the monophotonic process into its different parts.

A. SYSTEM DECOMPOSITION
The process can be divided as follows:

1.   A computer for CAD which also monitors the whole system
2.   A photochemical reactor
3.   A laser
4.   A motorized optomechanical system with several degrees of freedom which concentrates the laser beam on the monomer surface and shifts it on horizontally
5.   Another motorized mechanical system which monitors the object construction along the vertical axis

The critical parts of the system will briefly come under review.

1. CAD Softwares and the Monitoring Electronics
The representation of real objects according to their geometrical and not functional properties is known as geometrical modelization. Its final purpose is to represent objects which are either full or hollow solids. Therefore, the geometrical modelization has to respect

a certain amount of conditions due to their being manufactured. Of course, the modelization is all the better since it takes into account more of these conditions.

For instance, a bidimensional modelization, which only manages views made of segments and arcs without even trying to match the different views, will be considered as poor. On the contrary, a modelization, which controls the physical reality and manufacturing of the design solids, will be considered a high-quality one.

The conditions to be fulfilled to have a good modelization are (1) validation: any designed object corresponds with only one real object; (2) power: it can make the model of many objects, and (3) versatility: it can be used for different uses (numerical command, finite elements modelization).

However, most authors agree on the difficulty of transformation of the desired object into a mathematical modelization.

In practice, the system has to use softwares of computed assisted design and manufacturing (CADM) able to (1) verify the object feasibility; (2) choose the construction axis; (3) decompose the object into voxels; (4) foresee the object deformations due to the shrinkage; and (5) command the process.

As far as hardware and the amount of information to be processed are concerned, powerful microprocessors (32 bits) in a coherent and reliable system must be used.

2. Choice of Laser

Carboxyled compounds are the more usual photochemical initiators. They absorb light up to 400 nm with a maximum at about 350 nm. Therefore, we have to use UV lasers which are either (1) pulsed UV lasers — × excimer lasers (248,308,351 nm most common wavelengths), × nitrogen lasers (337 nm), or × tripled or quadrupled YAG lasers (266,355 nm) or (2) continuous UV lasers — argon laser (351,363) nm) or helium cadmium laser (325 nm).

We made our first objects with an experimental excimer laser and we are using now a tripled YAG laser. Only lasers which can be used by unskilled people and can work for months with a good stability should be chosen for any industrial process.

Damage to optical components submitted to high intensity pulsed flux must also be considered.

Furthermore, when lasers and in particular pulsed lasers are focused onto the monomer, the power density can be so high that there is a volatilization of the monomer instead of a polymerization, although it should be profitable to be able to focus the beam on surface from 0.01 to 4 mm².

In any case the pulse rate is the limiting factor for the motor speed.

We are led to choose continuous UV lasers rather than pulsed ones for practical reasons and, in particular, because of their reliability, stability, and ease of use, although pulsed lasers can initiate polymerization reactions much more rapidly.

The power of HeCd lasers is too limited (10 mW), so the conclusion is that argon lasers, which are more powerful, up to 5 W seem to be the best choice at the moment.

3. Shifting and Focusing of the Laser

Shifting systems with step-by-step motors or direct current motors can be used in laboratories or in industry systems so as to make polymer models a few centimeters large.

Whereas to make bigger objects (meters long), other systems should be developed in order to reduce the building time. We restate here evidence that the building is proportional to the cube of the object average radius, so to double the length leads to multiplying the time by eight at least. Therefore, it is of paramount importance to reduce the voxel average building time as much as possible, thanks to the weightlessness of light, which can be shifted by low inertia techniques (galvanometric mirrors).

4. Layer-Building System

Different systems have been patented. For example, one consists in covering the polymer by pumping up the liquid monomer. Another consists in shifting down in the liquid medium a plateau which supports the solid polymer. All these systems work if the liquid monomer can wet the polymer. However, in all cases, the monomer level must reach an equilibrium state which depends on the monomer viscosity and the polymer surface. This is a real limitation of the process.

B. DEFINITION OF THE TECHNICAL SPECIFICATIONS OF THE SYSTEM

Let us consider that the models can be set inside a cube with a L long side and that the precision to be reached is P. Let us notice that a useful parameter is d = L/P. For example, l = 100 mm and P = 0.1 mm. Therefore, there is a maximum of 10^9 voxels to polymerize. The time (T) it takes to polymerize the model can be divided into three stages:

1. t_1 is the time it takes the computer to prepare the numerical commands.
2. t_2 is time required to deposit a monomer layer.
3. t_3 is the average time needed to polymerize each layer, which depends on whether the system polymerizes the perimeter of each layer or the surface. In the first case, this time is about Δt_3 4d. In the second case, it is about $\Delta t_3 d^2$, where Δt_3 is the voxel average building time.

We may derive from these obvious calculations that the building time will be defined as

$$T = t_1 + dt_2 + t_3 \tag{14}$$

Therefore, in the case of perimeter polymerization

$$T = T_P = t_1 + dt_2 + 4d^2\Delta t_3 \tag{15}$$

and in the case of surface polymerization

$$T = T_V = t_1 + dt_2 + d^3\Delta t_3 \tag{16}$$

What is problematic in the industrialization of the process is to get a reasonable value for T_P or T_V, i.e., a few hours so as to be competitive with classical processes.

Let us take a motorized system made of step-by-step motors with a speed limited to 10 cm/s; then Δt_3 is about 1 ms.

Therefore, we have (1) t_3 # 1h, and T_P # 2 h and (2) t_3 # 280 h, and T_V # 280 h.

Let us take a motorized system made of scanning mirrors. Then Δt_3 is about 10 μs, so (1) $t_3 = 40$ s and T_P # t_2 # 1 h and (2) t_3 # 3 h, so $T_V = 4$h.

In conclusion, it is possible to make complex models in a few hours, and there is compromise to reach so to obtain precision, rapidity, and dimensions large enough for an industrial process of interest.

V. CONCLUSION

We have presented here a way to make three-dimensional objects by a space-resolved laser photopolymerization in its principle. The industrial development of this still very young process depends on the clearing up of several remaining difficulties.

In the case of the two-and-a-half processes, they arise from

1.   The material: reaction kinetics, volume shrinkage, and exothermicity
2.   The laser source
3.   The laser beam deflection system
4.   The way to make the layers

To solve these problems, specialists in several branches of science will have to work together. These branches are chemistry, photochemistry, polymer science, optics, computer science, and automation at least. In our opinion, it must be possible to solve these last remaining difficulties so that in the future this process becomes one among other CADM techniques.

The possibility of making large objects with precisely respected dimensions is still a long way ahead; nevertheless, we remain optimistic. One only has to remember that Nicephore Niepce's first two-dimensional photochemical objects, i.e., photographs, required a long time to be made. However, within a century this time has been reduced down from several minutes to a few microseconds. Therefore, we think that scientific researchers, as well as industrials convinced of the theoretical and applied interests of this process, will join forces in bringing its industrial development to a successful conclusion.

REFERENCES

1. **Luzy, E. and Dupuis, C.,** Procédé pour obtenir des projectionsen relief, French Patent 461,600,1912.
2. **Swainson, W. K.,** Method medium and apparatus for 3D figure production, U.S. Patent 4,041,476, 1975.
3. **Swainson, W. K. and Kremer, S. D.,** Three-dimensional systems, U.S. Patent 4,078,229, 1975.
4. **Swainson, W. K. and Kremer, S. D.,** Methods and apparatus for producing 3D objects and patterns, U.K. Patent 1,556,451, 1976.
5. **Adamson, A. W.,** Methods and apparatus for generating three-dimensional patterns, U.S. Patent 3,609,706, 1968.
6. **Lewis, J. D., Verber, C. M., and McChee, R. B.,** Computer-controlled 3D pattern generator, U.S. Patent 3,829,938, 1971.
7. **Schwerzel, R. E., Wood, E., McGinniss, V. D., and Weber, C. M.,** 3D photochemical with lasers, *Appl. Lasers Ind. Chem., SPIE*, 548, 90, 1984.
8. **Schwerzel, R. E., Ivancic, W. A., Johnson, D. R., McGinniss, V. D., Wood, V. E., Jenkins, J. A., Kramer, S., and Swainson, W. K.,** Three-dimensional photochemical machining with lasers, US Air Force Final Technical Report, 1984.
9. **André, J. C., Le Mehauté, A., and De Witte, O.,** Dispositif pour réaliser un modèle de pièce industrielle, French Patent 8,411,241, 1984.
10. **Hebert, A. J.,** Solid object generation, *J. Appl. Phot. Eng.,* 8,185, 1982.
11. **Nakai, T. and Marutani, Y.,** Shaping 3D solid objects using a UV laser and photopolymer, in Conf. Laser and Electrooptics, San Francisco, 1986.
12. **Corbet, A., André, J. C., and Grisoni, B.,** Dispositif de réalisation d'objets réels par photopolymérisation induite thermiquement, French Patent 8,602,327, 1986.
13. **André, J. C., Bouchy, M., Cabrera, M., Le Méhauté, A., and De Witte, O.,** Procédés et dispositif pour réaliser un modèle de piece industrielle, French Patent 8,509,055, 1985.
14. **André, J. C., Cabrera, M., Jezequel, J. Y., Le Méhauté, A., and De Witte, O.,** Procédé pour réaliser un modèle de pièce industrielle et dispositif de mise en oeuvre de ce procédé, French Patent 8,509,054, 1985.
15. **Cabrera, M.,** Formogénèse Laser 3D, Thesis, Institut National Polytechnique de Lorraine, Nancy, France, 1986.
16. **Grisoni, B.,** Morphosynthèse Laser, Thesis, Institut National Polytechnique de Lorraine, Nancy, France, 1986.
17. **Haas, C.,** Usinage et fabrication par lasers, DEA Génie des Procédés, Institut National Polytechnique de Lorraine, Nancy, France, 1987.
18. **Duclos, A. M.,** Modélisation des déformation de pièces en cours de photopolymérisation, DEA, Génie des Procédés, Institut National Polytechnique de Lorraine, Nancy, France, 1987.

19. **Cabrera, M., Andre, J. C., De Witte, O., and Le Méhauté, A.,** Photochimie industrielle IX: fabrication assistée par ordinateur, par Laser UV, *Entropie,* 133, 35, 1987.
20. **Ben Shaul, A., Haas, Y., Kompa, K. L., and Levine, R. D.,** Eds., *Laser and Chemical Change,* Springer-Verlag, Berlin, 1981, 10.
21. **Ambartzumian, R. V. and Letokhov, V. S.,** *Appl. Opt.,* 11(2), 354, 1972.

Chapter 4

POLYMERS FOR HIGH-POWER LASER APPLICATIONS

Robert M. O'Connell

TABLE OF CONTENTS

I. INTRODUCTION

Optical components fabricated from organic polymers (plastics) offer several advantages over their silicate glass counterparts. They can be injection molded or cast, which is less expensive than the grinding and polishing procedures usually required with glass. They are also lighter than glass components, which makes them attractive for portable systems and others where total weight is important.[1] Perhaps most importantly, some of them can easily be doped with certain organic dyes for use as passive Q-switches or as the active elements in dye lasers. In fact, many plastic optical components (lenses, windows, prisms, Q-switches) are now commonly available and are being used in low-power optical systems. Plastic Q-switches, for example, are used in commercially available laser range finders.

These and other appealing attributes make polymers attractive for use in high-power systems, too, e.g., laser fusion, but certain materials problems have thus far prevented it. Chief among them is a very low optical strength. As compared to glass, plastic tends to suffer permanent optically induced damage at much lower optical energy density levels. Researchers first became aware of the problem in 1966,[2] and have been working to solve it since then. In fact, some recent results[3,4] suggest that significant progress has already been achieved toward the goal of plastics with optical strength comparable to that of silicate glass.

There are other problems associated with the implanted organic dye molecules in plastic-host Q-switches and dye lasers that have further hindered the development of plastics for high-power applications. In plastic-host Q-switches, the problem is poor switching efficiency,[5] which is caused by the combination of relatively low absorption during the absorptive part of the switching cycle and relatively high residual absorption during the transmissive part of the switching cycle. Poor switching efficiency results in laser output pulses with less than optimum peak powers. In plastic-host dye lasers, the problem is the permanent degradation (bleaching) of the implanted dye molecules due mainly to the optical pump light, as was first reported in 1971.[6] This effect results in serious reductions in pump conversion efficiency,[7] which in turn leads to reduced output. Research has sought to solve these dye-related problems and, as with the optical strength problem, good progress is apparently being made.[4,8,9]

In this chapter, we review some of the progress that has been made in developing polymers for high-power laser applications. We begin in Section II with a summary discussion of some specific polymeric materials and their applications in high-power laser systems. In Section III we review research to understand and eliminate the causes of laser damage in polymers. In Section IV we consider the problems of low switching efficiency in polymer-host Q-switches and photobleaching in polymer-host dye lasers. We conclude the chapter with a brief summary of the progress that has been made in developing polymers for high-power laser systems and an indication of some of the problems yet to be solved.

II. POLYMERS IN HIGH-POWER LASER SYSTEMS

A. BULK COMPONENTS AND OTHER UNDYED APPLICATIONS

The most obvious potential application for polymers in high-power systems is in the various transparent bulk components, such as lenses, windows, and prisms, that are used in such systems. As the research summarized in Section III will show, in addition to being relatively inexpensive, safe, and easy to handle, polymethylmethacrylate (PMMA) has the highest optical strength of the polymers studied to date; thus, it is the material most likely to compete with glass in these components.

Other polymers have recently been studied for use in laser fusion systems as debris shields[10] and as edge claddings for large laser disk amplifiers.[11] In the debris shield application, a 0.025-mm thick sheet of FFP teflon, a relatively inexpensive copolymer of tetrafluoroethylene and hexafluoropropylene, is placed between the target and the expensive focusing lenses of the system and used to prevent the target debris from reaching the lenses.[10] Tests showed that, although the shield was susceptible to optical damage by the laser, it could still be used as long as the optical damage did not seriously perturb the laser beam impinging on the fusion target.

Edge claddings are used with the disk amplifiers in laser fusion systems to absorb spurious scatter and thereby eliminate edge reflections that could deplete the excited gain medium and reduce the useful gain of the amplifier.[11] In this application, a composite polymer-glass edge cladding, consisting of copper-doped phosphate glass and a highly cross-linked bis-phenol-based epoxy adhesive, was tested as a low-cost substitute for monolithic glass and found to perform acceptably.

B. DYE-DOPED POLYMERS AS PASSIVE Q-SWITCHES

Q-switches are used in lasers to produce high-power output pulses. They do this by greatly lowering the optical quality (the Q) of the laser resonator cavity during the pumping cycle so that lasing is prevented while the population inversion builds to a level far in excess of the normal lasing threshold. This is the low Q (high loss) state. When the cavity Q is switched back to its usual high Q (low loss) state, the energy stored in the highly inverted atomic population is released by stimulated emission. The laser thus emits a high energy optical pulse in a time comparable to the lifetime of the upper level of the laser transition, e.g., 5 to 50 ns.[12] The ideal Q-switch would be an alternately opaque/transparent optical shutter.

Both active and passive Q-switching schemes are available. In systems requiring maximum switching speed (<10 ns), precise timing and control of the laser pulse, good pulse-to-pulse stability, and active Q-switch devices based on electrooptic shutters must usually be used.[13] Such systems involve expensive crystals, additional cavity optics, and timing and control circuitry. They are rather elaborate. However, in those systems where speed, timing, and stability are not so critical, passive Q-switches, which do not require electrooptic crystals or additional optics and electronics, can be used.

In passive Q-switching, a relatively thin (~1 mm) cell is usually placed between the laser active medium and the totally reflecting mirror in the optical resonator. The thin cell contains an organic dye in liquid or solid solution that can alter the cavity Q properly if it has a saturable spectral transition at the laser wavelength. During the beginning of the optical pumping cycle, the spontaneous emission that would normally initiate laser oscillation is absorbed by the dye molecules of the Q-switch, thus allowing the necessary large excess population inversion of the active laser medium to develop.[13] When the populations of the upper and lower energy levels of the absorbing transition of the Q-switch dye become approximately equal, the dye is said to be saturated or temporarily bleached,[14] that is, it becomes transparent to the fluorescent emission that initiates oscillation and the emission of a Q-switched laser pulse.

The first passive Q-switches were developed in 1964. They consisted of liquid solutions of phthalocyanine[15] and cryptocyanine[16,17] dyes for Q-switching the ruby laser and a liquid solution of a polymethine dye for Q-switching the neodymium laser.[18] Polymer-based solid solution Q-switches were developed soon after the liquid solution devices. Once fabricated, which can usually be done by simply dissolving the Q-switch dye in the monomeric liquid prior to polymerization, they eliminate any further need to work with liquids. They are also small, inexpensive, disposable, and therefore ideally suited to portable applications.

Table 1 lists several polymer-host dye combinations that have been developed for Q-switching the ruby and neodymium lasers.[8,19-24] Note that whereas vanadium phthalocyanine (VOPc) has continued to be an acceptable Q-switch dye for ruby, the polymethine dyes were found to suffer premature deterioration or aging and have been replaced with the highly

TABLE 1

Some Polymer Host:Dye Combinations that have been Studied as Q-Switches for Ruby and Neodymium Lasers

Laser/ wavelength	Author/year	Polymer-host	Q-Switch dye	Comments	Ref.
Ruby/ 0.69 μm	Kusakawa/1968	PEMA and PMMA	VOPc	Low optical strength observed and studied	19
	Kusakawa/1969	PPMA, PBMA, and copolymers of MMA-nBMA	VOPc	Optical strength studied	20
	Sharlay/1970	PVA film	CuPc	Low optical strength	21
	Gromov/1982	Modified PMMA	VOPc	Similar performance to liquid-solution devices	8
Neodymium/ 1.06 μm	Abdeeva/1971	Cellulose Triacetate film	Polymethine	Poor switching efficiency, low optical strength, and rapid dye aging	22
	Kirk/1973	PMMA	bis-Nickel	Stable Q-switch dye	23
	Gromov/1982	Modified PMMA	bis-Nickel	Improved optical strength and switching efficiency	8
	Acharekar/1983	Cellulose acetate sheet	bis-Nickel	Improved optical strength but low switching efficiency	24

Note: (PEMA) polyethylmethacrylate, (PMMA) polymethylmethacrylate, (PPMA) polypropylmethacrylate, (PBMA) polybutylmethacrylate, (PVA) polyvinyl alcohol, (VOPc) vanadium phthalocyanine, (CuPc) copper phthalocyanine, and (bis-Nickel) bis-(4-dimethylaminodithiobenzil)-nickel.

stable bis-(4-dimethylaminodithiobenzil)-nickel (bis-nickel) compound for Q-switching neodymium.[25] Note also that, as can be seen from the comments given in the table, the primary problems hindering the development of high-power polymer-based Q-switches are low optical strength and poor switching efficiency. We will review research that has attempted to understand and solve these problems in Sections III and IV, respectively. It will be seen that the polymer-host plays significant roles in both problem areas.

C. POLYMER-HOST ORGANIC DYE LASERS

The most attractive and distinguishing feature of organic dye lasers is their tunability through the visible portion of the spectrum. This is due to the fact that there are many organic dyes with broad fluorescent spectra in the visible region. Thus, a laser system consisting of an appropriate collection of interchangeable dyes and a cavity resonance tuning mechanism constitutes a continuously tunable visible laser.

In most dye laser systems, the active dye is contained in a flowing liquid solution, which eliminates the need for cooling and the tendency to optical damage that are common with solid-state laser rods. It also automatically removes dye molecules that have been thermally or optically degraded (permanently bleached).[26] Thus, the first dye lasers, reported in 1966[27-30] and 1967,[31,32] were of the flowing liquid solution type. Due to the rather fragile nature of the dye molecules, optical pumping methods were and have continued to be used. (Pumping refers to the creation of the necessary population inversion in the active laser medium). Initially, the ruby laser was used to pump various cyanine dyes.[27-30] Shortly afterward, flashlamp pumping was used to excite several xanthene dyes.[31,32]

Along with the advantages of flowing liquid solution systems, there are certain drawbacks.[33] For example, the liquid dyes are subject to stagnation and/or evaporation, thereby requiring periodic replacement. Also, the flowing system requires pumping and recirculation equipment, which could fluctuate during operation, thereby affecting laser gain. Of course, such systems are elaborate, i.e., neither lightweight nor compact. These drawbacks can potentially be eliminated by replacing the flowing liquid with an inexpensive, lightweight, disposable dye-doped solid polymer-host. As in the case of plastic Q-switches, such devices offer the possibility of vastly simpler dye lasers. They are, however, subject to the thermal and optical problems that are eliminated in flowing liquid systems.

The first polymer-host dye laser was reported in 1967,[34] and many others have been reported since then. A representative listing is given in Table 2.[4,6,7,9,34-44] Note that, as the laser wavelength data shows, these polymer-host devices have already spanned the visible spectrum (400 to 700 nm) reasonably thoroughly. As indicated above, however, they all suffer from a variety of thermal and photochemical effects (see the comments in Table 2). In Section IV, we will review some of the research that has studied these effects. It will be seen that, as in the problems with plastic Q-switches, the polymer-host plays an important role. Note finally that, in contrast to Table 1, there is little mention of low optical strength in the comments of Table 2. This is simply because optical energy densities in the dye laser studies were kept well below the optical damage thresholds in order to study the dye-degradation effects.

III. THE OPTICAL STRENGTH OF POLYMERS

As we said in Section I, the main reason why polymers have not found widespread use in high-power laser systems is that they have relatively low optical strength (or, equivalently, low laser damage thresholds). In this section we review some of the research that has sought to characterize, understand, and improve the optical strength of polymers. We begin with definitions of laser damage and its threshold, and with a brief description of how laser damage thresholds are typically measured. We then present data comparing the bulk optical strength of polymers to their surface optical strength, the damage thresholds of PMMA and several readily available polymers, and the optical strengths of polymers and optical glasses. Next, we review research showing that optical damage in polymers is apparently always initiated extrinsically, i.e., by absorbing defects, and that its growth can be controlled by adjusting the thermoelastic properties of the polymer material.

A. LASER DAMAGE: DEFINITIONS AND MEASUREMENTS

The optical strength of a material is defined by its laser damage threshold. Traditionally, laser damage has been defined as "the irreversible, destructive alteration of a material substance by a sufficiently intense laser light wave."[45] In this definition laser damage and its threshold can be identified with "destructive material alteration" and "sufficient laser intensity", respectively. Unfortunately, there are certain difficulties involved in accurately identifying these two quantities.

The "destructive material alterations" that constitute laser damage occur as molten or carbonized regions, or as pits or fractures. Their occurrence is usually indicated by at least one of the following: scattered light from a nondamaging probe laser beam, emission of sparks or charged particles, or the results of a postillumination microscope examination.[46,47] The difficulty is that these damage indicators are somewhat subjective and not equally sensitive. Thus, thresholds are operator and indicator dependent and may decrease as the indicators become more sensitive, even though material improvement efforts may actually increase their optical strength.

It would appear that the difficulty in identifying laser damage could be eased by specifying a single damage indicator. However, the problem is compounded by the fact that optical materials often contain several damage-causing mechanisms that cause damage differently from each other. Thus, a damage indicator that is relatively sensitive to damage by one mechanism may be insensitive to damage by another. For example, one mechanism may produce sparks visible to the unaided eye while another one produces pits requiring

TABLE 2
Some Polymer-Host:Dye Combinations in which Lasing has been Observed

Author/year	Polymer host	Laser dye	Laser wavelength (nm)	Pump source	Comments	Ref.
Soffer/1967	PMMA	Rhodamine 6G	560—580	Nd laser at 532 nm	First observation	34
Peterson/1968	PMMA	Rhodamine 6G	601	Flashlamp	Low triplet state quenching and thermal lensing observed in both cases	35
		Rhodamine B	632			
Naboikin/1970	PMMA	α-NPO	396	Nd laser at 353 nm	Wide coverage of visible spectrum; poor photochemical stability (all cases)	36
		POPOP	415			
		Dimethyl POPOP	423			
		Eight oxazole derivatives	410, 414, 425, 428, 443, 444, 455, 480			
		Three oxadiazole derivatives	400, 416, 418			
		Two pyrazoline derivatives	450, 457			
		One dimethyl aniline derivative	478			
	Polystyrene	Fluoran	514			
		α-NPO	411			
		POPOP	420			
		Dimethyl POPOP	424			
	Polyvinyl xylene	Dimethyl POPOP	425			
Kaminow/1971	PMMA	Rhodamine 6G	570	Nd laser at 532 nm	Narrow bandwidth (0.1Å); excessive photobleaching observed	6
Ulrich/1972	Polyurethane	Rhodamine 6G	605	Nitrogen laser at 337 nm	Thin film; photobleaching observed	37
Chang/1972	Polyurethane	Rhodamine B	633	Nitrogen laser at 337 nm	Thin film	38
Drake/1972	PMMA	Acridine red	615	Flashlamp	Thermal and photochemical effects reduced with light-converter solutions (all cases)	7
		Rhodamine 6G	585			
		Fluorescein	540			
		4-Methylumbelliferone	405			

Author/year	Polymer	Dye	Wavelength	Pump source	Comments	Ref.
Reich/1974	Polyacrylonitrile	Rhodamine 6G	Not reported	Pulsed dye laser at 520 nm	Photobleaching studied	39
Heinsohn/1976	Two-component epoxy resin	Rhodamine B	600—620	Nd laser at 530 nm	Photobleaching studied	40
Wang/1976	Polyester PMMA	Rhodamine 6G Cresylviolet Rhodamine 6G	560—600 680—690 562	Nd laser at 532 nm	Photobleaching reduced by raster pumping	41
Itoh/1977	PMMA	Seven Coumarin dyes	500	Nitrogen laser at 337 nm	Thin film; low output power and dye photodegradation observed	42
Sriram/1980	Polyurethane	Rhodamine 110	643	CW argon ion laser at 514.5 nm	Thin film	43
Gromov/1982	Modified PMMA	Xanthene 11B	561	Nd laser at 530 nm	Precision machining technology and long-life laser dye xanthene 11B reported	9
Bezrodnyi/1982	Polyurethane	Triphenylmethane	750	Nd laser at 530 nm	Photobleaching, pump conversion efficiency, and optical strength studied	44
Gromov/1985	Modified PMMA	Rhodamine 6G chloride Rhodamine 6G perchlorate Rhodamine 111 Xanthene 11B Oxazine 17	560—570 560—570 560—570 560—570 620—640	Nd laser at 530 nm	Photobleaching, pump conversion efficiency, and optical strength studied (all cases)	4

100 × microscopy in order to be seen. Since the laser beams used in damage testing usually irradiate only a relatively small portion of the material, it cannot be assumed that identical material environments are irradiated from test to test. Hence, any one of the various damage mechanisms may be expected to cause damage in a given test, necessitating the use of several damage indicators, rather than one.

Another important consequence of the relatively small size of laser damage test beams (and their nonidentical irradiated volumes) is that the "sufficient laser intensity" required to cause damage must be determined by a statistical procedure that is subject to wide fluctuations. Typically, several separate locations (sites) on (for surface damage tests) or in (for bulk damage tests) the optical material under test are illuminated by the test laser with its output set at a fixed level. At each illuminated location various damage indicators are used to determine whether damage occurred during the irradiation. The ratio of damaged to illuminated sites thus determines the probability of damage by the laser at the given output level. The procedure is repeated at several such laser output levels, producing a range extending from the 0 to the 100% damage probability level. The laser damage threshold is then defined as the test laser output associated with a particular damage probability, which has traditionally been the 50% level. The difficulty arises because the optical materials being tested are usually relatively small, which necessitates the use of small spot size test beams and allows testing of relatively few sites at each laser output level. Since, as we stated, the irradiated volume is not usually identical from site to site, especially with small spot size beams, the measured damage probabilities can fluctuate widely.

The statistical problem is compounded by the fact that when discrete absorbing defects are responsible for the damage, they are more likely to be irradiated (and thus cause damage) by larger spot size beams than by smaller ones.[48] Therefore, the test laser output corresponding to any nonzero damage probability, e.g., 50%, will be spot size dependent. In contrast, the damage onset level,[49] which is the highest test laser output that never causes damage, is spot size independent,[50,51] and is thus a better choice of damage threshold in the defect-dominated case. As we will see, defects do appear to be the cause of laser damage in polymers.

The described damage testing procedure is valid with either continuous-wave (cw) or pulsed lasers. With pulsed lasers, it can be used to determine the threshold for damage in one laser pulse (single-pulse damage) or after any specified number of pulses (multiple-pulse damage).

A final difficulty involved in measuring a laser damage threshold concerns the accurate determination of the laser output in the damage statistics measurement. Although the statistics approach is valid with cw as well as pulsed lasers, we will consider only pulsed lasers with pulse widths (full width at half maximum) on the order of nanoseconds here because they are used in the vast majority of laser damage studies. In such pulsed measurements, the laser output is characterized either by its peak fluence in joules per square centimeter or by its peak intensity in watts per square centimeter, which is just the ratio of the peak fluence to the easily measured temporal pulse width. The peak fluence in a laser pulse can be determined either directly by measuring the peak energy transmitted through a pinhole of known area[49] or indirectly through a calculation that requires a measurement of the total energy in the pulse and its spatial fluence distribution (the beam profile).[52] Beam profiling methods include the use of photographic film,[53] video cameras,[54,55] and various masking schemes, such as the pinhole,[56] slit,[57] and knife-edge[52,58] methods. The difficulty here is that each of these methods for measuring peak fluence involves its own assumptions and/or approximations that can introduce significant uncertainty into the results. Consequently, peak fluences cannot usually be determined with less than 10% uncertainty. Although a discussion of beam profiling is beyond the scope of this chapter, the details of the various methods can be found in References 49 and 52 to 58.

In summary, the measurement of the optical strength of a material, defined as the laser damage threshold, involves the subjectivity of deciding whether damage has occurred on a given test, statistical fluctuations that result from finite test beam spot sizes and relatively small numbers of tested sites, and the experimental uncertainties inherent in obtaining beam profiles. Consequently, considerable caution is needed in interpreting and comparing published results.

B. PMMA IN PERSPECTIVE

It will be apparent in Section C that most polymer laser damage research has been devoted to the problem of bulk damage in PMMA. Before discussing some of that work, we review here a few characterization studies that enable one to compare the optical strength of PMMA to that of silicate glasses and other polymers and which at least partly explain the emphasis of research on the bulk (rather than surface) optical strength of PMMA (rather than some other polymer).

1. Single-Pulse Optical Damage Thresholds

In an early investigation of laser damage in transparent dielectrics, Ashkinadze et al.[2] reported that the single-pulse threshold for bulk damage by 20 to 30-ns wide pulses from either a ruby laser (operating at the wavelength 0.69 μm) or a neodymium laser (operating at 1.06 μm) was ten times higher in K-3 glass than in PMMA. They noted that this ratio matched the ratio of the elastic moduli of the two materials. Later, Bass and Barrett[59] measured the single-pulse surface optical damage thresholds of several materials with 12-ns wide pulses from a 1.06-μm neodymium laser. Included among the materials studied were fused quartz (a silicate glass) and PMMA, whose measured surface damage thresholds were 24.0 and 16.1 GW/cm,[2] respectively. When compared with Ashkinadze's result, this one suggests that the relative weakness of the bulk optical strength of PMMA is far more serious than that of its surface optical strength.

The seriousness of the bulk optical damage problem in PMMA is supported and clarified by the results of work by Fradin and Bass[60] and Bebchuk et al.,[61] all of whom measured ratios of bulk-to-surface single-pulse damage thresholds in various transparent dielectrics. Fradin and Bass studied fused quartz and BSC-2 glass with a 1.06-μm, 4.7-ns pulse-width Nd:YAG laser. They found that the ratio of the thresholds (in electric fields) for bulk and surface damage varied between 1.0 and 1.5 for fused quartz and 1.0 and 1.3 for BSC-2 glass. The larger ratios were obtained on conventionally polished samples whose surfaces were clean but which contained certain polishing imperfections. When special steps were taken to provide imperfection-free surfaces, the bulk-to-surface damage threshold ratios dropped to 1.0. They suggested that the observed results were due to electric field enhancement[62] at those imperfections found only on the conventionally polished surfaces. They also measured a bulk-to-surface electric field damage threshold ratio of 2.0 for conventionally polished sapphire. The significance of their work for our purposes is that the surface optical strength of the dielectrics studied is generally weaker than their bulk optical strength. This result has been observed in many other dielectric materials, and it explains why most studies of laser damage in dielectrics have been concerned primarily with surface rather than bulk damage.

Bebchuk et al.[61] used a 25-ns pulse-width ruby laser (at 0.69 μm) to make similar measurements on ruby and PMMA. Their results for ruby agreed with those of Fradin and Bass, that is, the ratio of the bulk-to-surface peak fluence damage thresholds varied between 7.7 and 20 (or between 2.7 and 4.5 in terms of electric fields), and depended on the "degree of imperfection" of the polished ruby surface. With PMMA, however, the bulk-to-surface peak fluence damage threshold ratio was only 0.5, and was relatively independent of the quality of the finish of the PMMA surface. Thus, PMMA has the unusual property of greater optical strength on the surface than in the bulk. To explain these results, they claimed that

TABLE 3
Measured Single-Pulse Threshold Peak Fluences (F_0) for Bulk Laser Damage in Several Commonly Available Transparent Polymers and Optical Silicate Glasses

Author	Polymer	Test wavelength (μm)	Laser pulse width (nsec)	F_0(J/cm²)	Ref.
Agranat	PMMA	0.69	20	17	64
	Polystyrene			8	
	Polycarbonate			24	
Milam	PMMA	1.06	0.125	1.6	46
	Polystyrene			0.8	
O'Connell	PMMA	1.06	8	41	65,66
	Polycarbonate			15	
	Cellulose acetate butyrate			15	
Fradin	Fused quartz	1.06	4.7	503	60
	BSC-2 Glass			413	
Merkle	Supracil 1 fused quartz	0.53	15	1350	67
		1.06	21	2310	
	Corning 7940 fused quartz	0.53	15	1650	
		1.06	21	2310	

in brittle (nonplastic) dielectric materials, the surface polishing action creates a capillary effect that causes some of the polishing abrasive to be drawn into the surface, thereby creating a defective surface layer with a relatively weak optical strength. They also stated that no such capillary effect occurred in PMMA and other plastic dielectrics, but they did not explain why. Aldoshin et al.[63] and Dyumaev et al.,[3] whose work we will discuss later, explained the effect in terms of thermoelastic properties.

The next point to be made here is that the optical strength of PMMA is generally greater than that of other commonly available polymers. This is illustrated in Table 3, which lists the results of measurements of the single-pulse bulk laser damage threshold peak fluence (F_0) of PMMA and other polymers.[46,60,64-67] The only exception in the table is Agranat's early result[64] for polycarbonate at 0.69 μm, which is higher than his result for PMMA. At 1.06 μm, however, PMMA apparently has significantly greater optical strength than polycarbonate, as shown by O'Connell's results[65,66] in the table.

We close this part of the discussion by including in Table 3 the results of measurements of single-pulse bulk laser damage thresholds in several optical quality silicate glasses. Although the data for the polymers and glasses were obtained in different investigations and should not therefore be compared too closely, the general statement that even the strongest polymer, PMMA, is much weaker than optical silicate glass, can clearly be made. By comparing the 1.06 μm, similar pulse-width data of O'Connell et al.,[65,66] Fradin and Bass,[60] and Merkle et al.,[67] it becomes evident that the single-pulse bulk optical strength of optical glass is 10 to 50 times as great as that of commonly available PMMA.

2. Multiple-Pulse Optical Strength

Thus far, we have compared only the single-pulse optical strengths of various polymeric and silicate glasses. Since many high-power laser systems are operated in a steady-state pulsed mode, the multiple-pulse optical strengths of these materials are also important. One popular method for evaluating the multiple-pulse optical strength of a material is to statistically determine the average number of pulses needed to damage it when the laser is operated at a specified pulse repetition frequency (PRF in pulses per second, pps) with its peak fluence set at each of several fractional levels below the previously determined threshold for single-pulse damage (fractional subthreshold fluences, F/F_0). To compare the results of several

TABLE 4
Fractional Subthreshold Fluences at which Laser Damage Occurred after an
Average of 100 Pulses [F/F$_0$(100)] in Several Multiple-Pulse Laser Damage Studies

Author	Material	Pulse repetition frequency (PRF) in pulses per second	F/F$_0$(100)	Ref.
Balitskas	K-8 glass	12.5	0.53	68
		50	0.45	
Merkle	Fused quartz	10	0.75	67
Butenin	PMMA	—[a]	0.15	69
Emelyanova	PMMA	—[a]	<0.1	70
Manenkov	PMMA	—[a]	<0.1	71
O'Connell	PMMA	1	<0.1	65

Note: All data were obtained with 1.06-μm wavelength, Q-switched pulses from neodymium lasers.

[a] PRF data not reported.

such measurements, Table 4 lists the fractional subthreshold fluence levels at which various optical quality silicate glasses and PMMA damaged after an average of 100 laser pulses.[65,67-71] In those cases which provided it, the laser PRF is given also. The data in the table illustrate the important point that the multiple-pulse optical strength of silicate glass is much greater than that of PMMA. For example, whereas Merkle et al.[67] required a fractional subthreshold fluence as high as 0.75 in order to damage fused quartz (Supracil-1 and Corning 7940) after 100 pulses, O'Connell et al.[65] needed an F/F$_0$ of only 0.10 to obtain the same result with PMMA (Plexiglas-G).

The PRF data given in Table 4 accentuate the seriousness of the relatively weak multiple-pulse optical strength of PMMA. To establish this point, we note first that multiple-pulse laser damage to silicate and organic glasses is PRF dependent, such that the higher the PRF, the smaller the number of laser pulses required to cause damage. Data from Balitskas' measurements[68] at PRFs of 12.5 and 50 pps (see Table 4) show this to be true of K-8 glass, and O'Connell et al.[65] observed that at 5 pps PMMA damaged in fewer pulses than at 1 pps, thus showing it for PMMA. Since multiple-pulse laser damage is PRF dependent, it can be seen that if the measurements of, for example, Merkle et al.[67] and O'Connell et al.[65] were made at equal PRFs, the discrepancy between the multiple-pulse optical strengths of fused quartz and PMMA indicated by the data in Table 4 would be even greater.

It should be clear from the research reviewed in this subsection that in polymers the bulk optical strength is weaker than the surface optical strength, that, compared to other commonly available tested polymers, PMMA has the highest optical strength, and that transparent polymers have less resistance to single- and multiple-pulse laser damage than silicate glass. Considering these results, it is understandable why research to understand and improve the optical strength of polymers has concentrated on bulk damage in PMMA. In addition, the methylmethacrylate (MMA) liquid monomer is relatively safe to handle, and it can be easily filtered, dyed, and polymerized,[72] making PMMA am excellent subject for research.

C. CAUSES OF POLYMER LASER DAMAGE AND THEIR ELIMINATION

Since Ashkinadze et al.[2] first reported their observations of laser damage in PMMA, a great deal of research has been conducted to understand and eliminate the causes of laser damage in polymers and to thereby improve their optical strength. We review some of that research here.

1. Two Major Themes

One of the earliest studies of the causes of laser damage in polymers was conducted by

Kusakawa et al. during their work with polymer-host Q-switches[19,20] (see Table 1). In their first report[19] they observed that after successfully Q-switching a ruby laser a number of times, their devices would cease to operate, due to the growth of laser-induced damage to the Q-switch. Using the number of Q-switch output pulses (laser oscillation times) as the indicator of damage, they measured the temperature dependence of the laser damage growth rate of several polyethylmethacrylate (PEMA) Q-switches and found that it was always slower at higher temperatures. Also, in fixed temperature measurements, they observed that devices based on PMMA as host damaged faster than those based on PEMA and that PEMA devices doped with 3-μm diameter VOPc-saturable absorber dye particles damaged faster than PEMA devices doped with 1-μm diameter VOPc particles.

To explain these observations, they proposed that the laser damage process begins with the absorption of laser photons and their conversion to thermal energy by the dye particles. Particles of 3-μm diameter absorb more efficiently than 1-μm particles. The thermal energy is then transferred from the dye particles to the surrounding polymer molecules, which may then experience damaging thermomechanical stresses, depending on the amount of heat transferred and the micro-Brownian motion of the polymer molecules. If this motion is vigorous, the stresses will be delocalized by the thermal relaxation of the polymer molecules, thereby either preventing damage or minimizing its extent. They noted that micro-Brownian motion generally increases as the temperature of a polymer approaches its glass transition temperature. (In their second report[20] (below) they associated this with increasing thermal conductivity). Thus, the laser damage growth rate should decrease with increasing temperature, as observed in their measurements. Also, since the glass transition temperature of PMMA and PEMA are, respectively, approximately 105 and 65°C[73], at a typical operating temperature, e.g., 25°C, PMMA is further from its glass transition temperature than PEMA is and should therefore damage faster. They observed this too, although it should be noted that for a fair comparison the dye concentration should have been the same in both devices. They did not say whether this were true.

In conclusion to their first study, Kusakawa et al. suggested that the growth of laser damage in polymers could be minimized by eliminating absorbing impurities, especially those larger than 3 μm in diameter, and by using the material at a temperature slightly below its glass transition temperature. In their second report[20] they extended their study of the relationship between the laser damage growth rate of a polymer and its glass transition temperature, seen earlier with PMMA and PEMA Q-switches, to the copolymer system of MMA and n-butylmethacrylate (n-BMA), i.e., P[MMA-nBMA]. They used thermal polymerization at 80°C without an initiator to synthesize both VOPc-dyed and undyed samples of the copolymer with MMA:n-BMA weight percent ratios of 80:20, 60:40, 40:60, and 20:80. The dyed samples had identical transmittances (40%) at the 0.69-μm ruby laser wavelength, and their glass transition temperatures, measured with the dilatometric method,[74] decreased steadily with increasing weight percent of n-BMA. The laser damage growth rates of the copolymers at 27°C, measured as in the earlier report,[19] were found (as with PMMA and PEMA) to be proportional to their glass transition temperatures. This result was obtained separately with both the undyed samples and the equal transmittance dyed samples, although the dyed sample of each copolymer damaged much faster than its undyed counterpart. The results of Kusakawa's second report thus support the proposal made in the first report concerning the roles of impurities such as dye particles in initiating damage and of micro-Brownian motion (which increases with the proximity of the ambient and glass transition temperatures) in preventing or slowing its growth.

They concluded their second study by noting that, since in their measurements (both reports) the growth of laser damage decreased with either increasing ambient temperature or decreasing glass transition temperature, the two processes might be considered as equivalent. Then, since the thermal conductivity of an amorphous polymer generally increases with increasing ambient temperature,[75] it might also increase with decreasing glass transition temperature. Thus, the damage mechanism proposed in the first report could be modified to say that the thermomechanical stresses imparted to the polymer matrix by the heated dye particles are delocalized in proportion to the thermal conductivity (or, equivalently, micro-Brownian macromolecular motion) of the material, which in turn depends on the proximity of it glass transition and ambient temperatures.

Whereas Kusakawa et al. observed the role of dye particles in initiating laser damage to polymers, Butenin and Kogan[69,76] showed that micron-sized atmospheric soot particles also play a significant role. In an early study[76] they noted that the atmosphere contains significant concentrations of dust that could easily contaminate a monomeric liquid at any of several stages prior to polymerization. They argued that, upon exposure to optical fluence levels on the order of 1 J/cm², such a particle could easily be heated to a high enough temperature (10,000 K) to cause a microexplosion and subsequent fracture of the surrounding polymer matrix. It should thus be possible to raise the single-pulse optical strength of a polymer by removing dust from the monomeric liquid before polymerization.

To test their idea, Butenin and Kogan used a Q-switched ruby laser to measure the single-pulse laser damage thresholds of samples of PMMA which they had synthesized from as-supplied (unpurified) and purified MMA monomers. Their purification procedure consisted of repeated distillation through an FPP-15 aerosol filter (with 0.1-μm pores) and freezing, a process which concentrates dust particles in one of two adjoined vessels and leaves a relatively dust-free liquid in the other. They used scattered helium-neon laser light to monitor the procedure. The measured single-pulse optical strength of the polymer made from the purified monomer was ''several times larger'' than that of the polymer made from the unpurified monomer. They did not provide any polymerization details.

In a later study, they investigated the role of atmospheric dust in multiple-pulse laser damage.[69] They used a simple calculation to show that at small fractional subthreshold fluence levels, a dust particle as small as 0.02 μm in diameter could still reach high enough temperatures (2000°C) to cause thermolysis of surrounding polymer material and the formation of additional absorbing products such as carbon. Under multiple-pulse irradiation, the absorbing impurity would therefore grow in size, thereby absorbing increasingly more light, and reach higher temperatures with each laser pulse, until eventually it became hot enough to cause single-pulse-like damage (microexplosion and fracture).[76] Thus, polymers made from purified monomers should have improved multiple-pulse optical strength, as well as improved single-pulse optical strength. Measurements with a Q-switched neodymium laser on samples of PMMA made from unpurified and purified (as described above)[76] monomers showed this to be true. For illustration, some of their data are given in Table 5. As shown, purification doubled the single-pulse damage threshold, in agreement with their earlier result.[76] In addition, with the laser operating at the fractional subthreshold fluence of 0.25 (relative to the purified example), it took 60 times as many pulses (approximately 300 vs. 5) to cause multiple-pulse damage in the purified sample as in the control sample. The laser PRF was not reported.

To investigate the hypothesis that the light-absorbing dust particles cause thermolysis and carbon formation during multiple-pulse irradiation, Butenin and Kogan also measured the multiple-pulse optical strength of copolymers of MMA and styrene.[69] They argued that thermolysis and carbon formation should occur more easily in aromatic rather than aliphatic compounds. Therefore, since the styrene molecule contains an aromatic fragment, the multiple-pulse optical strength of the copolymer should decrease with increasing styrene content. Their samples behaved accordingly, which supports the thermolysis idea and shows that polymers intended for high-power optical applications should not contain aromatic components.

In the same report[69] they also noted that they could increase the optical strength of PMMA by not carrying the polymerization reaction to completion, i.e., by producing a

TABLE 5
Results of Various Efforts to Improve the Bulk Optical Strength of PMMA

Author	Material improvement effort	Test laser wavelength (μm)	F_f/F_0[a]	F/F_1[b]	Pulse repetition frequency (pps)	N_1[c]	N_0[c]	Ref.
Butenin	Distill, freeze, and filter monomer with 0.1-μm pore filter	1.06	2.0	0.25	Not reported	300	5	69
Agranat	Distill, freeze, and filter monomer with 0.1-μm pore filter	1.06	2.5	—	—	—	—	77
Emelyanova	Copolymerize BA and MMA, i.e., form P[MMA(.98)-BA(.02)]	1.06	2.0	0.25	Not reported	45	9	70
Manenkov	Purify monomer with 0.22-μm pore filter	0.69	3.5	0.10	Not reported	50	—	71
O'Connell	Distill monomer	1.06	2.2	0.25	1	>200	10	84
				0.19	1	>1000	—	
				0.19	10	35	—	
Aldoshin	Plasticize with 20% ethyl alcohol	0.69	1.4	—[d]	—[d]	>10^4	20	63
	Filter monomer with 0.22-μm pore filter		3.5	—[d]	—[d]	90	20	
Aldoshin	Plasticize with 30% dibutyl phthalate	0.69	1.4	—[d]	—[d]	>10^3	20	87
	Copolymerize decylmethacrylate (DMA) and MMA; i.e., form P[MMA(.80)-DMA(.20)]		1.4	—[d]	—[d]	>10^3	20	
	Copolymerize monomethacrylic ester of ethylene glycol (MMEG) and MMA, i.e., form P[MMA(.71)-MMEG(.29)]		1.8	—[d]	—[d]		70	20

Note: All data were obtained with Q-switched ruby or neodymium lasers.

[a] F_f/F_0 = ratio of single-pulse damage thresholds of improved (F_f) and unimproved (F_0) material.

[b] F/F_1 = fractional subthreshold fluence level for multiple-pulse measurements, in terms of the single-pulse damage threshold of the improved material.

[c] N_1, N_0 = number of laser pulses needed to cause multiple-pulse damage to improved and unimproved material, respectively, at the given fractional subthreshold fluence F/F_1.

[d] Not specified, but the same in all data from References 63 and 87.

relatively soft material. They asserted that the softer material reduced the efficiency of thermolysis and carbonization, but they did not explain how it did so.

As a general conclusion to both of their reports, Butenin and Kogan suggested that high optical strength polymers would result from choosing macromolecular structures that ensure the absence of thermolysis/carbonization and by carefully purifying the liquid monomers to remove atmospheric soot before polymerization.

In a study similar to those of Butenin and Kogan, Agranat et al.[77] used 10-ns wide 1.06-μm laser pulses to measure the single-pulse laser damage threshold of both purified (using Butenin's procedure, described above)[76] and unpurified (as-supplied) liquid MMA and solid PMMA. They observed that the damage threshold of the unpurified MMA was 50% greater than that of the unpurified PMMA and that purification improved the MMA and PMMA by factors of 8.0 and 2.5, respectively. (The PMMA result is given in Table 5.) To explain the superior behavior of the monomer, they suggested that defects in the microstructure of the PMMA might be responsible for its lower optical strength.

To study the effect further, they measured the damage threshold during the polymerization of a sample of purified MMA and found that, when the polymerization (via a thermal procedure described in Reference 78) reaction was 90% complete, the damage threshold suddenly dropped by a factor of 4. They claimed that the microstructure of a polymer is essentially completely formed at the 90% point of the polymerization reaction, so that the reduction in damage threshold cannot be due to the formation of microstructure. It might, however, be a consequence of defects in the microstructure, which are known to form during the last stages of polymerization.[78] Agranat et al. suggested that these microstructure defects might "somehow affect the thermal processes in the course of absorption of laser radiation by fine (<0.1μm) foreign inclusions," but they did not explain how it might happen.

Instead of removing impurities to improve the optical strength of a polymer (as suggested by Kusakawa et al.[19,20] and experimentally done by Butenin and Kogan[69] and Agranat et al.[77]), Emelyanova et al.[70] suggested that the optical strength might also be improved by modifying the structure of the polymer in such a way as to (among other things) "intensify the thermal micro-Brownian motion of the macromolecules." (Recall that Kusakawa et al.[19,20] had shown that similar ambient and glass transition temperatures could accomplish the same thing.) They proposed that macromolecular thermal micro-Brownian motion could be intensified by copolymerizing into the main polymeric chains relatively long linear side radicals that would increase the flexibility of the main chains, allowing them to absorb light-induced elastic stresses from absorbing inclusions without becoming damaged. The added side radicals would thus act as a sort of intramolecular plasticizer. (Recall that Butenin and Kogan[69] had shown that aromatic (ring) copolymeric components have the opposite effect, i.e., they reduce optical strength).

To test their hypothesis, Emelyanova et al. copolymerized small amounts of butylacrylate (BA) into PMMA, thus forming the composition P[MMA-BA]. Since the C_4H_9 radical in each BA molecule is longer than the $COOCH_3$ and CH_3 radicals in each MMA molecule that it would replace, it should provide the required enhanced flexibility to the macromolecular chains. Also, P[MMA-BA] is easily synthesized using radical block polymerization with thermal initiation. In laser damage measurements (made with 40 to 50-ns, 1.06-μm wavelength pulses from a Nd:glass laser) on samples of PMMA and P[MMA-BA] thus fabricated from as-supplied monomers, the largest improvement in optical strength was seen in the copolymer containing 2% BA. As shown in Table 5, its single-pulse damage threshold was twice as large as that of PMMA and, at the fractional subthreshold fluence level of 0.25 (relative to the 2% BA copolymer), it took 5 times as many pulses (45 vs. 9) to damage the copolymer as it did to damage the pure PMMA. The laser PRF was not reported.

It should be noted here that Emelyanova's results are consistent with Kusakawa's idea concerning the proximity of ambient and glass transition temperatures, enhanced micro-

Brownian motion, and improved optical strength. According to Emelyanova, the glass transition temperature of pure polybutylacrylate is $-56°C$. The glass transition temperature of P[MMA(.98)-BA(.02)] should therefore be slightly lower than that of PMMA and, according to Kusakawa, have the greater optical strength that Emelyanova observed.

Considered together, the reviewed results of Kusakawa, Butenin, Agranat, and Emelyanova and co-workers suggest two very important themes concerning the mechanism of laser damage in polymers. The first one is that micron- and submicron-sized absorbing foreign impurities and microstructural defects play a significant initiating role. This is unusual because it says that even in highly purified polymers the cause of laser damage is extrinsic. In most other transparent dielectrics, e.g., alkali halide crystals and silicate glasses, intrinsic mechanisms such as avalanche or multiphoton ionization are often the cause of damage.[45] The second theme is that the thermoelastic properties of the polymer matrix have a significant effect on whether and how efficiently the light-absorbing impurity or defect will ultimately damage the polymer.

2. An Extrinsic Damage Initiation Mechanism

Manenkov, Nechitailo, and co-workers have studied laser damage in polymers extensively, and their efforts have contributed significantly to a further understanding of the two above-mentioned themes. Therefore, in this and the next subsection, we summarize some of their research (and a few others') concerning, respectively, the roles of absorbing defects and material thermoelastic properties in the polymer laser damage process.

Aldoshin et al.[79] studied the dependence of the single-pulse laser damage threshold of PMMA on the spot size (see Section A) and wavelength of the test laser using nanosecond pulses at the wavelengths of 0.53 and 1.06 μm from a Nd:glass laser and at 0.69 μm from a ruby laser. The PMMA was synthesized from MMA that had been filtered with 0.22-μm pore filters. (No other synthesis details were reported.) In their spot size measurements, the damage threshold was reduced by factors of approximately 18 and 3, respectively, as the spot size diameter was increased from 30 to 640 μm at the 0.69-μm wavelength and from 60 to 640 μm at the 1.06-μm wavelength. Because they defined the damage threshold as the peak fluence corresponding to the 50% damage probability[3] (rather than the damage onset level), this result implies that isolated absorbing defects smaller than 0.22 μm in size were responsible for the damage. This follows because, as we explained earlier (see Section A), when laser damage is defect dominated or extrinsic, the 50% damage probability fluence is spot size dependent.[50] For spot sizes greater than 275 μm, their data showed no further decrease in the 50% damage thresholds, that is, they became spot size independent and thus identical with the more desirable onset damage thresholds. They reported their results not in absolute fluence levels, but only in arbitrary relative units.

For their wavelength-dependence measurements, they used a laser spot size of 110 μm and obtained single-pulse damage thresholds of 1.0, 1.4, and 2.0 (in arbitrary relative units) at wavelengths of 0.53, 0.69, and 1.06 μm, respectively. To explain these results, they noted that, according to their own theory of laser damage by absorbing impurities,[80] when absorbing particles much smaller than the laser wavelength are responsible for the damage, and in the absence of any resonant absorption, the absorption coefficient varies inversely with wavelength, and the damage threshold is therefore proportional to the wavelength. In their measurments, the ratios of wavelength to threshold were 1.9, 2.0 and 1.9, which can be considered as approximately equal, given the uncertainties (discussed earlier) in measuring damage thresholds. Thus, their measurements behaved according to the theory, suggesting that absorbing particles smaller than 0.1 μm in size caused the damage.

In the same report,[79] Aldoshin et al. also pointed out that when the absorbing defects do exhibit resonant absorption, the damage threshold should be reduced, so that the proportionality between threshold and wavelength would not be obeyed. To illustrate this, they used the 0.69-μm ruby laser to damage-test samples of PMMA doped with increasing concentrations of VOPc, which is the same well-known ruby laser saturable (resonant) absorber Q-switch dye that Kusakawa et al. had studied earlier.[19,20] Samples with low dye concentrations had the same damage threshold as the previously measured undyed material, but those with higher dye concentrations had lower damage thresholds. They suggested that at higher concentrations the dye molecules form conglomerates that act as resonant absorbing defects, which require less energy to cause damage than do the nonresonant 0.1-μm particles. This result confirms Kusakawa's earlier observation[20] that VOPc-dyed polymers damaged faster than their undyed counterparts.

Manenkov and Nechitailo[71] studied the optical strength of PMMA in terms of the purity of the starting monomer, the morphology of the laser-damaged polymer, and the cumulative multiple-pulse effect. (Their results were expanded and clarified by Dyumaev et al.[3]) In the monomer purity experiment, they filtered the starting monomers with a range of pore sizes from 16 to 0.22 μm and measured the single-pulse laser damage thresholds of the resulting polymers (they reported no polymerization details) with 0.69-μm wavelength, 10-ns pulses from a Q-switched ruby laser. The laser spot size was a relatively wide 110 μm. In general agreement with the results of Butenin and Kogan[69] and Agranat et al.,[77] the damage threshold improved monotonically with decreasing filter pore size so that, as shown in Table 5, the material made from 0.22-μm filtered monomer had 3.5 times the optical strength of "technical grade" PMMA. The fact that the data showed no sign of saturation, even after 0.22-μm monomer filtration, suggests that finer filtration would have raised the damage threshold even further and that absorbing particles smaller than 0.22 μm were responsible for the damage.

One difficulty with the above results is that they do not include a value for the single-pulse damage threshold of technical grade PMMA. In Reference 3, the same result is presented as 0.28 times the unprovided damage threshold of K-8 glass. Balitskas' multiple-pulse data on K-8 glass, discussed earlier (see Table 4), is referenced to the single-pulse damage threshold of K-8 glass but, again, that quantity is not reported in the useful form of either a peak fluence or peak intensity. Consequently, it is difficult to compare the optical strength of Manenkov's 0.22-μm filtered PMMA to that of optical quality glass.

To study damage morphology, Manenkov et al. examined their laser-damaged samples microscopically.[3,71] In the materials made from all except the 0.22-μm filtered monomers, the damage consisted of star-shaped cracks. In the 0.22-μm filtered material, the damage consisted instead of opaque ellipsoidal zones. Both the star-shaped cracks and the opaque ellipsoids varied in size between 50 and 100 μm and were distributed randomly in the focal volume of the laser. The authors claimed that the variations in size and the randomness in positions of the observed damage showed that it could not be explained by an intrinsic mechanism, so it must be due to absorbing defects.

To explain the observed change in morphology from star-shaped cracks to opaque ellipsoids in the 0.22-μm filtered material, Manenkov et al. applied the theory of References 80 and 81 (an extended version of Reference 80), which ascribes different damage-causing processes to different-sized absorbing defects. Accordingly, defects larger than approximately 0.1 μm are associated with lower damage thresholds, i.e., lower laser fluences (as shown in their monomer purity measurements), and, thus, with lower temperatures. Such defects induce damage via the low temperature "triboelectric process" or "triboprocess", in which the heated defect places thermoelastic stresses on the surrounding polymer matrix (just as Kusakawa had proposed),[19] causing microscopic cracks to appear. Surface charge associated with the cracks then absorbs laser energy (from the same pulse), causing the cracks to propagate. Smaller defects (<0.1 μm), on the other hand, are associated with higher threshold fluences (as shown in the monomer purity measurements) and, therefore, with higher temperatures. These heated defects behave like blackbodies with UV spectral maxima. The UV light preionizes some of the surrounding polymer macromolecules, providing electrons that absorb laser light (from the same pulse), leading to an explosion that results in the observed opaque ellipsoid.

The explanation given for the observed damage morphology in Manenkov and Nechitailo's report[71] also accounts for the monotonic increase in the single-pulse damage threshold with decreasing filter pore size observed in the same report. That is, increasingly finer filtration leaves only increasingly smaller particles in the monomer, which require higher temperatures, i.e., higher laser peak fluences, in order to damage the polymer.

The multiple-pulse cumulative effect is considered a test for the influence of absorbing defects in laser damage because it essentially involves the accumulation from pulse to pulse of thermal stresses in the vicinity of microstructural defects[82] until microcracks appear and grow via the triboprocess.[81] Manenkov and Nechitailo[71] observed a pronounced cumulative effect, even in 0.22-μm filtered PMMA. For example, at the fractional subthreshold fluence of 0.10, the material suffered macrodamage after only "a few dozen" (50) pulses, as shown in Table 5. The damage morphology consisted of star-shaped cracks, characteristic of the expected low fluence, low defect temperature triboprocess. (Like Butenin and Kogan[69] and Emelyanova et al.,[70] they did not state the laser PRF in their multiple-pulse measurements). Dyumaev et al.[3] noted that the proposed triboprocess is significantly different from thermolysis, the multiple-pulse damage mechanism suggested earlier by Butenin and Kogan.[69] Whereas in thermolysis strongly absorbing soot products are formed early in the microdamage growth process, in the triboprocess, they are not.

Aldoshin et al.[63] offered an explanation for such a pronounced cumulative effect in polymers. They suggested that as heated defects induce accumulating thermal stresses on the surrounding polymer material during the triboprocess, the associated microcracks will form when the breaking stress or brittle fracture limit of the polymer is reached. The breaking stress is proportional to the product of defect temperature and the efficiency β with which laser energy absorbed by the defect is converted into elastic deformation (stress) of the surrounding matrix. This efficiency can be expressed as

$$\beta = \frac{T_0}{C_v} \left(\frac{1 + \nu}{1 - \nu}\right)^2 \alpha^2 C_l^2 \tag{1}$$

where T_0 and C_l are the initial temperature and the speed of sound in the surrounding material, respectively, and C_v, ν and α are its specific heat, Poisson coefficient, and coefficient of linear thermal expansion. Using typical values for these quantities, they showed that β is 10 to 100 times larger in relatively soft materials like polymers than in relatively hard ones like fused quartz. Thus, breaking stresses can be reached at substantially lower defect temperatures, i.e., at lower peak fluence levels with the same number of laser pulses, or alternatively, in fewer pulses at the same peak fluence level.

To determine whether intrinsic damage could be observed in their 0.22-μm filtered material, Manenkov et al. also studied the single-pulse damage morphology of samples damaged with the much smaller laser spot size of 3.6 μm.[83] Such a small spot size would hopefully miss the absorbing defects on at least some pulses and produce intrinsic damage. The resulting damage morphology was identical to that observed with the 110-μm spot size, i.e., it consisted of opaque ellipsoidal zones of various sizes randomly located in the focal volume of the laser. They thus concluded that absorbing defects were the cause of damage with the narrowly focused laser beam, too.

In summary, then, Manenkov and co-workers' studies of the dependence of the single-pulse damage threshold on both the spot size and wavelength of the laser and on monomer purity, and their studies and of damage morphology and the multiple-pulse cumulative effect all support the important theme suggested by the earlier works of Kusakawa et al.,[19,20]

Butenin and Kogan,[69] and Agranat et al.[77] that laser damage in polymers is always caused by absorbing impurities and defects of various kinds, that is, it is always extrinsic.

Besides showing the extrinsic nature of laser damage initiation in polymers, Butenin and Kogan,[69] Agranat et al.,[77] and Manenkov et al.,[71] also showed that the optical strength of a material could be improved by purifying the starting monomer so as to reduce the impurity content before polymerization. The question thus arises as to whether monomer purification alone can produce polymers with optical strength comparable to that of optical quality silicate glass. Concerning single-pulse optical strength, the question can be answered by recalling the point made earlier in conjunction with the data in Table 3, i.e., that the single-pulse optical strength of PMMA is at least 10 to 50 times weaker than that of glass. The factors of 2.0 to 3.5 in improvement listed in Table 5 strongly suggest that monomer purification alone is inadequate.

Butenin's[69] and Manenkov's[71] reports cannot be used to compare the multiple-pulse performance of purified PMMA with that of optical glass because they do not include PRF information, which (as we noted earlier) has a significant influence on multiple-pulse behavior. We refer instead to a recent study of laser damage in monomer-purified PMMA by O'Connell et al.,[84] which does include enough PRF information to make the required comparison. The monomer in O'Connell's study was purified with a simple drying/distillation procedure and polymerized thermally with azobisisobutyronitrile as the initiator. Scattered, depolarized helium-neon laser light[85] was used to verify that the material made from the purified monomer contained vastly fewer micron-sized impurities than the material made from unpurified (as-supplied) monomer. Damage tests with 8-ns, 1.06-μm wavelength pulses from a Q-switched Nd:YAG laser showed that, in agreement with previous results, monomer purification could improve both single- and multiple-pulse optical strength. Some of their results are given in Table 5. As shown, the single-pulse damage threshold improved by a factor of 2.2 (from 21 to 47 J/cm^2). Also, with the laser operating at 1 pps with its fractional subthreshold fluence set at 0.25 (relative to the purified material), the average number of pulses required for damage increased from 10 to more than 200.

The same argument used to explain the improvement in the single-pulse optical strength of Manenkov's purified material[71] can be applied here to O'Connell's multiple-pulse data. That is, since the damage process involves the accumulation of thermally induced stresses,[63] and the defect heating is size-dependent,[80,81] smaller defects (in the purified material) simply require more laser pulses in order to reach the required higher temperatures.

O'Connell et al.[84] also showed that the multiple-pulse optical strength of the purified material has a severe PRF dependence. For example, as shown in Table 5, at the fractional subthreshold fluence level of 0.19, the purified material did not damage after 1000 laser pulses at 1 pps, but did so after an average of only 35 pulses at 10 pps. This last result can be compared with Merkle's 10-pps result[67] for fused quartz given in Table 4. Thus, whereas it required the large fractional subthreshold fluence of 0.75 in order to damage the glass in 100 pulses at 10 pps in Merkle's work, it would require fractional subthreshold fluence levels of only 0.19 or smaller to produce the same result in O'Connell's purified PMMA. Therefore, as in the case of single-pulse behavior, monomer purification can improve the multiple-pulse optical strength of PMMA, but evidently not enough to make it comparable to optical quality silicate glass.

3. The Influence of Material Thermoelastic Properties

Manenkov and co-workers also conducted an in-depth study of the second theme developed initially by Kusakawa et al.[19,20] and by Emelyanova et al.,[70] i.e., that the thermoelastic properties of the polymer matrix have a strong influence on the laser damage process. Aldoshin et al.[63] defined the thermoelastic (or viscoelastic) nature of a polymer in terms of its induced (or forced) elastic limit, which could be varied (reduced) either by plasticizing the material or by heating it.

To study the effects of lowering the induced elastic limit of a polymer on its optical strength, they used 20-ns, 0.69-μm ruby laser pulses to measure the optical strength of plasticized PMMA as a function of the concentration of the plasticizer (ethyl alcohol) and as a function of material temperature in the case of PMMA plasticized with 20% ethyl alcohol. In the variable-plasticizer-concentration measurements, as the concentration of ethyl alcohol was increased from 0 to 20%, the single-pulse damage threshold improved by 40% and the multiple-pulse damage threshold, defined as the number of laser pulses required to cause macroscopic (>100-μm size) damage at some unspecified fractional subthreshold fluence and unspecified PRF, increased from 20 to more than 10^4 (see Table 5). In the variable-temperature measurement, only multiple-pulse data were taken, and the number of pulses needed to damage the material increased from 10^3 at $-60°C$ to more than 10^4 at 20°C. These results showed that lowering the induced elastic limit of PMMA produces modest improvement in its single-pulse optical strength and substantial improvement in its multiple-pulse optical strength.

In the same report, they also compared the effects of plasticization on optical strength with those of monomer purification by measuring the single- and multiple-pulse damage thresholds in unplasticized PMMA made from 0.22-μm filtered monomers.[71] As shown in Table 5, the single-pulse damage threshold of the purified (but unplasticized) material improved by a factor of 3.5 vs. only 1.4 for the 20% plasticized (but unpurified) material. On the other hand, it took only 90 subthreshold pulses to damage the purified material, vs. over 10^4 for the plasticized material. The fact that the multiple-pulse threshold of the purified material improved from 20 to 90 pulses is due, as discussed earlier (see Section 2) to the slower heating of smaller defects.

To explain these multiple-pulse results, they noted that the induced elastic limit of the 20% plasticized material was 3.9 kg/mm², which is below the brittle fracture limit of PMMA of 8.0 kg/mm². On the other hand, the induced elastic limit of the purified but unplasticized material was 12 kg/mm². (They did not say how the induced elastic limits were determined.) They thus proposed that, when a heated defect (shown earlier to be the cause of damage initiation)[71] places accumulating elastic stresses on the surrounding polymer material during the triboprocess (as described earlier),[63] if the induced elastic limit of the material is below its brittle fracture limit, the accumulating stresses will reach the elastic limit first. Then, instead of forming microcracks, which (as described earlier)[63] would happen if the brittle fracture limit was reached first, "the material around the defect undergoes induced elastic deformations characterized by appreciable relative elongation and acquires a mechanical strength several times greater than the brittle fracture limit." The multiple-pulse optical strength is thus greatly improved. They did not explain why, in contrast, the improvement in the single-pulse optical strength of the plasticized material was so modest. In a later report Manenkov et al.[86] derived an expression which showed that the damage threshold varies inversely with Young's modulus, which, of course, decreases with plasticization.[74] Also, in their review article, Dyumaev et al.[3] suggested that the plasticizer molecules improve optical strength by deactivating those excited electrons that appear on the surfaces of microcracks during the triboprocess, absorb laser radiation, and stimulate further cracking.

In support of the idea concerning the relationship between the induced elastic and brittle fracture limits of a polymer and its multiple-pulse optical strength, Aldoshin et al. presented additional data in another report.[87] They measured the single- and multiple-pulse damage thresholds (exactly as in Reference 63) of PMMA with various other plasticizers and of various copolymers consisting mainly of MMA. In all cases, the results, some of which are given in Table 5, confirmed the previous observation. Thus, in PMMA plasticized with 30% dibutyl phthalate and MMA copolymerized with 20% decylmethacrylate, the multiple-pulse damage threshold pulse numbers were more than 1000 as compared to 20 in unpurified PMMA, and their induced elastic limits (2.0 and 6.6 kg/mm², respectively) were below the

PMMA brittle fracture limit (8.0 kg/mm^2). In contrast, in MMA copolymerized with 29% monomethacrylic ester of ethylene glycol, the multiple-pulse damage threshold was only 70 pulses, and its induced elastic limit (20 kg/mm^2) was above the PMMA brittle fracture limit. (They explained the difference in multiple-pulse optical strengths of this latter material and PMMA as due to differences in monomer purity.) Thus, the results presented in References 63 and 86 show that orders of magnitude improvement in multiple-pulse optical strength can be produced by modifying the material in such a way as to reduce its induced elastic limit to a level below its brittle fracture limit.

Dyumaev et al.[88] studied the temperature dependence of multiple-pulse laser damage more thoroughly than had been done by Aldoshin et al.[63] In measurements on six different materials, including PMMA, PBMA, P-iBMA, and PMMA plasticized with 20% of either ethyl alcohol, dimethyl phthalate, or dibutyl phthalate, the multiple-pulse damage threshold (defined and measured exactly as in Reference 63) increased sharply in a narrow temperature range near the glass transition temperature of each material. For example, as the temperature of PMMA was increased from 20 to 100°C, the number of pulses required for damage increased from approximately 50 to only approximately 100, but between 100 and 120°C the pulse number increased from 100 to 1000. They reported a glass transition temperature for PMMA of 119°C and concluded that their results showed that polymers undergo a transition to a damage-resistant state at temperatures near the glass transition temperature, where they become highly elastic, as characterized by a substantially reduced Young's modulus, and have greatly altered relaxation properties. (Recall that Kusakawa et al.[19,20] had earlier described this in terms of enhanced micro-Brownian macromolecular motion.)

The damage morphology in Reference 88 confirmed the observed results. Using scattered helium-neon laser light as a probe, Dyumaev et al. observed scattering centers of approximately 1-μm size that did not change during multiple-pulse irradiation at temperatures near the glass transition temperature, where they had observed greatly improved optical strength. At temperatures well below the glass transition temperature, however, where damage thresholds were lower, the same scattering centers were unstable, that is, they grew in size from pulse to pulse and quickly formed into star-shaped cracks, characteristic of the triboprocess.[81,86]

Taken together, Manenkov and co-workers' studies of the influence of thermoelastic properties showed that, as expected, they do have a very significant effect on the growth of laser damage in polymers. By plasticizing a polymer either internally (via copolymerization) or externally (via an additive plasticizer) such that the induced elastic limit of the material is lower than its brittle fracture limit, the triboprocess can be greatly slowed or even stopped completely.

Aldoshin et al.[63] also claimed that thermoelastic properties could be used to explain the anomalous fact that the optical strength of polymers is greater on the surface than in the bulk. As we pointed out earlier, Bebchuk et al.[61] had shown that this unusual property of polymers is true regardless of the quality of the surface finish. Aldoshin et al.[63] stated that the anomaly is due to a difference between the thermoelastic stresses found near absorbing defects located on the surface and those in the bulk of the material, but they did not elaborate. In their review article, Dyumaev et al.[3] added that in polymers the induced elastic limit is lower on the surface than in the bulk because the surface is plasticized by monomer and oligomer fragments that appear in the surface layer as an unavoidable result of mechanical processing (e.g., polishing). Recall that it had been shown by Butenin and Kogan[69] and Agranat et al.[77] (also by Aldoshin et al.[63]) that incompletely polymerized material has greater bulk optical strength than more completely polymerized material. Thus, as Aldoshin[63] noted, the polymer can be plasticized by its own monomer, and this evidently occurs to a greater degree on the surface than in the bulk.

The research reviewed in this section has shown that monomer purification and plasticization can each separately improve the optical strength of PMMA. Material that is both purified and plasticized has been named modified PMMA (MPMMA) by Dyumaev et al.[3] and is the strongest optical form of PMMA. While we have been able to carefully compare the performance of purified, but unplasticized, PMMA with that of optical quality glass, the same cannot be done with MPMMA because there does not appear to be enough data to do so. Dyumaev et al.[3] reported a single-pulse damage threshold for MPMMA (measured with 10-ns, 0.69-μm laser pulses) equal to 59% of that of K-8 glass. As we stated earlier, they did not report a numerical value for the damage threshold of K-8 glass, so this result cannot be compared to those in Table 3. In a study of laser dyes in MPMMA (see Section IV and Table 2), Gromov et al.[4] reported that the single-pulse damage threshold of their dye-impregnated MPMMA (measured with 50-ns, 1.06-μm wavelength pulses) was more than 13 J/cm^2, and the material suffered no multiple-pulse damage after 10^4 pulses with the laser fluence set at 2 J/cm^2. The stated single-pulse threshold is clearly much smaller than the corresponding values given for glass in Table 3, but since the PRF of the laser was not given, the multiple-pulse data cannot be properly compared to the data on glass in Table 4.

IV. PROBLEMS IN DYE-DOPED POLYMER DEVICES

In the last section we considered the optical strength of polymers as an important limiting factor in their use in the optical components of high-power laser systems. This is true of both the dyed and the undyed components. As we stated in the Section I, however, the dye-doped Q-switches and lasers have other problems, in which the polymer-host plays a significant role, that restrict their use in high-power systems. We consider those problems in this section.

A. SWITCHING EFFICIENCY OF POLYMER-HOST Q-SWITCHES

In order to produce high-power laser pulses (and thereby exhibit good switching efficiency), the ideal passive Q-switch should alternate between opacity (to allow the population inversion to grow) and transparency (to allow the laser to emit the entire pulse). This requirement can be described as the combination of a small optical transmittance T_0 in the initial opaque or absorptive state and a large transmittance T_F in the final or transparent state. Both of these quantities (and therefore the laser output) have been shown to be influenced by the host material in the Q-switch.

In an early study of passive Q-switching, Dovger et al.[89] measured the transmittance vs. incident optical intensity of several ruby Q-switch dyes in liquid solutions. In every case, the transmittance saturated at a level T_F in proportion to the initial transmittance T_0. For example, two solutions of VOPc in nitrobenzene with different dye concentrations had (T_0, T_F) values of (41%, 90%) and (9%, 63%), respectively. In both cases, the optical intensities required for saturation were approximately equal ($\sim$7 MW/cm^2). Dovger et al. explained the effect as due to nonsaturable residual absorption by the upper level of the saturable transition. This is consistent with the proportional behavior of T_0 and T_F because a small T_0 implies a large initial absorption, which provides more excited-stated dye molecules for residual absorption. Thus, T_F will also be small.

In the same study, Dovger et al. showed that residual loss is host-dependent and limits Q-switched laser output (switching efficiency). To do this they measured T_F and the relative laser output pulse energy of each of several Q-switches with identical initial transmittances. Some of their results are listed in Table 6. Thus, for example, when toluene is replaced with nitrobenzene in a VOPc Q-switch, T_F drops from 92 to 83%, and the Q-switched laser output drops by 13%. The table shows similar behavior for phthalocyanine in different hosts. They suggested, in conclusion, that residual loss could affect switching efficiency by causing local heating of the host and subsequent perturbation of its optical quality.

TABLE 6
Initial (T_0) and Saturation (T_F) Transmittances and Pulsed Laser Outputs for Several Passive Q-Switches Discussed in the Text

Author	Dye/host	T_0 (%)	T_F (%)	Laser output	Ref.
Dovger	VOPc/toluene	39	92	1	89
	VOPc/nitrobenzene	39	83	0.87[a]	
	Pc/ethanol	39	74	1	
	Pc/water	39	65	0.34[b]	
Szabo	VOPc/PMMA	55	—	200 KW	5
	VOPc/nitrobenzene	55	—	50 KW	
Abdeeva	Polymethine/nitrobenzene	15, 50	—	80J,13J	22
	Polymethine/cellulose triacetate film	15, 50	—	50J,1J	
Emelyanova	Polymethine 3955/PMMA	40	85	—	92
	Polymethine 3955/P[MMA(0.58) - styrene (0.42)]	40	65	—	
	Polymethine 3955/polystyrene	40	60	—	

Note: VOPc: vanadium phthalocyanine and Pc:phthalocyanine.

[a] Energy relative to VOPc/toluene.
[b] Energy relative to Pc/ethanol.

Shortly after the work of Dovger et al., Szabo and Erickson[5] measured the switching efficiency of VOPc in two different hosts, the solid polymer PMMA and liquid nitrobenzene. Their data, given in Table 6, show that even though the initial transmittances of the Q-switches were identical (55%), the laser output was four times greater with the polymer-host Q-switch. They did not report the bleached-state transmittance of the two Q-switches, but they claimed that the difference in performance was due to a difference in residual losses. In discussing possible reasons for the difference in residual loss between liquids and polymers, they cited other research that showed that internal conversion among excited states in liquids is very efficient and nonsaturable[90] and that internal conversion between the first excited and ground state singlets depends on the viscosity of the host.[91] Thus, residual losses might be expected to be more serious in liquids.

While the work of Dovger et al.[89] and Szabo and Erickson[5] showed that the switching efficiency of passive Q-switches is related to residual loss, which in turn is host dependent, it did not prove the assertion made in both papers that residual loss is the dominant factor in Q-switch efficiency. To do so would require showing that differences in absorptive-state transmittance T_0, with identical bleached-state transmittance T_F, resulted in no significant laser output differences. This was not done in either study.

Abdeeva et al.[22] studied the neodymium laser Q-switching behavior of an unspecified polymethine dye in hosts of liquid nitrobenzene and thin films of the polymer cellulose triacetate (see Table 1). The cellulose triacetate films, 125-μm thick, were formed by dissolving the dye in liquid solutions of cellulose triacetate in methylene chloride and allowing the methylene chloride to evaporate. Measured laser output pulse energies of the Q-switches varied inversely with T_0 and were host dependent. Some of their data are given in Table 6. Thus, whereas the laser output dropped from 80 to 13 J as T_0 increased from 15 to 50% in the liquid devices, the corresponding output energies with the polymer devices were only 50 and 1 J. The observed variation of laser output is as expected because, as stated earlier, smaller values of T_0 represent greater opacity, which is needed for laser population inversion buildup in the low Q, high-loss portion of the switching cycle. Note also that, in contrast to Szabo's result, the polymer-host device in Abdeeva's work was less efficient than the liquid-host device. Thus, Szabo's suggestions concerning the differences in residual losses of liquids and polymers are insufficient. Abdeeva et al. also noted that polymer film devices with T_0 less than 35% suffered optical damage in one pulse. This observation is consistent with those of Kusakawa et al.[20] and Aldoshin et al.,[79] discussed in Sections III.C.1 and 2, respectively, in which increasing dye concentrations correlate with decreasing laser damage thresholds because of absorption by conglomerates of the dye molecules. Thus, efforts to improve device switching efficiency by increasing the dye concentration to reduce T_0 would entail two important tradeoffs: reduced optical strength as observed by Abdeeva and reduced T_F as observed by Dovger et al.[89]

In the same report, Abdeeva et al. studied the host dependence of T_0 by measuring the absorption cross-section spectra of their polymethine dye in the nitrobenzene and cellulose triacetate hosts. The observed spectra were host dependent; for example, the resonant (near 1.06 μm) portion of the polymer-host spectrum was broader and weaker (less absorptive) than the liquid-host spectrum and shifted slightly toward a smaller resonant wavelength. They thus suggested that the performance of the polymer-host devices could be improved by altering the chemical structure of the polymer film in such a way as to modify its absorption spectrum.

Taken together, the results of Dovger et al.,[89] Szabo and Erickson,[5] and Abdeeva et al.[22] show the need in efficient passive Q-switching for a small T_0 and a large T_F, and they illustrate the host dependence of each. Thus, in studies of the potential use of polymethine-doped polymers and copolymers for Q-switching the neodymium laser, Emelyanova et al.[92,93] measured both absorptive- and transmissive-state properties. The measurements were made on platelets of PMMA and polystyrene, and on 50-μm-thick copolymer films of MMA with styrene, i.e., P[MMA-styrene], all of which were doped with various concentrations of four different polymethine dyes. The platelets were synthesized using radical polymerization at 60°C without an initiator in an argon atmosphere, and the films were formed by solvent evaporation, as in Abdeeva's report.[22]

In agreement with Abdeeva's result, the measured absorption spectra of Emelyanova's[92,93] samples were all host dependent. In particular, the position λ, width Δλ, and intensity (absorption cross-section) of the spectral resonance varied from host to host. For example, in the case of each dye, in spectra obtained with constant dye concentration, the resonant absorption was strongest in the pure PMMA sample, weakest in the polystyrene sample, and intermediate and in proportion to the MMA content in the copolymer samples. Also, spectra measured on samples of polystyrene[92] and PMMA[93] with increasing concentrations of the dyes polymethine 3955 and thyropyrylohepamethene, respectively, did not obey the Lambert-Beer absorption law. For example, the intensity of a secondary resonance near 0.8 μm grew relative to the primary one at 1.06 μm. They suggested that this was evidence of the host-dependent formation of complicated aggregates of the dye molecules, which absorb more strongly at 0.8 than at 1.06 μm. To further investigate the nature of the influence of the solid host on the dye spectra, they dissolved the dye-doped samples in the solvents from which they were originally synthesized and remeasured their absorption spectra. For each dye, the spectra were identical and independent of the dissolved monomers, and identical to spectra measured before polymerization. They claimed that this shows that the interaction between dye and host in solid solution is totally reversible and therefore nonvalent.

In the same reports,[92,93] Emelyanova et al. measured the transmittances of their dye-doped samples as a function of the energy density of a Q-switched neodymium laser. The results both agreed with Dovger's results with liquids[89] and revealed an effect not seen by him. The point in common is that a saturation level was reached in every case (as in Dovger's work), revealing the presence of residual loss. As shown in Table 6, different hosts, doped with polymethine 3955 and having initial transmittances of 40%, had different T_F values, showing that the residual loss was host dependent and increased with the content of styrene in the host. Together with their measurements of the host dependence of T_0, these measurements show that for Q-switching PMMA has the advantage over polystyrene of greater absorptive-state absorption and less transmissive-state residual loss, i.e., a smaller T_0 and a larger T_F.

The effect observed by Emelyanova and Ivanova and co-workers, but not by Dovger, is that, in the samples consisting of any fraction of styrene and doped with either polymethine 3955[92] or thiopyrylohepamethene II,[93] the transmittance decreased when the energy density of the neodymium laser was increased beyond the level required to measure maximum T_F. The effect was most pronounced in pure polystyrene and not present at all in pure PMMA. It was also accompanied by a darkening of the sample, which was most likely optical damage of the polymer host due, as we have said before, to increased absorption by conglomerates of dye molecules. This result illustrates again that, as switching efficiencies in polymer-host Q-switches are improved and the devices handle larger optical powers, their optical strength will become their performance-limiting factor.

The research discussed thus far established the importance of both T_0 and T_F in efficient Q-switching and showed that the polymer-host has an influence on each. In a recent study of Q-switches based on modified PMMA (MPMMA,[3] Section III.C.3) as host, Gromov et al.[8] (see Table 1) recognized the importance of both T_0 and T_F by defining a parameter ϕ as the ratio of the absorptive- and transmissive-state optical densities. (Optical density is defined as the base 10 logarithm of the reciprocal of transmittance.) Thus, an efficient Q-switch would be characterized by a relatively large value of ϕ. They found that ϕ depended significantly on the extent of dye decomposition during polymerization, which in turn was related to the purity of the dye, the monomer, and the modifier used to fabricate the Q-switch. For example, in devices using bis-(4-dimethylaminodithiobenzil)-nickel (to Q-switch the neodymium laser), a careful purification procedure (not described therein) resulted in the improvement of ϕ from 3.7 to 5.5. In life tests, such devices could be used in over 10^4 Q-switching operations with laser energy densities up to 1 J/cm².

B. PHOTOBLEACHING IN POLYMER-HOST DYE LASERS

In Section II we noted that, while polymer-host dye lasers offer several advantages over flowing liquid systems, they are subject to a variety of thermal and optical problems. The most serious of these is photobleaching, i.e., the light-induced fragmentation of the active laser dye molecules into components that no longer absorb at the pump wavelength. Photobleaching is irreversible and should not be confused with the reversible (and desirable) bleaching that occurs in saturable absorber Q-switch dyes. In this section we review some of the research that has sought to characterize, understand, and reduce photobleaching in polymer-host dye lasers.

Photobleaching and its severity in dye lasers were first observed by Ippen et al.[94] in their work with nonflowing capillary-guide liquid solution dye lasers. Having visually observed the effect in devices longitudinally pumped with a cw argon laser at 514.5 nm, they measured the absorption spectra of larger samples of dye solutions both before and after argon laser irradiation. Dyes, their concentrations, and pump laser power levels were varied. With the exception of a weak but characteristic new absorption peak in the UV portion of the spectrum,[95] the absorption spectra of the irradiated solutions were identical in shape (but weaker, of course) to the spectra taken before irradiation. This result showed that the bleached dye molecular fragments did not absorb in the visible portion of the spectrum. They also observed that the rate of photobleaching in their measurements varied linearly with the power of the argon laser.

To quantify their measurements Ippen et al. combined the change in extinction coefficients (from the measured absorption spectra) with the concentration of the dye, the total laser energy absorbed, and the energy of the 514.5-nm photons to calculate the number of molecules bleached per absorbed photon in each experiment (defined by a specific dye, its concentration, and argon laser power). These so-called bleaching quantum efficiencies were found to be independent of laser power levels or dye concentrations for wide ranges of both. They explained slight increases in the bleaching quantum efficiency at either high laser power levels or high dye concentrations as probably due to heating effects rather than nonlinear processes. Among four dyes tested, all in ethanol solutions, rhodamine 6G had the lowest photobleaching quantum efficiency, 5×10^{-7} molecules per photon ($\pm 50\%$). Using this number, they showed, with a simple calculation, that under typical cw operation, a sample of the dye would have a useful lifetime of only 40 ms.

Shortly after Ippen's report, the first reports of the observation of photobleaching in polymer-host dye lasers were made by Kaminow et al.[6] and by Ulrich and Weber.[37] In their work on the development of a narrow band ($\Delta\lambda = 0.1$ Å) polymer-host dye laser (see Table 2), Kaminow used a cw argon laser to measure photobleaching of rhodamine 6G in PMMA. They reported obtaining a bleaching number (the reciprocal of bleaching quantum efficiency) "comparable" to Ippen's result[94] for rhodamine 6G in ethanol. They, too, used a simple calculation to show that under pulsed operation, their rhodamine 6G-doped devices could be used for only 3×10^4 pulses.

Ulrich and Weber[37] observed photobleaching in their study of dye lasers fabricated from rhodamine 6G-doped thin films of polyurethane (see Table 2). They used the pulsed outputs of a nitrogen laser at 337.1 nm and a neon laser at 540.0 nm to pump (create the population inversion required for lasing) their thin film devices and observed that the dye was bleached "considerably more slowly" by the 540-nm (green) source than by the 337.1-nm (UV) source.

Kaminow et al.[6] and Ulrich and Weber[37] simply observed the severity of photobleaching in polymer hosts. Early studies to understand and reduce it were made by Fork and Kaplan,[96] Kaminow et al.,[97] and Reich and Neumann.[39] Fork and Kaplan studied the effect of reduced temperatures on the photobleaching of rhodamine 6G in PMMA using both cw argon and pulsed neodymium lasers at 515 and 531 nm, respectively. Using Tomlinson's theory of photobleaching,[98] which says that the transmittance T of a photobleaching beam varies with time t according to the expression

$$T(t) = \{1 + [T(0)^{-1} - 1]e^{-(\sigma n\phi)t}\}^{-1} \tag{2}$$

they measured the temporal behavior of the transmitted intensity of the argon or neodymium laser. In Equation 2 σ is the absorption cross section of an isotropic dye molecule at the wavelength of the probe beam, n is the incident photon flux density, and ϕ is the bleaching quantum efficiency in molecules per photon. The resulting data, together with Tomlinson's theory, showed that the quantum efficiency for photobleaching of the dye was significantly reduced with cooling of the PMMA host. For example, at saturation, defined as the point where the bleaching quantum efficiency was reduced to the measuring accuracy of the system (10^{-7}), virtually all of the dye in the room-temperature PMMA samples was bleached. By contrast, in samples cooled to 193°K, only 22% of the dye was bleached at saturation.

To explain their results, Fork and Kaplan proposed that cooling either reduces the mobility of impurities that "catalyze the dye photodegradation process" or "increases the microscopic rigidity" of the polymer macromolecules surrounding the dye molecules. These rigid macromolecules would act as a protective "polymer cage" that becomes more rigid with lower temperatures.

Fork and Kaplan also observed that if the intensity of the bleaching laser in their measurements exceeded some maximum value, "sudden onset of irreversible blackening" of the sample would occur after partial photobleaching. In light of the discussions in Sections III and IV.A, it is likely that the observed effect was just the occurrence of laser damage. Thus, just as in the case of polymer-host Q-switches, as resistance to photobleaching in

TABLE 7
Measured Photobleaching Numbers (B) for Rhodamine 6G in Various Polymer-Hosts Discussed in the Text

Author	Polymer-host	Bleaching number B ($\times 10^5$ photons/molecule)	Bleaching light source	Ref.
Kaminow	Purified PMMA	25	CW argon laser at 514.5 nm	97
	Peroxidized PMMA	13		
	Epoxy resin	5.5		
Reich	PMMA	0.049	CW argon laser at 514.5 nm	39
	PMMA	>30[a]		
	PAN	2.45		
	PAN	3.1[a]		
Higuchi	P[MMA(0.99)—MA(.01)]	7.1	High-pressure mercury lamp	99
	P[MMA(0.90)—MA(.10)]	14.3		
	P[MMA(0.90)—MA(.10)]	33.3[b]		

Note: All measurements at 300 K unless otherwise noted. PAN: polyacrylonitrile and MA: methylacrylate.

[a] Measured at 100 K.
[b] After heat treatment at 80°C for 100 h.

polymer-host dye lasers is improved, the optical strength of the devices will become their performance-limiting factor.

Kaminow et al.[97] extended Tomlinson's photobleaching model[98] to include the anisotropic nature of the laser dye molecules. They also showed (by simple differentiation) that Equation 2 could be written as

$$B = \sigma nT(0)[1 - T(0)]/T'(0) \qquad (3)$$

where $B = 1/\phi$ is the bleaching number in photons per molecule. Equation 3 says that bleaching numbers can be determined from measured anisotropic dye absorption cross sections, probe laser power, and the initial value and rate of change of the transmittance of the probe laser. To examine the roles of impurities and polymer matrix rigidity in photobleaching (as proposed by Fork and Kaplan) they used a cw argon laser at 514.5 nm to study photobleaching of rhodamine 6G in PMMA and epoxy resin. The PMMA was synthesized from both purified and peroxidized (via UV-irradiated in oxygen) monomers, which would provide information on the role of impurities. Epoxy resin is much harder than PMMA and would provide information on the role of matrix rigidity.

Some of the results of Kaminow's measurements are shown in Table 7. The smaller measured values of B for peroxidized vs. purified PMMA and for epoxy vs. PMMA led them to conclude that impurities play a greater role than matrix rigidity in photobleaching. They also observed that B was relatively independent of laser power density, which indicates that photobleaching is a one-photon (linear) process. From this they concluded that bleaching is due to a photochemical reaction between the dye molecule in either its first excited singlet or lowest triplet state and either an impurity or the polymer-host.

As part of a study of polyacrylonitrile (PAN) as a dye-laser host for rhodamine 6G (see Table 2), Reich and Neumann[39] used a cw argon laser at 514.5 nm and Kaminow's method[97] to measure photobleaching of rhodamine 6G in PAN and PMMA at room temperature and at 100 K. Some of their data are shown in Table 7. The large increase in B with reduced temperature for rhodamine 6G in PMMA verifies the observation of Fork and Kaplan [96] concerning the advantage of cooled operation. However, as the data in the table also show, no such substantial improvement in B occurs for rhodamine 6G in PAN. Thus, the benefit is host dependent. To explain this, Reich and Neumann suggested that, if the polymer caging process proposed by Fork and Kaplan for rhodamine 6G in PMMA is valid, the dye molecule might cause a steric hindrance of the effect in PAN. They did not consider the role of impurities in their results.

Although inconclusive, the relatively early studies of Fork and Kaplan,[96] Kaminow et al.,[97] and Neumann and Reich[39] showed that the extent of photobleaching in dyed polymers depends on both the choice of polymer host and on such embedded impurities as oxygen and that it can be reduced (to varying extent) by cooling. Soon afterward, Heinsohn and Weber[40] and Wang and Gamper[41] illustrated some simple ways of increasing the photolimited lifetimes of polymer-host dye lasers. Heinsohn and Weber[40] measured the effect of cooling by comparing the outputs of pulsed rhodamine B- and rhodamine 6G-doped epoxy dye lasers (see Table 2) at 25 (room temperature) and $-140°C$. After 1000 pump pulses, the outputs of the room temperature rhodamine 6G and rhodamine B devices were reduced to approximately 55 and 10% of their initial values, respectively. In contrast, the cooled devices still produced approximately 90% of their initial energies after 1000 pulses. In all cases, the decrease in laser output began within the first 100 pump pulses and continued monotonically. Note that this work, although not explicitly labeled as such, was essentially a measurement of the pump conversion efficiency of the devices, as defined by Wang and Gampel[41] (below).

Wang and Gampel[41] showed that the lifetime of a rhodamine 6G-doped PMMA dye laser (see Table 2) could be increased by scanning the pump beam over a large area of the device to reduce the average power (and thus the photobleaching) in any one area. They achieved variable-area scanning by mounting the dye-doped polymer host to one of the laser cavity mirrors and rotating the assembly. For devices with different scanned areas they measured the decrease in pump conversion efficiency (defined as the ratio of the output energy of the dye laser to the input energy of the pump laser) as a function of the number of pump pulses. In one device, with a scanned area of 0.3 cm^2, the pump conversion efficiency dropped from 47 to 25% after 3.2×10^4 pump pulses. In another device, with a scanned area of 2 cm^2, the efficiency dropped from 35 to 25% after 10^6 pulses. As in Heinsohn and Weber's measurements[40] the decrease in pump conversion efficiency began early and continued monotonically. They claimed, in conclusion, that 25% efficiency is useful and suggested that more sophisticated scanning schemes would lead to devices competitive with flowing systems.

Higuchi and Muto[99] essentially continued the work begun by Fork and Kaplan,[96] Kaminow et al.[97] and Reich and Neumann[39] concerning the role of the polymer-host in photobleaching. Using a high-pressure mercury lamp and Kaminow's measurement method,[97] they measured photobleaching quantum efficiencies in rhodamine 6G-doped samples of copolymers of MMA and methylacrylate (MA), i.e., P[MMA + MA]. Some of their data (inverted to give bleaching numbers) are shown in Table 7. It can be seen that as the content of MA was increased from 1 to 10% in the copolymer, its resistance to photobleaching essentially doubled. Applying Fork and Kaplan's idea,[96] Higuchi and Muto suggested that their result was due to protective caging of the dye molecules by the MA in the copolymer chains.

In another experiment, they measured photobleaching quantum efficiencies in rhodamine 6G-doped P[MMA + MA] samples that had been heat treated at 80°C for 4 d. They found that the heat treatment inproved resistance to photobleaching. For example, in a 10% MA sample (see Table 7), the heat treatment increased the bleaching number from 14 to 33. They explained this effect as the result of further polymerization of residual monomers and annealing of the sample, but they did not explain how those processes influence the photodegradation of the dye molecules. In light of Bezrodnyi's work,[44] to be discussed below, this latter result might be better explained in terms of the removal (by completion of polymerization) of the initiator molecules that generate free radicals, which could in turn attack the photoexcited dye molecules in either their first excited or lowest triplet states.[97] Higuchi and Muto used azobisisobutyronitrile to initiate their polymerization reactions.

Concerning the heat treatment of dye-doped polymers, Higuchi and Muto made another important observation.[100] If the treatment temperature was close to the glass transition temperature of the material, significant thermal bleaching of the dye occurred. They observed this in samples of rhodamine 6G-doped PMMA which were heated at various temperatures for at least 4 d. From absorption measurements taken before and after the heat treatment, they determined bleaching rates for the dye-doped samples. Samples heated at 70°C and above bleached approximately 10 times as fast as those heated at 65°C. To explain this result, they argued that near the glass transition temperature some of the polymer bonds become free to rotate. Such enhanced micro-Brownian motion could damage the impregnated dye molecules. No such thermal bleaching was observed in the heat-treated P[MMA + MA] sample because the glass transition temperature of P[MMA + MA] is higher than that of PMMA. Considering the earlier discussion of polymer optical strength, this result suggests that in materials with lower glass transition temperatures there exists an inherent tradeoff between increased resistance to optical damage and reduced resistance to thermal bleaching of impregnated dyes.

The most recent investigations[4,9,44] of photobleaching in dye-doped polymers have concentrated on the role of impurities, particularly free radicals. We complete this section with reviews of such reports by Bezrodnyi et al.[44] and Gromov et al.[4,9] Bezrodnyi et al.[44] (see Table 2) studied the roles of free radicals and atmospheric oxygen in the photobleaching of several polymethine dyes in polyurethane. They found that, depending on the structure of the specific organic polymethine dye molecule, both could be very important. To examine the role of free radicals, they compared the absorption spectra of different laser dyes both in the urethane monomer before polymerization and in the polymer after radical photopolymerization. Radical photopolymerization was used because it involves the photoexcitation of an initiator and subsequent release of free radicals that initiate the chain polymerization reaction. If the embedded laser dye molecules were sensitive to the free radicals, some of them would be attacked and degraded. Photobleaching would thus accompany the polymerization process and the resonant absorption would be weaker in the polymer than in the monomer.

Their results showed that for laser dyes with an extended π-electron system (and thus very sensitive to the action of free radicals), this was indeed the case. For example, in spectra from samples of polyurethane doped with polymethine dye PK-890 and photopolymerized with each of four different initiators, the resonant absorption at 930 nm decreased relative to its value in the monomer by at least 33 and usually by more than 50%, indicating that excessive photobleaching had occurred during polymerization. On the other hand, samples of polyurethane doped with rhodamine 6G, which has a much shorter π-electron chain than PK-890, showed no such spectral evidence of bleaching during photopolymerization.

In the same report, Bezrodnyi et al. also suggested that any residual initiator molecules that did not participate in the polymerization process could generate free radicals later during, for example, normal device operation and cause photobleaching then. Considering this point, it seems likely that in Higuchi and Muto's work,[99] where heat treating the polymer reduced photobleaching, most of the residual initiator molecules were thermally induced to participate in polymerization, leaving fewer of them for subsequent bleaching activity.

Concerning the role of atmospheric molecular oxygen in the photobleaching of polymethine-doped polyurethane, Bezrodnyi et al. measured the time dependence of the resonant absorption of polymethine dye PK-686 in samples of polyurethane both with and without an unspecified concentration of dissolved oxygen. Their results agreed with those of Kaminow et al.,[97] discussed earlier. That is, whereas the resonant absorption in the oxygen-free samples was virtually unchanged after 50 h, it decreased by 50% after 40 h in the oxygenated samples, indicating the occurrence of excessive photobleaching. They described this effect as the "photoreaction of attachment of molecular oxygen" and argued that the quantum efficiency of the process depends on the structure of the dye molecule. This is evidenced by changes in the absorption spectra of polymethine dyes with different conjugated chain lengths but identical otherwise; the longer the conjugated chain, the greater the proportion of molecules that photodegrade.

Gromov et al.[4,9] studied the role of free radicals in the photobleaching of xanthene laser dyes in modified PMMA[3] (see Table 2). Instead of the polymerization initiator being the source of the destructive free radicals (Bezrodnyi's idea), they proposed that radical products are formed from polymer molecules when the dye and polymer molecules interact during vibrational relaxation of the excited electronic states of the dye. Photobleaching occurs when the radical product induces the excited dye molecule to lose an electron, thereby producing a reduced and inactive form of the dye. (Note that in this mechanism the polymer-host is involved, interestingly, in producing the impurity that causes photobleaching). To reduce photobleaching, they suggested the use of additives in the polymer matrix that would retard the generation and branching of free radical chains by inducing cross-relaxation between the vibrational levels of the polymer molecules and those of the additive.

To test this hypothesis, they measured the variation with the number of pump pulses (at 530 nm) of the pump conversion efficiency of rhodamine 6G perchlorate-doped PMMA that contained ethanol and either carbonic acid ether or ethylene glycol ether as the modifying additive. Whereas it took only 100 pump pulses for the sample containing ethylene glycol to lose 20% of its pump conversion efficiency, it took 650 pulses for the sample with carbonic acid ether. The initial pump conversion efficiency was the same in both samples. Although the difference in performance of the two samples is obvious, Gromov et al. did not explain how the carbonic acid and ethylene glycol additives did and did not, respectively, induce cross-relaxation between their vibrational levels and those of the PMMA.

One interesting feature of Gromov's pump conversion efficiency data is that it remained constant at its initial value for a certain number of pump pulses, then dropped rapidly. This was observed in the two rhodamine 6G perchlorate samples discussed above, as well as in a rhodamine 6G chloride sample with ethanol plus carbonic acid ether as the modifying additive. The interesting point is that no such threshold pulse number was observed by either Heinsohn and Weber[40] or Wang and Gampel[41] in their photobleaching measurements, described above. Since the primary difference in these experiments is the modifying additive in Gromov's work, it seems likely that the additive was responsible for the observed threshold. Presumably, while performing the role of inhibitor to the generation and branching of free radicals, the additive molecules become consumed or occupied in some permanent way. After enough pump pulses, essentially all the additive is occupied, allowing normal photobleaching to proceed.

Using knowledge gained in their experiments, Gromov et al. synthesized a new xanthene dye, which they called 11B.[9] They provided no other information about the dye, but they claimed that laser devices based on it in modified PMMA had the best performance of all the dyes they considered (they are listed in Table 2). For example, whereas the threshold pump laser fluence for single-pulse photobleaching of a rhodamine 6G chloride device was 1.6 J/cm^2, it was 10 J/cm^2 in an 11B device. Also, with a pump laser fluence of 1 J/cm^2, the laser conversion efficiency and critical number of laser pump pulses for photobleaching in the 11B device were 52% and 2000, respectively, both of which were better than the corresponding data for the other dyes.

V. SUMMARY AND CONCLUSIONS

Because they are lightweight, simply processed, and dye impregnable, polymers are an attractive potential substitute for crystals and glasses in a variety of bulk optical components for high-power laser systems. Examples include lenses, windows, prisms, passive Q-switches, and active dye laser elements. Polymers are uniquely suited to low-cost disposable components for portable systems.

Before the advantages that polymers offer can be realized in high-power systems, however, certain problems need to be solved. These include relatively low single-pulse and multiple-pulse optical strength and, for Q-switches and dye laser elements, the additional problems of low switching efficiency and photobleaching, respectively.

Most of the research on laser damage in polymers has been concerned with bulk rather than surface optical strength in PMMA and its derivatives because the surface optical strength of polymers is anomalously higher than that of the bulk and because PMMA has the highest optical strength of those commonly available polymers whose optical strengths have been measured. The associated monomers are also relatively safe to handle and easily processed.

Early investigations of the optical strength of polymers showed that micron- and sub-micron-sized absorbing impurities (e.g., dye particles and atmospheric soot) and micro-structural defects initiate laser damage by imparting thermal stresses to the surrounding polymer molecules. The extent (growth) of damage depends on the thermoelastic properties of the polymer matrix; the more efficiently the matrix can delocalize thermal stresses, the less extensive the damage. Those investigations showed that both single- and multiple-pulse damage thresholds could be raised by removing micron-sized impurities from the monomeric liquid (e.g., by distillation or filtration) prior to polymerization and/or by modifying the polymer structure (e.g., by internal plasticization) to enhance the thermal micro-Brownian macromolecular motion that is responsible for thermal stress delocalization. This process is characterized by either an increased thermal conductivity or a reduced glass transition temperature.

Later investigations studied the roles of absorbing defects and material thermoelastic properties in laser damage more extensively. The damage-initiating role of absorbing impurities and defects was supported by the results of the following: (1) measurements of the dependence of the single-pulse damage threshold on the spot size and wavelength of the test laser and on the purity (via filtration) of the starting monomer; (2) examination of damage morphology; and (3) the observation of a well-pronounced multiple-pulse cumulative damage effect. It thus appears that laser damage in polymers is always caused by an extrinsic mechanism, i.e., either foreign-body microinclusions and/or microstructural defects.

When relatively small (<0.1 μm) defects and impurities cause single-pulse damage, they require large laser fluence levels and thus reach high enough temperatures to explode, producing opaque ellipsoidal zones. When larger defects cause single-pulse damage or when defects of any size cause multiple-pulse damage, laser fluence levels and thus defect temperatures are lower. In that case, damage occurs via the triboprocess, in which heated defects place thermoelastic stresses on the surrounding polymer matrix, causing cracks to form when the thermal stresses exceed the breaking stress or brittle fracture limit of the polymer. Multiple-pulse damage is so much more serious in polymers than in silicate glasses because the efficiency of conversion of absorbed laser energy into elastic stresses is 10 to 100 times higher in polymers.

Investigations of the role of material thermoelastic properties in laser damage showed that orders of magnitude improvement in multiple-pulse optical strength can be achieved by either plasticizing a material such that its induced elastic limit is below its brittle fracture limit or by raising its temperature to a point near its glass transition temperature. Both of these steps are known to reduce the Young's modulus of the material and greatly increase its relaxation properties, i.e., its macromolecular micro-Brownian motion. The result is that the triboprocess is greatly hindered or even stopped completely, evidently due to deactivation of excited electrons that normally appear on the faces of microcracks during the process, absorb laser radiation, and stimulate further cracking. These results support the proposal made in the early investigations that thermoelastic properties primarily control the growth of laser damage.

Although limited, available data suggests that the optical strength of polymers (e.g., modified PMMA) is still inferior to that of silicate glass. Multiple-pulse improvements have apparently outpaced single-pulse improvements, and it appears that polymers (e.g., modified PMMA) can now be used as a low-cost compromise material in relatively low-fluence multiple-pulse applications. Before polymers can be used in place of crystals and glasses in the truly high-power applications, however, much more progress needs to be made on the single-pulse optical strength problem. Perhaps the solution will lie with an entirely different class of polymers.

Besides low optical strength, polymer-host Q-switches and dye laser elements have the respective additional problems of poor switching efficiency and rapid photobleaching. Optimum switching efficiency requires, respectively, small and large absorptive-state and bleached-state transmittances T_0 and T_F. Higher dye concentrations can be used to reduce T_0, but that also reduces T_F (by increasing nonsaturable excited-state residual losses) and lowers the optical strength of the device by encouraging the formation of resonantly absorptive dye conglomerates. The choice of polymer host and its degree of purity also affect switching efficiency. Different hosts affect the absorption properties of the Q-switch dyes in different nonvalent ways so that, for example, PMMA is a more efficient polymethine host than polystyrene. As in the optical strength problem, the biggest gains in polymer-host switching efficiency will probably accompany the use of some as yet untried host material.

Investigations of the permanently destructive photobleaching of laser dyes in polymer-hosts showed that the process consists of a photochemical reaction between the dye molecule and either an impurity or the molecules of the host itself. The process is thus impurity and host dependent. The use of host materials with relatively low glass transition temperatures entails an inherent tradeoff between improved optical strength and accelerated degradation of laser dye molecules, both due to the enhanced micro-Brownian motion that accompanies lower glass transition temperatures. The most important impurities in photobleaching appear to be dissolved atmospheric molecular oxygen and free radicals that are generated either by residual polymerization initiator molecules or in reactions between polymer-host molecules and excited dye molecules. Removal of oxygen, polymerization without free radical initiators, and hosts that do not generate free radicals will obviously minimize that problem. Thus, problems with impurities, dyes, and hosts remain before long-lived high-power polymer-host dye lasers will be available.

REFERENCES

1. **Kreidl, N. J. and Rood, J. L.**, Optical materials, in *Applied Optics and Optical Engineering*, Vol. 1, Kingslake, R., Ed., Academic Press, New York, 1965, 181.
2. **Ashkinadze, B. M., Vladimirov, V. I., Likhachev, V. A., Ryvkin, S. M., Salmanov, V. M., and Yaroshetskii, I. D.**, Breakdown in transparent dielectrics caused by intense laser radiation, *Sov. Phys. JETP,* 23, 788, 1966.
3. **Dyumaev, K. M., Manenkov, A. A., Maslyukov, A. P., Matyushin, G. A., Nechitailo, V. S., and Prohkorov, A. M.**, Transparent polymers: a new class of optical materials for lasers, *Sov. J. Quantum Electron.*, 13, 503, 1983.
4. **Gromov, D. A., Dyumaev, K. M., Manenkov, A. A., Maslyukov, A. P., Matyushin, G. A., Nechitailo, V. S., and Prokhorov, A. M.**, Efficient plastic-host dye lasers, *J. Opt. Soc. Am. B.*, 2, 1028, 1985.
5. **Szabo, A. and Erickson, L. E.**, Study of saturable absorber switching efficiencies, *J. Appl. Phys.*, 40, 3574, 1969.
6. **Kaminow, I. P., Weber, H. P., and Chandross, E. A.**, Poly(methyl methacrylate) dye laser with internal diffraction grating resonator, *Appl. Phys. Lett.*, 18, 497, 1971.

7. **Drake, J. M., Tam, E. M., and Morse, R. I.,** The use of light converters to increase the power of flashlamp-pumped dye lasers, *IEEE J. Quantum Electron,* QE-8, 92, 1972.
8. **Gromov, D. A., Dyumaev, K. M., Manenkov, A. A., Maslyukov, A. P., Matyushin, G. A., Nichitailo, V. S., and Prokhorov, A. M.,** *Izv. Akad. Nauk SSSR Ser. Fiz.,* 46, 1959, 1982.
9. **Gromov, D. A., Dyumaev, K. M., Manenkov, A. A., Maslyukov, A. P., Matyushin, G. A., Nichitailo, V. S., and Prokhorov, A. M.,** *Izv. Akad. Nauk SSSR Ser. Fiz.,* 46, 1956, 1982.
10. **Staggs, M. C. and Rainer, F.,** Damage thresholds of fused silica, plastics and KDP crystals measured with 0.6-ns 355-nm pulses, in *Laser Induced Damage in Optical Materials: 1983,* Bennett, H. E., Guenther, A. H., Milam, D., and Newnam, B. E., Eds., Spec. Publ. No 688, National Bureau of Standards U.S. Government Printing Office, Washington, D.C., 1983, 84.
11. **Campbell, J. H., Edwards, G., Frick, F. A., Gemmell, D. S., Gim, B. M., Jancaitis, K. S., Jessop, E. S., Kong, M. K., Lyon, R. E., Murray, J. E., Patton, H. G., Powell, H. T., Riley, M. O., Wallerstein, E. P., Wolfe, C. R., and Woods, B. W.,** Development of composite polymer-glass edge claddings for nova laser disks, in *Laser Induced Damage in Optical Materials: 1986,* Bennett, H. E., Guenther, A. H., Milam, D., and Newnam, B. E., Eds., Spec. Publ. No. 752, National Bureau of Standards, U.S. Government Printing Office, Washington, D.C., 1986.
12. **Koechner, W.,** *Solid-State Laser Engineering,* Springer-Verlag, Berlin, 1976, chap. 8.
13. **Siegman, A. E.,** *Lasers,* University Science Books, Mill Valley, CA, 1986, chap. 26.
14. **Smith, W. V. and Sorokin, P. P.,** *The Laser,* McGraw-Hill, New York, 1966, chap. 4.
15. **Sorokin, P. P., Luzzi, J. J., Lankard, J. R., and Pettit, G. D.,** Ruby laser Q-switching elements using phthalocyanine molecules in solution, *IBM J. Res. Dev.,* 8, 182, 1964.
16. **Soffer, B. H.,** Giant pulse laser operation by a passive, reversibly bleachable absorber, *J. Appl. Phys.,* 35, 2551, 1964.
17. **Kafalas, P., Masters, J. I., and Murray, E. M. E.,** Photosensitive liquid used as a nondestructive passive Q-switch in a ruby laser, *J. Appl. Phys.,* 35, 2349, 1964.
18. **Vanyukov, M. P., Dmitrievskii, O. D., Isaenko, V. I., and Serebryakov, V. A.,** Fast working liquid shutter for modulation of cavity laser of a neodymium glass laser, *Sov. Phys. Dokl.,* 11, 233, 1966.
19. **Kusakawa, H., Takahashi, K., and Ito, K.,** Temperature dependence of laser-induced damage in polyethylmethacrylate with Q-switching dye, *J. Appl. Phys.,* 39, 6116, 1968.
20. **Kusakawa, H., Takahashi, K., and Ito, K.,** Relationship between the growth rate of laser-induced damage in polyalkylmethacrylates and their glass transition temperature, *J. Appl. Phys.,* 40, 3954, 1969.
21. **Sharlay, S. F.,** Phototropic film shutter for passive Q-modulation of a ruby laser, *Zh. Prikl. Spektrosk.,* 13, 730, 1970.
22. **Abdeeva, V. I., Alperovich, M. A., Vanyukov, M. P., Isaenko, V. I., Levkoev, I. I., Serebryakov, V. A., and Starikov, A. D.,** Utilization of liquid and film bleachable switches in GOS-1000 lasers, *Sov. J. Quantum Electron,* 1, 166, 1971.
23. **Kirk, J. P., Lo, H. M., Farmer, G. I., Wright, R., and Buser, R. G.,** Passive Q-switching of a Nd:YAG laser compared with electrooptic switching, *IEEE J. Quantum Electron.,* QE-9, 633, 1973.
24. **Acharekar, M. A.,** Increase in the optical damage threshold of cellulose acetate, in *Laser Induced Damage In Optical Materials: 1983,* Bennett, H. E., Guenther, A. H., Milam, D., and Newnam, B. E., Eds., Spec. Publ. No. 688, National Bureau of Standards, U.S. Government Printing Office, Washington, D.C., 1983, 70.
25. **Drexhage, K. H. and Muller-Westerhoff, U. T.,** New Q-switch compounds for infrared lasers, *IEEE J. Quantum Electron.,* QE-8, 759, 1972.
26. **Snavely, B. B.,** Flashlamp-excited organic dye lasers, *Proc. IEEE,* 57, 1374, 1969.
27. **Sorokin, P. P. and Lankard, J. R.,** Stimulated emission observed from an organic dye, chloroaluminum phthalocyanine, *IBM J. Res. Dev.* 10, 162, 1966.
28. **Sorokin, P. P., Culver, W. H., Hammond, E. C., and Lankard, J. R.,** End-pumped stimulated emission from a thiacarbocyanine dye, *IBM J. Res. Dev.,* 10, 401, 1966.
29. **Schafer, F. P., Schmidt, W., and Volze, J.,** Organic dye solution laser, *Appl. Phys. Lett.,* 9, 306, 1966.
30. **Spaeth, M. L. and Bortfield, D. P.,** Stimulated emission from polymethine dyes, *Appl. Phys. Lett.,* 9, 179, 1966.
31. **Sorokin, P. P. and Lankard, J. R.,** Flashlamp excitation of organic dye lasers: a short communication, *IBM J. Res. Dev.* 11, 148, 1967.
32. **Schmidt, W. and Schafer, F. P.,** Blitzlampengepumpte farbstoff-laser, *Z. Naturforsch.,* 22a, 1563, 1967.
33. **O'Connell, R. M. and Saito, T. T.,** Plastics for high-power laser applications: a review, *Opt. Eng.,* 22, 393, 1983.
34. **Soffer, B. H. and McFarland, B. B.,** Continuously tunable, narrow-band organic dye lasers, *Appl. Phys. Lett.,* 10, 266, 1967.
35. **Peterson, O. G. and Snavely, B. B.,** Stimulated emission from flashlamp-excited organic dyes in polymethyl methacrylate, *Appl. Phys. Lett.,* 12, 238, 1968.

36. **Naboikin, Y. V., Ogurtsova, L. A., Podgornyi, A. P., Pokrovskaya, F. S., Grigoryeva, V. I., Krasovitskii, B. M., Kutsyna, L. M., and Tishchenko, V. G.,** Spectral and energy characteristics of organic molecular lasers in polymers and toluene, *Opt. Spectrosc. (USSR)*, 28, 528, 1970.
37. **Ulrich, R. and Weber, H. P.,** Solution-deposited thin films as passive and active light-guides, *Appl. Opt.*, 11, 428, 1972.
38. **Chang, M. S., Burlamacchi, P., Hu, C., and Whinnery, J. R.,** Light amplification in a thin film, *Appl. Phys. Lett.*, 20, 313, 1972.
39. **Reich, S. and Neumann, G.,** Photobleaching of Rhodamine 6G in polyacrylonitrile matrix, *Appl. Phys. Lett.*, 25, 119, 1974.
40. **Heinsohn, C. and Weber, J.,** Cooled dye laser, *Z. Naturforsch.*, 30a, 606, 1975.
41. **Wang, H. H. L. and Gampel, L.,** A simple, efficient plastic dye laser, *Opt. Commun.*, 18, 444, 1976.
42. **Itoh, U., Takakusa, M., Moriya, T., and Saito, S.,** Optical gain of coumarin dye-doped thin film lasers, *Jpn. J. Appl. Phys.*, 16, 1059, 1977.
43. **Sriram, S., Jackson, H. E., and Boyd, J. T.,** Distributed-feedback dye laser integrated with a channel waveguide formed on silicon, *Appl. Phys. Lett.*, 36, 721, 1980.
44. **Bezrodnyi, V. I., Przhonskaya, O. V., Tikhonov, E. A., Bondar, M. V., and Shpak, M. T.,** Polymer active and passive laser elements made of organic dyes, *Sov. J. Quantum Electron.*, 12, 1602, 1982.
45. **Smith, W. L.,** Laser-induced breakdown in optical materials, *Opt. Eng.*, 17, 489, 1978.
46. **Milam, D.,** Laser-induced damage at 1064 nm, 125 psec, *Appl. Opt.*, 16, 1204, 1977.
47. **Domann, F. E., Becker, M. F., Guenther, A. H., and Stewart, A. F.,** Charged particle emission related to laser damage, *Appl. Opt.*, 25, 1371, 1986.
48. **Aleshin, I. V., Bonch-Bruevich, A. M., Zinchenko, V. I., Imas, Y. A., and Komolov, V. L.,** Effect of absorbing inhomogeneities on optical breakdown of transparent dielectrics, *Sov. Phys. Tech. Phys.*, 18, 1648, 1974.
49. **Foltyn, S. R. and Newnam, B. E.,** Ultraviolet damage resistance of dielectric reflectors under multiple-shot irradiation, *IEEE J. Quantum Electron.*, QE-17, 2092, 1981.
50. **Foltyn, S. R.,** Spot size effects in laser damage testing, in *Laser Induced Damage in Optical Materials: 1984*, Bennett, H. E., Guenther, A. H., Milam, D., and Newnam, B. E., Eds., Spec. Publ. No. 727, National Bureau of Standards, U.S. Government Printing Office, Washington, D.C., 1984, 368.
51. **Porteus, J. O. and Seitel, S. C.,** Absolute onset of optical surface damage using distributed defect ensembles, *Appl. Opt.*, 23, 3796, 1984.
52. **O'Connell, R. M. and Vogel, R. A.,** Abel inversion of knife-edge data from radially symmetric pulsed laser beams, *Appl. Opt.*, 26, 2528, 1987.
53. **Milam, D.,** Fluence in 1064-nm laser beams: its determination by photography with Polaroid film, *Appl. Opt.*, 20, 169, 1981.
54. **Smith, W. L., DeGroot, A. J., and Weber, M. J.,** Silicon vidicon system for measuring laser intensity profiles, *Appl. Opt.*, 17, 3938, 1978.
55. **O'Connell, R. M. and Stewart, A. F.,** Performance characteristics of a beam profiling system consisting of various solid state imaging devices and an 8-bit image processor, in *Laser Induced Damage in Optical Materials: 1986*, Bennett, H. E., Guenther, A. H., Milam, D., and Newnam, B. E., Eds., Spec. Publ. No. 752, National Bureau of Standards, U.S. Government Printing Office, Washington, D.C., 1986.
56. **Shayler, P. J.,** Laser beam distribution in the focal region, *Appl. Opt.*, 17, 2673, 1978.
57. **McCally, R. L.,** Measurement of gaussian beam parameters, *Appl. Opt.*, 23, 2227, 1984.
58. **Mauck, M.,** Knife-edge profiling of Q-switched Nd:YAG laser beam and waist, *Appl. Opt.*, 18, 599, 1979.
59. **Bass, M. and Barrett, H. H.,** Avalanche breakdown and the probabilistic nature of laser-induced damage, *IEEE J. Quantum Electron.* QE-8, 338, 1972.
60. **Fradin, D. W. and Bass, M.,** Comparison of laser-induced surface and bulk damage, *Appl. Phys. Lett.*, 22, 157, 1973.
61. **Bebchuk, A. S., Gromov, D. A., and Nechitailo, V. S.,** Measure of surface defectiveness and optical strength of transparent dielectrics, *Sov. J. Quantum Electron.*, 6, 986, 1976.
62. **Bloembergen, N.,** Role of cracks, pores, and absorbing inclusions on laser induced damage threshold at surfaces of transparent dielectrics, *Appl. Opt.*, 12, 661, 1973.
63. **Aldoshin, M. I., Gerasimov, B. G., Manenkov, A. A., and Nechitailo, V. S.,** Decisive importance of the viscoelastic properties of polymers in their laser damage mechanism, *Sov. J. Quantum Electron.*, 9, 1102, 1979.
64. **Agranat, M. B., Krasyuk, I. K., Novikov, N. P., Perminov, V. P., Yudin, Y. I., and Yampolskii, P. A.,** Destruction of transparent dielectrics by laser radiation, *Sov. Phys. JETP*, 33, 944, 1971.
65. **O'Connell, R. M., Deaton, T. F., and Saito, T. T.,** Single- and multiple-shot laser-damage properties of commercial grade PMMA, *Appl. Opt.*, 23, 682, 1984.

66. **O'Connell, R. M., Saito, T. T., Deaton, T. F., Siegenthaler, K. E., McNally, J. J., and Shaffer, A. A.,** Laser damage in plastics at the Frank J. Seiler Research Laboratory (FJSRL), in *Laser Induced Damage in Optical Materials: 1983*, Bennett, H. E., Guenther, A. H., Milam, D., and Newnam, B. E., Eds., Spec. Publ. No. 688, National Bureau of Standards, U.S. Government Printing Office, Washington, D.C., 1983, 59.

67. **Merkle, L. D., Bass, M., and Swimm, R. T.,** Multiple pulse laser-induced bulk damage in crystalline and fused quartz at 1.064 and 0.532 μm, *Opt. Eng.*, 22, 405, 1983.

68. **Balitskas, S. K. and Maldutis, E. K.,** Bulk damage to optical glasses by repeated laser irradiation, *Sov. J. Quantum Electron.*, 11, 541, 1981.

69. **Butenin, A. V. and Kogan, B. Y.,** Mechanism of damage of transparent polymer materials due to multiple exposure to laser radiation pulses, *Sov. J. Quantum Electron.*, 6, 611, 1976.

70. **Emelyanova, G. M., Ivanova, T. F., Votinov, M. P., Ovchinnikov, V. M., Piterkin, V. D., and Smirnova, Z. A.,** Optical strength of copolymers of methyl methacrylate and butylacrylate, *Sov. Tech. Phys. Lett.*, 3, 280, 1977.

71. **Manenkov, A. A. and Nechitailo, V. S.,** Role of absorbing defects in laser damage to transparent polymers, *Sov. J. Quantum Electron.*, 10, 347, 1980.

72. **Brydson, J. A.,** *Plastic Materials*, Butterworths, London, 1982, chap. 15.

73. **Rogers, S. S. and Mandelkern, L.,** Glass formation in polymers. I. The glass transitions of the poly-(*n*-alkyl methacrylates), *J. Chem. Phys.*, 61, 985, 1957.

74. **Allcock, H. R. and Lampe, F. W.,** *Contemporary Polymer Chemistry*, Prentice-Hall, Englewood Cliffs, NJ, 1981, chap. 17.

75. **Anderson, D. R.,** Thermal conductivity of polymers, *Chem. Rev.*, 66, 677, 1966.

76. **Butenin, A. V. and Kogan, B. Y.,** Mechanism of optical breakdown in transparent dielectrics, *Sov. J. Quantum Electron*, 1, 561, 1972.

77. **Agranat, M. B., Novikov, N. P., Perminov, V. P., and Yampolskii, P. A.,** Some aspects of the initial stage of the development of laser damage in polymethylmethacrylate, *Sov. J. Quantum Electron.*, 6, 1240, 1976.

78. **Simonin, R. A., Kasaikin, V. A., Lachinov, M. B., Zubov, V. P., and Kabanov, V. A.,** Structure formation during radical polymerization of methyl methacrylate, *Dokl. Akad. Nauk. SSSR*, 217, 631, 1974.

79. **Aldoshin, M. I., Manenkov, A. A., Nechitailo, V. S., and Pogonin, V. I.,** Frequency and size dependence of the threshold for laser damage in transparent polymers, *Sov. Phys. Tech. Phys.*, 49, 1412, 1979.

80. **Danileiko, Y. K., Manenkov, A. A., Nechitailo, V. S., Prokhorov, A. M., and Khaimov-Malkov, V. Y.,** The role of absorbing impurities in laser-induced damage of transparent dielectrics, *Sov. Phys. JETP*, 36, 541, 1973.

81. **Danileiko, Y. K., Manenkov, A. A., and Nechitailo, V. S.,** The mechanism of laser-induced damage in transparent materials, caused by thermal explosion of absorbing inhomogeneities, *Sov. J. Quantum Electron.*, 8, 116, 1978.

82. **Danileiko, Y. K., Manenkov, A. A., and Nechitailo, V. S.,** Prethreshold phenomena in laser damage of optical materials, *Sov. J. Quantum Electron.*, 6, 236, 1976.

83. **Manenkov, A. A., Nechitailo, V. S., and Tsaprilov, A. S.,** Laser damage to transparent polymers in sharply focused single-mode beams, *Izv. Akad. Nauk SSSR. Ser. Fiz.*, 44, 1770, 1980.

84. **O'Connell, R. M., Romberger, A. B., Shaffer, A. A., Saito, T. T., Deaton, T. F., and Seigenthaler, K. E.,** Improved laser-damage-resistant polymethyl methacrylate, *J. Opt. Soc. Am. B*, 1, 853, 1984.

85. **O'Connell, R. M., Ellis, R. V., Romberger, A. B., Deaton, T. F., Siegenthaler, K. E., Shaffer, A. A., Mullins, B. W., and Saito, T. T.,** Laser damage studies of several methacrylate polymeric materials, in *Laser Induced Damage in Optical Materials: 1984*, Bennett, H. E., Guenther, A. H., Milam, D., and Newnam, B. E., Eds., Spec. Publ. No. 727, National Bureau of Standards, U.S. Government Printing Office, Washington, D.C., 1984, 49.

86. **Manenkov, A. A., Nechitailo, V. S., and Tsaprilov, A. S.,** Analysis of a mechanism of laser damage to transparent polymers associated with their viscoelastic properties, *Sov. J. Quantum Electron.*, 11, 502, 1981.

87. **Aldoshin, M. I., Gerasimov, B. G., Manenkov, A. A., Maslyukov, A. P., Nechitailo, V. S., and Ponomarenko, E. P.,** Laser damage in transparent polymers of various atomic compositions, *Sov. Phys. Tech. Phys.*, 49, 1411, 1979.

88. **Dyumaev, K. M., Manenkov, A. A., Maslyukov, A. P., Matyushin, G. A., Nechitailo, V. S., and Tsaprilov, A. S.,** Influence of viscoelastic properties of the matrix and of the type of plasticizer on the optical strength of transparent polymers, *Sov. J. Quantum Electron.*, 12, 838, 1982.

89. **Dovger, L. S., Ermakov, B. A., and Lukin, A. V.,** Characteristics of the optical transmission of phototropic shutters and their effect on laser parameters, *Opt. Spectrosc. (USSR)*, 25, 346, 1968.

90. **Giuliano, C. R. and Hess, L. D.,** Nonlinear absorption of light: optical saturation of electronic transitions in organic molecules with high intensity laser radiation, *IEEE J. Quantum Electron.*, 3, 358, 1967.

91. **Spaeth, M. L. and Sooy, W. R.,** Fluorescence and bleaching of dyes for a passive Q-switch laser, *J. Chem. Phys.,* 48, 2315, 1968.

92. **Emelyanova, G. M., Avdeeva, V. I., Alperovich, M. A., Votinov, M. P., Eremeeva, E. P., and Ivanova, T. F.,** Degree of bleaching and spectral properties of phototropic polymer media, *Zh. Prikl. Spektrosk.,* 31, 1116, 1978.

93. **Ivanova, T. F., Votinov, M. P., Dokukina, A. F., Piterkin, B. D., Smirnova, Z. A., and Emelyanova, G. M.,** Dyed copolymers as laser materials, *Izv. Akad. Nauk SSSR Ser. Fiz.,* 45, 662, 1981.

94. **Ippen, E. P., Shank, C. V., and Dienes, A.,** Rapid photobleaching of organic laser dyes in continuously operated devices, *IEEE J. Quantum Electron.,* QE-7, 178, 1971.

95. **Turro, N. J.,** *Molecular Photochemistry,* Benjamin Cummings, Menlo Park, CA, 1965, chap. 6.

96. **Fork, R. L. and Kaplan, Z.,** Increased resistance to photodegradation of rhodamine 6G in cooled solid matrices, *Appl. Phys. Lett.,* 20, 472, 1972.

97. **Kaminow, I. P., Stulz, L. W., Chandross, E. A., and Pryde, C. A.,** Photobleaching of organic laser dyes in solid matrices, *Appl. Opt.,* 11, 1563, 1972.

98. **Tomlinson, W. J.,** Phase holograms in photochromic materials, *Appl. Opt.,* 11, 823, 1972.

99. **Higuchi, F. and Muto, J.,** Photo and thermal bleaching of rhodamine 6G(Rh6G) in the copolymer of methylmethacrylate (MMA) with methacrylic acid (MA), *Phys. Lett.,* 81A, 51, 1981.

100. **Higuchi, F. and Muto, J.,** Thermal bleaching of rhodamine 6G in polymethylmethacrylate (PMMA), *Phys. Lett.,* 81A, 95, 1981.

Chapter 5

ABLATIVE PHOTODECOMPOSITION OF POLYMERS BY UV LASER RADIATION

R. Srinivasan and B. Braren

TABLE OF CONTENTS

I. INTRODUCTION AND HISTORY

During the 14-year period from 1961, when the very first laser was described, to 1974, when the first excimer laser was built, the principal sources of laser radiation produced visible or IR light. The output of some of these lasers could be multiplied in frequency to emit pulses of UV radiation, but the energy output of these sources was too little (a few millijoules) to be useful in the study of the interaction of UV laser radiation with solid organic matter.

In the mid-1970s, excimer lasers were invented almost simultaneously by several research groups. These lasers are powerful sources of pulsed, monochromatic, UV laser radiation. Depending upon the gas fill, their output spans the entire UV wavelength range from 350 to 150 nm (Table 1), although the intensity is very variable at the different wavelengths. The principle of operation of an excimer laser is discussed in detail in Section II. Within a period of 3 to 5 years of their invention, excimer lasers with output pulses of 100 to 500 mJ, and repetition rates of 200 to 500 Hz have become available commercially.

It was first reported in 1982[1,2] that when pulsed UV laser radiation falls on the surface of an organic polymer, the material at the surface is spontaneously etched away to a depth of 0.1 to several microns. The principal features of this phenomenon which drew attention to it were the control which can be exercised over the depth of etching by controlling the number of pulses, the fluence of the laser, and the lack of detectable thermal damage to the substrate. The result is an etch pattern in the solid with a geometry that is defined by the light beam. Within a period of a year after the first report, other groups confirmed these observations in several other polymers and at different UV wavelengths. The possibility of extending the process to biological tissue was reported in 1983.[3,4]

Research devoted to understanding the science and developing the technology behind UV laser ablation of polymers has grown at a surprising rate over the past 3 years. This chapter summarizes the state of our knowledge at the beginning of 1987. The organization of the text can be described best in terms of the schematic representation of UV laser ablation shown in Figure 1. The process requires the absorption by the substrate of the pulse of UV photons which penetrate to a depth ℓ_a (Figure 1A). It results in an etch pit that is ℓ_f deep (Figure 1B). The science of UV laser ablation is concerned with the nature of the reactions that take place in the depth ℓ_f of the material that is no longer in the substrate at the end of the pulse. These reactions, in turn, depend upon the absorption characteristics of the material, the bond-breaking steps, the ejection of the products, and the time profile of all these steps. These scientific aspects are discussed in Sections II to IV. The application of laser ablation as a technology is more concerned with the substrate that is left behind after ablation. Some practical aspects of this are considered in Section V.

The reader should bear in mind that, since this field is active and fast-changing, there are no "absolute truths" that have been unequivocally established. One hopes that at least a major portion of the data will survive the test of time. As for the interpretations, they are bound to evolve as our understanding of these phenomena grows in sophistication.

II. ELEMENTS OF UV LASER ABLATION

In this section we will describe the nature of the two basic elements that interact to give rise to UV laser ablation. These are the radiation in the form of laser pulses and the organic solid whose UV absorption characteristics make it susceptible to the laser pulse. The characteristics of the resulting ablation process will also be discussed.

A. EXCIMER LASER

The excimer laser is the source of choice for pulsed UV radiation. It consists of a laser cavity in which a long, narrow discharge tube is set. It is filled with a halogen, a noble gas,

TABLE 1
Different Gas Mixtures

Active medium	F_2	ArF	KrCl	KrF	XeCl	N_2	XeF
Wavelength (nm)	157	193	222	248	308	337	351
Helium (mbar)	2650	1410	1720	2300		960	2570
Krypton (mbar)			200	120			
Xenon (mbar)					60		10
Argon (mbar)		270					
Neon (mbar)					2760		
He/F_2 (5%) (mbar)	50	120		80			120
He/HCL (5%) (mbar)			80		80		
Nitrogen (mbar)						40	
Total pressure[a] (mbar)	2700	1800	2000	2500	2900	1000	2700
Pulse width (ns)	19	23	21	34	28	7	30
Beam dimensions	horizontal, 22 mm; vertical, 6—10 mm						

[a] The compositions are for The Lambda-Physik EMG 201-203 MSC excimer lasers.

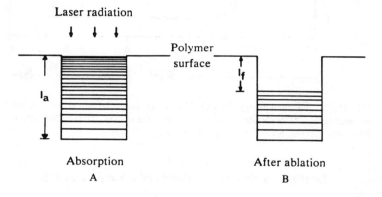

FIGURE 1. Schematic impact of laser pulse on polymer surface. (From Srini-vasan, R., *Science,* 234, 559, 1986. With permission.)

and a buffer gas to a total pressure of 2 to 3 atm. When a high-voltage discharge of 20 to 30 kV is fired through this mixture, laser radiation is produced at an UV wavelength. Table 1 lists the combination of halogen, noble gas, and buffer gas that is used to produce various UV wavelengths.

The compositions used are those recommended by one manufacturer and do vary slightly with the make of the laser. The working of the excimer laser can be understood by referring to the scheme shown in Figure 2.[5] Several characteristics of the excimer laser in their current state of development are worth noting. Although the reaction scheme described above suggests that there are no net reactions in the laser cavity, there are unwanted side reactions that result in the gradual degradation of the gas fill. This problem is increasingly more important with decreasing wavelength and causes a progressive decrease in the output of the laser. The problem can be formulated differently by stating that side reactions and the consequent degradation of the gas fill are more serious in those formulations which involve fluorine than with those that involve chlorine, which is usually introduced as hydrogen chloride. Table 2 lists the lifetime of a single fill for various wavelengths given by one manufacturer. The lifetime is prolonged by the use of an auxiliary reservoir for the gas fill, recirculation of the gases between the discharge tube and the reservoir, some kind of pro-prietary purifying equipment in the gas circulation, injection of a small amount of fresh

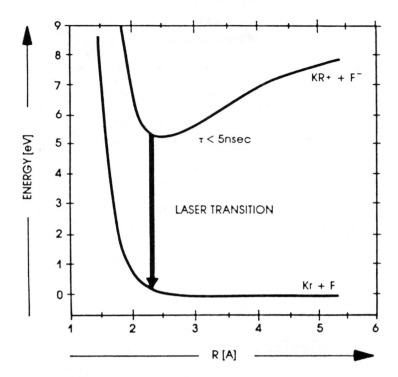

FIGURE 2.   Potential energy diagram showing the molecular structure of the KrF*
excimer laser. (*) Molecule exists only in an excited state. (From Pummer, H.,
Photonics Spectra, May, 1985. With permission.)

TABLE 2
Lifetime in Number of Pulses of a Single Gas Fill

Laser	F_2	ArF	KrCl	KrF	XeCl	N_2	XeF
Wavelength (nm)	157	193	222	248	308	337	351
EMG 101-104MSC	n.m	4×10^5	2×10^5	2×10^6	2×10^7	10^7	4×10^6
EMG 201-204MSC	n.m.	4×10^5	10^6	10^6	10^7	10^7	2×10^6

Note: The compositions are for The Lambda-Physik EMG 201-203 MSC excimer lasers. n.m.: not measured.

gases, ramping the discharge voltage as the gas mixture ages, and the use of a high-capacity
external processor. Obviously, the number of such options will be decided by the purpose
for which the laser is designed. A laser that is meant for use in a research laboratory may
be designed to be more adaptable to several wavelengths (i.e., use with fluorine or chlorine).
A laser meant only for industrial use will be designed for level output over a long period
of time (or number of pulses).

The output beam of a typical excimer laser is larger in cross-section than a visible or
IR laser (see Table 1). The central area of nearly uniform intensity which can have a cross-
section of square millimeters to several square centimeters can be made more homogeneous
and to be of any desired fluence by passage through suitable optics. Commercial lasers
usually have pulse widths of a few tens of nanoseconds. The stretching of a pulse to hundreds
of nanoseconds or its compression to a few picoseconds has been achieved only with specially
designed lasers. In view of the impact that changes in pulse width has been expected and
found to have on the efficiency, as well as the thermal damage, from UV laser ablation, it
can be foreseen that this will be an attractive area for future development in excimer lasers.

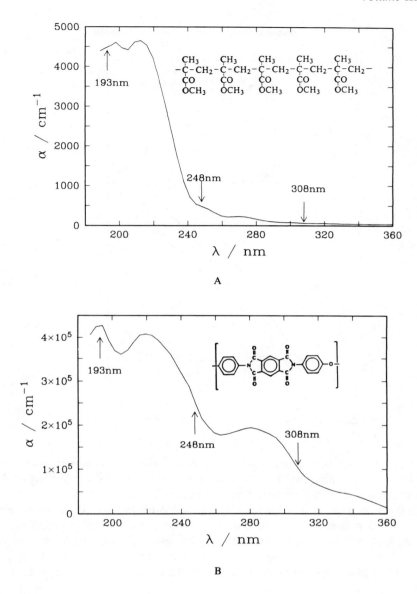

FIGURE 3. Formulas and UV absorption spectra of (A) PMMA and (B) polyimide. (From Sutcliffe, E. and Srinivasan, R., *J. Appl. Phys.*, 60, 3315, 1986. With permission.)

B. UV ABSORPTION OF POLYMERS

Two typical absorption spectra in the UV region for polymers are shown in Figure 3. It can be readily calculated that in a weak absorber (PMMA at 248 nm) the depth of penetration of the radiation (to 95% absorption) is of the order of 30 μm. In contrast, in a strong absorber (polyimide at 193 nm), the radiation will penetrate to only a fraction of a micron before 95% of it is absorbed. In the UV region of the spectrum, the absorption of a photon by an organic molecule leads to an electronically excited state. It is possible for the first excited state to absorb another photon to give a doubly excited state. The excimer wavelengths at 193, 248, and 308 nm correspond to 6.4, 5.0, and 4.0 eV, respectively. Successive absorption of two such photons by any one chromophore in a polymer will lead

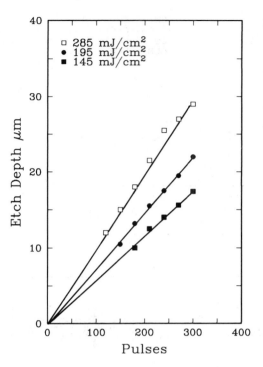

FIGURE 4. Etch depth as a function of number of pulses. Polyimide at 193 nm.

to an excited state which may undergo ionization. Sequential multiphoton excitation (which should be carefully distinguished from simultaneous multiphoton absorption) is of fundamental importance at the laser powers that are used in UV laser ablation. The efficiency with which the absorption of a single photon can lead to the rupture of a chemical bond is expressed in terms of a quantum yield. For one-photon excitation, the quantum yields for bondbreak in the condensed phase are invariably far less than unity. However, secondary reactions can follow the primary bond rupture and contribute to an increased yield. Reactions in the condensed phase following multiphoton excitation in the UV have scarcely been investigated.

C. LASER ABLATION CHARACTERISTICS

The ablation of the surface of a polymer by a UV laser pulse is a function of the energy deposited in the solid in unit time. If a typical UV pulse has a half-width of 20 ns, an energy of 450 mJ, and the size of the beam at the polymer surface is 1.5 cm^2, the fluence at the surface will be 300 mJ/cm^2, and the power density will be 1.5 $\times$ 10^7 W/cm^2. When this pulse strikes the surface, a loud audible report will be heard and, depending upon the wavelength, 0.01 to 0.1 μm of the material would have been etched away with a geometry that is defined by the light beam. If this experiment is performed in air, a bright plume will be ejected from the surface and will extend to a few millimeters. Typically, UV laser ablation is carried out with a succession of pulses. The etching of the surface with a train of uniform pulses is shown in Figure 4. The depth etched is a linear function of the number of pulses but note that there is a very long extrapolation between the origin (zero pulses) and the first data point. One can expect that, at the first few pulses, the uniformity does not exist because

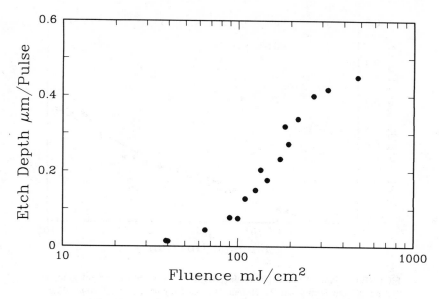

FIGURE 5.   Plot of etch depth vs. log fluence in laser ablation of PMMA at 193 nm.

the first pulse sees a virgin material, whereas each subsequent pulse sees a sample which has already been modified in part by the preceding pulse. The slopes of the lines in Figure 4 give an average value for the etch depth per pulse at that wavelength and fluence for that material. These values are reproducible within the uncertainties in the measurement of the fluence of the laser and the depth of the etching.

Figure 5 shows a plot of the depth of etching per pulse vs. the fluence at 193 nm for PMMA. In this and all subsequent discussion, although the fluence (number of joules per pulse per unit surface area of sample) is quoted, one should keep in mind that it is the power density (power per unit area) that is important, and discussions based on fluence are acceptable only so long as the pulse width is constant. Since the major portion of the data that are currently published and quoted in this volume is based on pulse widths of 15 to 30 ns, this imprecision is not serious but nevertheless worth bearing in mind. The profile of the etch plot in Figure 5 is the typical "lazy S" form of a polymer with a moderate or weak absorption at the laser wavelength. There is always a threshold value for the fluence for the onset of etching, and it is difficult to pinpoint this exactly because the etch curve approaches the abscissa asymptotically. This initial region of slight etching is followed by a steeply rising region in which the etch depth/pulse is linear with the fluence when the latter is plotted on a logarithmic scale. This linear region terminates in a third region in which the sensitivity of the polymer to etching rises more gradually or even decreases with increasing fluence. Usually, the absorption of the polymer decreases with increasing wavelength. The threshold fluence tends to increase with increasing wavelength and the linear portion increases in slope. Both the onset of etching and the flattening at high fluences become quite abrupt. It should be strongly emphasized that at a wavelength at which a polymer has no reported absorption (e.g., PMMA at 308 nm), etching does not decrease to zero. Instead, as the fluence is increased steadily, etching does set in, but the two characteristics that are readily observable in UV laser ablation are no longer observed. These are the control that can be exercised over the depth of etching in a reproducible manner and the lack of thermal damage to the substrate. It is reasonable to say that at 308 nm, the characteristics of the etching pass over from photoablation to the thermal ablation which is observed at visible and IR wavelengths. The transition is *not* sharply defined, as will be discussed in a later section. Figure 6 shows a plot of etch depth per pulse vs. log fluence for a polymer at a wavelength

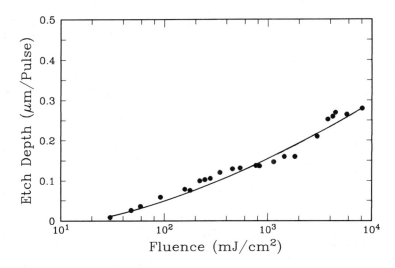

FIGURE 6.   Plot of etch depth vs. log fluence in laser ablation of polyimide at 193 nm. (From Srinivasan, R., Braren, B., and Dreyfus, R. W., *J. Appl. Phys.,* 61, 372, 1987. With permission.)

at which the absorption is very strong ($\alpha > 10^5$ cm^{-1}). The existence of a threshold for etching is evident, but there is no extended region over which the etch depth/pulse is a linear function of log fluence. The absence of thermal damage to the substrate which was earlier stated as a characteristic of UV laser ablation is really a function of the conditions under which the experiment is performed. Let us consider the response of the substrate to a single pulse after the material has already been exposed to a few hundred pulses. A depth ℓ_a(Figure 1) can be supposed to absorb the photons and a depth ℓ_f can be assumed to ablate. The photon energy in the pulse serves to break the chemical bonds, but there is always an excess of energy over that needed for bondbreaking, which serves to propel the products from the etched hole and also excites the fragments vibrationally and rotationally, i.e., the fragments are heated. The thermal damage to the substrate from the hot fragments are minimized because the expulsion process takes place in the same order of time as the laser pulse, as will be discussed in the next chapter. The small amount of energy that is dissipated in the substrate is shown by the shading in the figure. This heat does not cause any observable damage in polymers, which are quite heat resistant. In polymers which are heat sensitive, the heating becomes sufficient to cause melting and heat damage is evident especially when the pulses come in rapid succession. Thus, if the repetition rate of the laser exceeds 10 to 20 Hz, nearly all polymeric materials will show some heat damage. This is especially true if the etched hole has a high aspect ratio (ratio of depth to width) because the exiting fragments tend to heat the sides of the etched hole.

III. PHYSICS AND CHEMISTRY OF UV LASER ABLATION

A. CHRONOLOGY OF THE ABLATION PROCESS
The pictorial representation of the interaction of a laser pulse with a polymer surface that was shown in Figure 1 does not take into account the time frame in which the process occurs. The cartoon on the left in that figure assumes that there is a hypothetical moment at which all the photons in the laser pulse have been absorbed but the surface has not ablated, whereas the cartoon on the right is a representation of the surface long after the laser pulse has etched the surface.

A knowledge of the timing of the ablation process is fundamental to an understanding of the chemical physics of the phenomenon. Early attempts to study it were based on a spectroscopic investigation of the light emission which accompanies the impact of a UV laser pulse on a polymer surface. Koren and Yeh[6,7] timed the intensity of the emission at various distances from the polymer surface and concluded that, in the etching of polyimide films by 193-nm pulses (15 ns FWHM), the emission had a fast component which appeared simultaneously with the laser excitation taking into account the ~30-ns response time of the photomultiplier used in the detection, but a slower component lasted 10- to 100-fold longer than the laser pulse itself. The emission from the plume in the ablation of PMMA at 193 nm (20 ns FWHM) was studied spectroscopically by Davis and co-workers,[8] who timed the peak in the intensity of the emission of CH radicals at various distances from the surface in order to calibrate its velocity of ablation. Their data also placed the beginning of the emission signal at times of the order of the width of the laser pulse. They pointed out that "it is possible that the photodissociation processes responsible for creating the emission in the plume are separate and subsequent to the breaking of the polymeric bonds which cause ablation." These two studies (which are discussed further in Sections III to V) showed that the polymeric structure could begin to ablate on a time scale that is even shorter than the width of a pulse from the laser beam.

At fluences which are smaller than the threshold for ablation, it can be expected that the energy of the laser photons will mostly be degraded to heat. This will cause a rapid, local temperature rise which will be dissipated as the heat is conducted away to the region surrounding the irradiated volume. Dyer and Sidhu[9] used a thermocouple that was attached to the back of a 50-μm-thick film of a polymer to measure directly the transient temperature rise in the irradiated region. Temperature increases of $\geq 5 \times 10^{-2}$ K were detected. The total thermal energy deposited in the film by the laser pulse was measured by Gorodetsky et al.[10] by the use of a pyroelectric crystal of $LiTaO_3$ which was thermally bonded to a 75-μm-thick film of polyimide. Both these investigations showed that the temperature rise in the irradiated area was directly proportional to the photon energy that was deposited, but the linearity of the relationship broke off at the threshold for ablation (Figure 7). Neither method was of significance in the study of the time profile of ablation, but they pointed the way to other studies which achieved this purpose.

The local heating produced by a laser pulse at a fluence that is well below the threshold value at that pulse width causes a stress wave which is sinusoidal in character (Figure 8a). The compression wave caused by the heating is rapidly followed by a rarefaction wave caused by the cooling. As the fluence is increased, the onset of ablation is marked by a transition to an initially structured (Figure 8b and c) and then a relatively smooth compressive signal (Figure 8d). The last frame shows that when the fluence is sufficient to initiate strong ablation, the temporal width of the stress wave is about equal to that of the laser pulse and the point of initiation of both signals is nearly coincident. Dyer and Srinivasan[11] used the arrangement shown in Figure 9A.[12] The film of PMMA (12-μm thick) was on a wide bandwidth polyvinylidene fluoride (PVDF) piezoelectric transducer which, in turn, was pressure contacted using cyanoacrylate adhesive to a 4-mm-thick impedance matching stub of PVDF or PMMA. The latter minimized the acoustic reflection at the interface and gave a rise time limited by the transit time of the longitudinal wave in the transducer. When the electrical circuitry for the detection of the transducer signal was taken into account, the overall display system had a rise time which was estimated to be ≤ 5 ns. There is a time delay in the detection of the stress wave which is caused by its passage through the PMMA layer before it strikes the surface of the PVDF film. This delay can be measured in a separate experiment by measuring the transit time of an acoustic signal. The optical path of the laser beam to the silicon photodiode detector (Figure 9B) was adjusted so that the synchronization in the oscilloscope display of the start of the light pulse and the stress pulse referred to the front surface of the transducer was achieved to better than ± 1 ns.

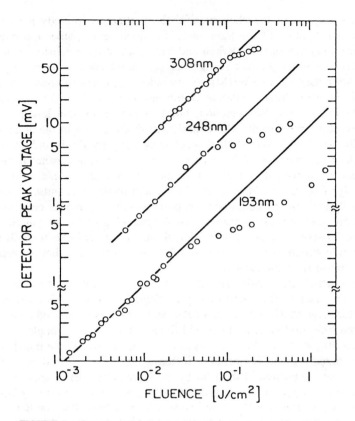

FIGURE 7.   Pyroelectric detector peak voltage vs. laser fluence for various wavelengths. Below the "threshold" kink in each set of data, all the incident laser fluence is detected as thermal energy. Above the threshold only a relatively small fraction is converted to thermal energy. (From Gorodetsky, G., Kazyaka, T. G., Melcher, R. L., and Srinivasan, R., *Appl. Phys. Lett.*, 46, 828, 1985. With permission.)

Compressive stress signals from the irradiation of polyimide films with 193- or 308-nm laser pulses are shown in Figure 10a and b. In this instance the stress wave due to ablation was observed to begin with a delay of ~4 to 6 ns after the start of the laser pulse. For the 16-ns (FWHM) 193-nm pulse, the duration of ablation (FWHM) as indicated by the stress pulse was 16 to 20 ns, whereas for the longer and more structured 308-nm pulse (~24 ns, the duration was 50 to 70 ns, but decreased to ≤25 ns as the fluence increased above 200 mJ/cm². These compressive stress pulses persisted down to fluences well below the "threshold" fluence for etching which was determined from etch depth measurements.

The precision of these photoacoustic experiments depends upon the minimization of the rise of the acoustic detector relative to the width of the laser pulse. For the same reason, the thickness of the polymer sample should also be small in order to keep the transit time of the stress wave to a minimum.

Particular note should be taken of the large magnitudes of the stress pulses which are generated in the ablation of both polymers by UV laser pulses. Figure 11 is a plot of the peak amplitude of the stress wave as a function of the fluence for the ablation of a polyimide film. The depth of the polymer that is etched is less than 1 μm per pulse at 193-nm wavelength. The ablated material is known to be ejected at two to five times the velocity of sound in air (see Section III). Therefore, the pressure generated on the substrate is consistent with these measurements. It is interesting that this stress pulse extends even below

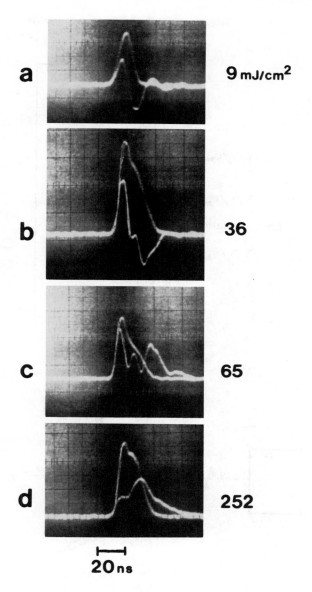

FIGURE 8. Time-synchronized laser and stress waveforms for laser-irradiated PMMA at 193 nm. The larger amplitude signal is the laser pulse which varies slightly in shape due to different operating voltages being used. (From Dyer, P. G. and Srinivasan, R., *Appl. Phys. Lett.*, 48, 445, 1986. With permission.)

the fluence at which significant etching is observed. It will be discussed at length in Section III that radicals and some ions are emitted at high velocity (which can be as large as 5×10^5 cm/s) at fluences at which the etch depth per pulse is below the limit for detection (about 500 Å even after 1000 pulses). The ablation of $\leqslant 5 \times 10^{-8}$ g of material at a velocity of 2×10^4 cm/s on a 20-ns time scale will account for the signal at 0.003 J/cm² in the interaction of 193-nm laser radiation with a polyimide film.

Acoustic signals from polymer films from the impact of UV laser radiation at even lower fluences (less than 0.0001 J/cm²) have been recorded by Gorodetsky et al.[10] In these experiments, the aim was not to resolve the time profile of the acoustic signal, but merely to

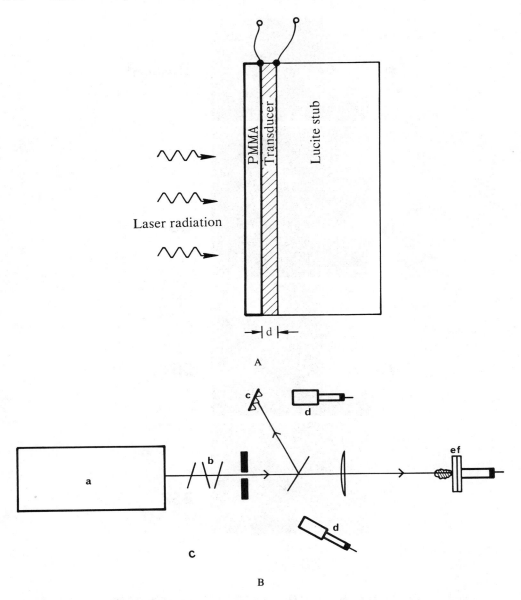

FIGURE 9. Photoacoustic measurements. (A) Arrangement of PMMA, transducer film, and lucite stub and (B) experimental arrangement for recording stress-wave and laser pulse in synchrony. (a) Excimer laser; (b) attenuators; (c) scatterer; (d, d′) silicon photodiodes; (e) PMMA; and (f) PVDF transducer. (From Srinivasan, R., Dyer, P. E., and Braren, B., *Lasers Surg. Med.*, 6, 514, 1987. With permission.)

measure its magnitude as a function of the fluence of the beam. There is an increase in the acoustic signal (Figure 12) at the fluence at which etching can be first observed, but the exact value of the so-called threshold fluence is seen to be as difficult to determine from acoustic data as from data on etching. The polymer that was used in this study was also polyimide. There appears to be a definite relationship between the chemical structure of a polymer and its acoustic behavior in the fluence range that lies well below the threshold for etching. For example, in the case of PMMA, which is a weakly absorbing polymer, Figure 8a shows that a laser pulse at 0.009 J/cm² (193 nm) gives a sinusoidal response which is in contrast to the behavior of polyimide. In order to elicit a acoustic signal that is due to

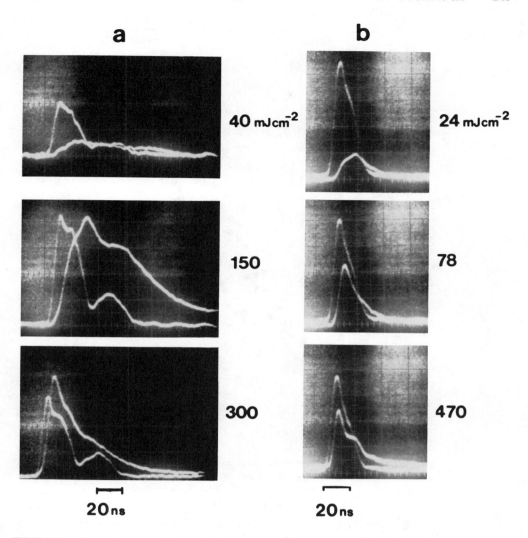

FIGURE 10.   Time-synchronized laser and stress waveforms for polyimide film irradiated at various fluences. (a) 308-nm and (b) 193-nm laser irradiation. (From Dyer, P. E. and Srinivasan, R., *Appl. Phys. Lett.,* 48, 445, 1986. With permission.)

thermoelastic origin from polyimide, a different experimental approach has been taken by von Gutfeld et al.[13] These workers followed the deformation of the surface of the film which corresponds to the rapid heating followed by rapid cooling that is induced by the laser pulse by reflecting a cw beam of a He-Ne laser off the polymer surface and monitoring the reflection with a position-sensitive detector. At the small values of the laser fluence that they used, these experiments do not really involve UV ablation, but they give important information about the magnitude and decay time of the temperature rise that is caused by the UV photons. In a film of polyimide of 125-μm thickness, for a laser pulse (193 nm) of 20-ns FWHM and ≤ 0.001 J/cm^2 fluence, the deformation rises to a maximum in about 2 μs. A thinner film (50 μm) shows the same response time, but the distortion of the surface is fourfold greater. The conclusions from this study with reference to the temperature rise in the film at the surface are pertinent to the mechanism of ablation which will be discussed in the next section.

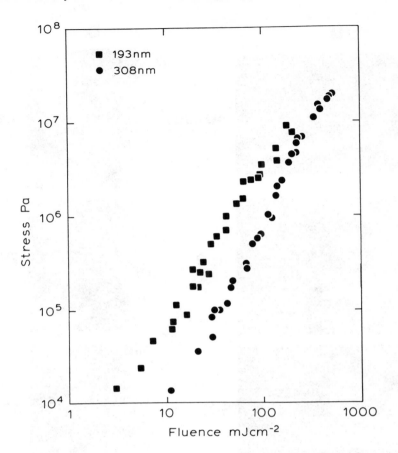

FIGURE 11.   Peak amplitude of the stress wave as a function of fluence for polyimide film irradiated using the 308 and 193 nm lasers. (From Dyer, P. E. and Srinivasan, R., *Appl. Phys. Lett.*, 48, 445, 1986. With permission.)

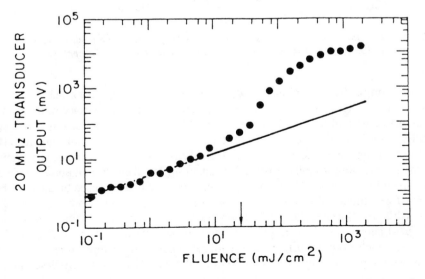

FIGURE 12.   Acoustic signal vs. laser fluence at 193 nm. An abrupt increase is detected at the ablation threshold. (From Gorodetsky, G., Kazyaka, T. G., Melcher, R. L., and Srinivasan, R., *Appl. Phys. Lett.*, 46, 828, 1985. With permission.)

B. CHEMICAL COMPOSITION OF ABLATED MATTER

A knowledge of the composition of the material that is ablated from the surface of a polymer is of importance in understanding the chemistry of the process. It is also of potential use in the design of a UV laser tool for technology that can ablate a given surface most economically and with the least damage to the sample.

Chemical analysis of the ablation products has not received as much attention as physical aspects of ablation. This can be attributed to a number of difficulties which include the following:

1. Systems in which the ablation products have been collected and analyzed show that the products range from atoms and diatomics to volatile, polyatomic molecules of low molecular weight (<200 mass units) to fragments of the polymer (up to 10,000 mol wt). Such a complex mixture of materials cannot be readily analyzed by a single analytical technique.
2. There is evidence that the composition of the products changes with the UV wavelength used, the repetition rate of the pulses (since this will affect the dissipation of any heat energy that is left in the substrate from one pulse to the next), the chronology of the pulses (since the first 1 to 100 pulses may not etch in the same manner as when the system settles down to a constant etch depth per pulse at a given fluence), and the absolute fluence value (since with increasing fluence the initial products of ablation can be decomposed further by incoming photons).
3. While the etch depth per pulse may not be sensitive to the pressure of air at the polymer surface, the products almost certainly undergo reaction with the oxygen and the nitrogen in the air.
4. The net quantity of products that is formed is of the order of only milligrams in typical experiments on a laboratory scale. This, when combined with the complexity of the products referred to in (1) above, increases the difficulty in the analysis.

Commercial polymers contain various additives which are meant to improve their physical and chemical properties. They also contain residual monomers, catalyst, and impurities. While polymers which are free from these contaminants can be prepared and used in research, the data that are so obtained may not be easy to relate to data that are obtained from commercial samples, that is, the impurities in the commercial samples may contribute to their laser ablation behavior.

The products of ablation can be classified into several categories according to their morphology and molecular weight. The analytical methods to detect them have been selected in relation to these properties. It is obvious that certain methods (e.g., IR spectroscopy) are more widely applicable than others.

The first category consists of volatile stable compounds usually of molecular weight less than 200. They have been analyzed by mass spectroscopy. A detailed description of the apparatus is given in the next section. While the mass spectral technique is capable of high sensitivity (picogram levels), a quantitative estimate of one component in a mixture of many products all of which are not fully characterized is not easy. It is more practical to use a combination of a gas chromatograph and a mass spectrometer to conduct a separation of the product mixture into its components and then measure each component quantitatively. This has been done in the cases of two polymers. Not only is this technique tedious, but it also requires considerable effort to obtain the maximum sensitivity.

The second category which is also the major weight fraction of the ablation products in many instances is found as solid material. Following its expulsion into the atmosphere, a part of this fraction settles on the sides of the etched surface. If the ablation is carried out in a closed vessel under a vacuum, this product can travel as much as 10 cm before settling

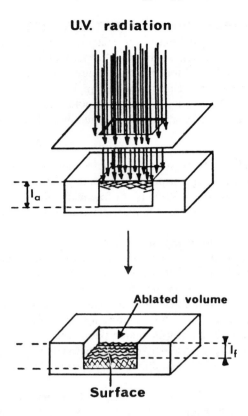

FIGURE 13.   Schematic impact of several laser pulses on a polymer surface.

down on the window and on the inner walls of the irradiation cell. This material can be collected and analyzed by standard chemical methods. To make the collection more exhaustive, the irradiation of polyimide film was carried out under a millimeter layer of water by Srinivasan et al.[14] The solid product, which in this case is carbon, formed a colloidal suspension in the aqueous medium, while HCN which is another significant product dissolved in the water.

Optical spectroscopy has been used by several groups for the detection of transient species such as atoms and diatomics, which form the third category of products. Initial experiments were based on the emission spectrum from the species. It has been pointed out that the excited diatomic species that is the source of light emission may be formed by a secondary photolysis of an initially formed product.[7] Thus, absortion spectroscopy would be a more fundamental method to probe for transient species in the plume. Laser-induced fluorescence spectroscopy (which is discussed in the next section), which combines absorption spectroscopy with fluorescence detection, has been successfully used to probe for transient species such as C_2 and CN in the laser ablation of several polymers.

While at first sight it may seem as if information about the chemistry of the ablated material can be gathered only by analyzing the plume itself, information can also be obtained indirectly by a study of the surface that is created by ablation, which is the fourth category of products. This is shown by the shaded area in Figure 13. The shading represents a section of the volume that has already been transformed by the last one or few pulses and will be ablated by the next pulse. Its chemistry is therefore a precursor to the processes that will occur in the plume on actual ablation. At a resolution down to 1 μm, this surface has been examined by scanning electron microscopy (SEM) and optical microscopy. This helps to see if the bottom of the etched pit is smooth or convoluted and whether it is marred by melting and/or gas evolution or not. It is also useful for observing the deposition of solid debris in and around the etch pit. For information at an atomic level, X-ray photoelectron spectroscopy has been used with considerable success. This gives information on the chemical composition in the first 100-Å depth of the etch pit. This technique has served to demonstrate that after the first laser pulse the composition of the polymer surface is permanently altered at least to a depth of 50 to 100 Å. This information is corroborated by the measurement of the contact angle for water at the new surface. Rutherford back scattering is another technique of use in the study of the composition of the etched surface.

Table 3 summarizes the data that are currently available on the products of the ablation of several polymers by UV lasers. While there is a great deal of qualitative information, the amount of quantitative information which correlates the yield of a given product to the mass of polymer that was removed by ablation is sparse. The three polymers about which some detailed data are available are discussed here, as they may be prototypes of the response of all polymers to UV laser pulses.

1. PMMA

The most important stable products in the gas phase are CO, CO_2, and the monomer, methyl methacrylate (MMA). The quantum yields at 193 and 248 nm are shown in Figures 14A and B. Since the absorption coefficient for this polymer at 248 nm, compared to 193 nm, drops by one order of magnitude, the quantum yields which are calculated from the total number of photons absorbed are not so useful as the chemical yield which relates the yield of MMA at a particular fluence to the mass of the polymer that was ablated at the same fluence. These values are 18 and 1%, respectively, at 193 and 248 nm. There are no quantitative data on the yield of solid polymer that was ejected. Solubility tests placed the average molecular weight of the product obtained at 193 nm to be smaller than the value of $\overline{M}_n = 2500$ for the product obtained at 248 nm. The morphology of the ablated polymer and of the etched surfaces are shown in Figures 15 and 16. They convincingly demonstrate that the local temperature in the etched volume is greater than the melting point of PMMA (about 150°C) during ablation with 248-nm but not with 193-nm laser pulses, both at a fluence that is close to the ablation threshold. The data on the transient species, C_2, in this system will be discussed in the next section.

An interesting perspective on the photochemistry of this material is brought out by the study of Estler and Nogar.[15] These workers analyzed the volatile products of the interaction of PMMA with UV laser pulses by mass spectroscopy. At 266 nm and a fluence of 0.05 J/cm², they observed CO, CO_2, and MMA. However, at a laser wavelength of 240 nm, at a comparable fluence, they found only CO and CO_2, but not MMA. Instead, they observed the formation of methyl formate, which is a major product of the one-photon photochemistry of PMMA.[16] They conclude that they are irradiating the polymer at wavelengths at which the fluences that they used are insufficient to lead to two-photon excitation. This work demonstrates that in polymer:UV laser pulse interactions, there can be very different decomposition pathways which are not only wavelength dependent (a concept that is easily understood) but also dependent upon the photon multiplicity.

2. Polyimide (PMDA-ODA)

Although many studies have been conducted on this polymer, there is no datum which correlates the yield of a given product with the disappearance of the polymer. A major experimental problem is the intractability of the polymer, which does not lend itself readily to molecular weight analysis or spectroscopic studies. The major products that have been identified suggest a stoichiometry according to Equation 1, but this is speculative.

$$\rightarrow 2HCN + 4CO + 3H_2 + 15C + H_2O \tag{1}$$

TABLE 3
Chemical Composition of Ablated Material

Polymer	Sample characteristics	Laser wavelength	Experimental conditions	Ablated products		Substrate analysis and morphology	Ref.
				Monomer or smaller	Larger than monomer		
PMMA (= polymethylmethacrylate)	Free-standing; various molecular weights from 13,700 to 10^6 Spun film on neutral substrate	193 nm 222 nm 248 nm 308 nm FWHM = 5—30 ns	0.04—18.0 J/cm² 0.05—? J/cm² 0.15—10.0 J/cm² 0.50—3.0 J/cm²[21] Air atmosphere or vacuum Repetition rate <10 Hz 90—293 K (external temperature)	C^a $C_2^{a,d}$ CN^a $CN^{a,f}$ CO^g CO_2^g MMA (= methyl methacrylate)[g,k] Methyl formate (c)	Polymer[b,c] $\overline{M_n}$ = 2500 248 nm $\overline{M_m}$ = 4500 248 nm No sign of melting at 193 nm at 10 W fluence[j] Melting at high fluence at 193 nm; at all fluences at 248 nm	No change in atomic composition of first 100Å[e] Change in UV absorption[h] No thermal damage (melting, charring) at 193 nm at 10 W fluence (see Figure 15) Melting bubbling at 248 nm[i,l]	5,8,15,21 26,44—47
Polysterene	Free-standing, low density (0.04—0.14 g/cm³)	193 nm	<1 J/cm²	Styrene[g]	Carbon[m] Polystyrene[a,e]		48,49,59
Deuterated polystyrene	High density	248 nm	Air atmosphere				
Poly α-methyl styrene	Spun on neutral substrate	193 nm	<0.5 J/cm² Air atmosphere			Partial loss of aromatic rings in surface[e]	5
Nitrocellulose	Nitrocellulose linen fiber—13.2% N Spun film on neutral substrate	193 nm[n] 351 nm FWHM = 5—30 ns	0.02—0.20 J/cm²	Gaseous products[g]	None	<3-nm residue at 193 nm; more residue at longer wavelengths[j]	31,50,51

Material		Conditions	Fluence	Gaseous products	Solid products	Effect	Ref.
Polyimide (Kapton®)	PMDA-ODA[o] spin coated as polyamic acid solution and cured / Free-standing film (Kapton®)	193 nm 248 nm 308 nm 351 nm FWHM = 7—30 ns 'Air atmosphere, vacuum, or water repetition rate <10 Hz	0.02—9.0 J/cm² 0.07—5.0 J/cm² 0.05—2.0 J/cm² 0.10—15 J/cm²	C I[a] C II[a] C_2[a,d] CN[a,d] CO[c] CO_2[c] H_2O[c] HCN[c] Benzene[g,k]	Carbon solid[c,j] Polymer[c,j,i]	Blackening[l] O/C ratio decreased by 12%[g] Carboxylic group decreased by 40%[e]	3,6,22,32 52—55
Mylar® PET (= poly ethylene terephthalate)	Free-standing film 1—7 mil thick	193 nm 248 nm 308 nm FWHM = 7—30 ns	0.03—0.50 J/cm² 0.20—2.00 J/cm² 0.10—3.00 J/cm² Repetition rate <10 Hz	H_2[g] CO[g] CO_2[g] C_2–C_{12} Numerous products:[g] benzene[g] toluene[g] benzaldehyde[g]	Solid polymer[i,l]	O/C ratio drops from 1.2 to 0.7 at surface[e] Surface hydrophilic[m] Corrugated surface[j]	12,30,32,54, 56,57
Polyorganosilane	Spincoated $\overline{M}$ = 9000	Multiphoton excitation with 532- or 626-nm laser irradiation	Power density >2.5 GW/cm² FWHM = 4—6 ns			Change in intensity of UV absorption but no change in spectrum	58

a By emission spectroscopy.
b By gel permeation chromatography.
c By IR spectroscopy.
d By laser-induced fluorescence.
e By X-ray photoelectron spectroscopy (XPS).
f An artifact produced by reaction of ablation products with atmospheric nitrogen.[g]
g By mass spectroscopy.
h By UV spectroscopy.
i No etching except at high (20 Hz) repetition rate.
j By scanning electron microscopy.
k By gas chromatography.
l By optical microscopy.
m By Raman spectroscopy.
n Also 157 laser exposure for lithography using this material.[42]
o Condensate of pyromellitic dianhydride and oxydianiline.

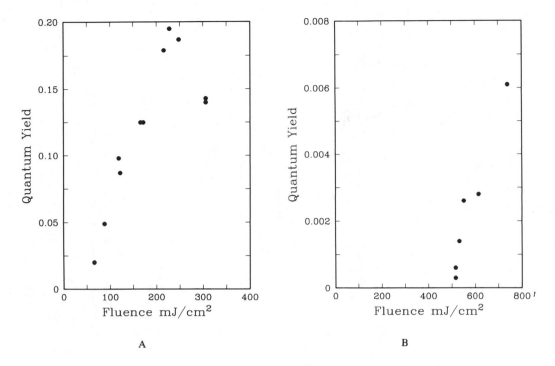

FIGURE 14.   Quantum yields of monomer in laser ablation of PMMA. (A) at 193 nm and (B) at 248 nm. (From Srinivasan, R., Braren, B., Seeger, D. E., and Dreyfus, R. W., *Macromolecules,* 19, 916, 1986. With permission.)

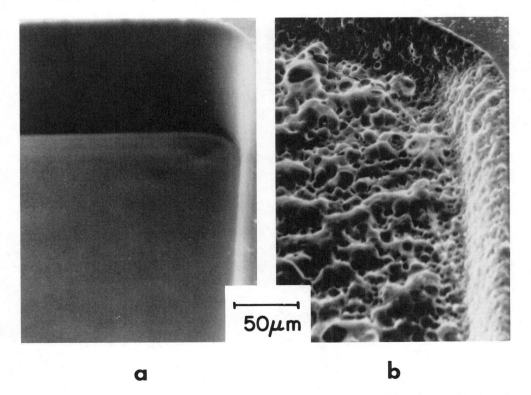

FIGURE 15.   SEM photographs of the inside surface of etch pits of PMMA. (a) 193 nm, fluence of 0.70 J/cm². (From Srinivasan, R., Braren, B., Seeger, D. E., and Dreyfus, R. W., *Macromolecules,* 19, 916, 1986. With permission.)

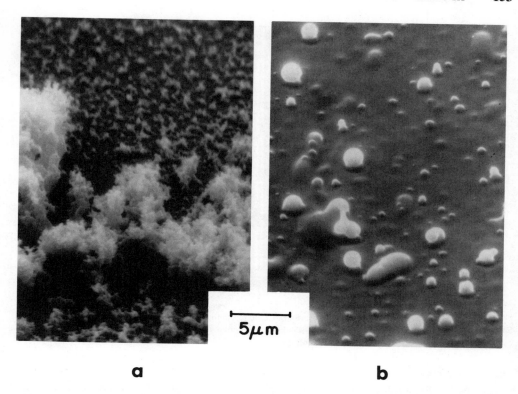

FIGURE 16.   SEM of solid material ejected during laser irradiation of PMMA. (a) 193 nm, fluence 0.15 J/cm^2 and (b) 248 nm, fluence of 2.70 J/cm^2. (From Srinivasan, R., Braren, B., Seeger, D. E., and Dreyfus, R. W., *Macromolecules*, 19, 916, 1986. With permission.)

Very little mechanistic information can be gained from the chemical analysis which have been published.

3. Poly(Ethylene Terephthalate)

The variety of volatile products from this polymer shows that the chemistry that must take place in the ablating volume during laser ablation is complex. A major volatile product is benzene, and its yield has been measured quantitatively.[1] A study of the structures of benzene and PET shows that to form a benzene molecule, out of the latter, two bonds must be broken and two atoms of hydrogen attached to these carbons. The quantum yield plot shows that the formation of benzene initially depends upon the second power of the fluence, but with increasing fluence it is also destroyed photochemically. Hence, the quantum yield starts to level off. The destruction of benzene in the vapor phase following multiphoton excitation by UV lasers is well documented.[17-19] Attempts to quench the formation of benzene by the addition of oxygen[1] showed that more than 20 torr of oxygen was needed to show any quenching effect. It is remarkable that in the presence of oxygen at a pressure of 200 torr, the formation of benzene can still be detected, since oxygen is known to destroy free radicals by chemical reaction at every collision.

The chemistry and morphology of the PET substrate after a surface layer has been etched away have been studied in detail.[20] XPS analysis of the surface (Figure 17) showed that there is a change in the oxygen/carbon ratio after exposure to even one pulse of 193-nm laser radiation at a fluence that is less than the threshold value. This change increases with increasing fluence until the threshold for etching by ablation (0.03 J/cm^2) is exceeded. Thereafter, while the average depth that is etched per pulse increases with increasing fluence,

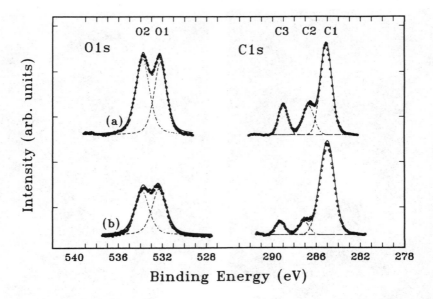

FIGURE 17.   XPS spectrum of poly(ethylene terephthalate). (a) untreated and (b) after irradiation with one laser pulse at 193 nm (100 mJ/cm²). (From Srinivasan, R. and Lazare, S., *Polymer*, 26, 1297, 1985. With permission.)

the oxygen/carbon ratio holds steady at a value of 0.7, which is considerably less than the value of 1.2 for an unirradiated sample. The chemistry of the new surface that is formed by ablation is complicated by the fact that commercial PET film contains crystalline and amorphous region which seem to etch at slightly different rates; the film may also contain an inorganic material which shades the polymer underneath it from further exposure so that a laser-ablated film has a very rough appearance (Figure 18).

C. DYNAMICS OF ABLATED MATERIAL

When a UV laser pulse impinges upon a polymer surface, the etched material is ejected from the surface with considerable velocity and directionality. The dynamics of this plume has been the subject of much research. The velocity of specific molecules in the stream and their angular distribution have been determined. From the distribution of the velocities, a translational temperature has been estimated for the species. This temperature has been compared to the vibrational-rotational temperature, which can be calculated for the same or other specific molecules in the plume from their spectra. The total momentum of the ejected material has been estimated. The velocity and angular distribution have been estimated for a hypothetical polymer whose ablation behavior has been modeled.

In Section III.A, mention has already been made of the measurement of the velocities of diatomic species such as C_2 and CH by timing the peak in the amplitude of the emission spectrum. The precision of this method was greatly improved by the use of laser-induced fluorescence (LIF) as the detection method for a specific diatomic species. The geometry that was used in this experiment[21-23] is shown in Figure 19. The plume was probed by a dye laser beam at a distance from the surface of the polymer which was large enough to avoid any interference from the background luminescence at the film surface during ablation. The laser wavelength was 248 nm. The UV pulses were repeated at 5 Hz and the polymer was slowly translated past the UV beam. The electronic detection system is shown schematically in Figure 20. Following a UV laser pulse, after a preset time delay, a probe pulse was fired from the dye laser at the plume. The wavelength of the probe pulse was sufficiently narrow to excite only one predetermined transition in the diatomic product that was being monitored.

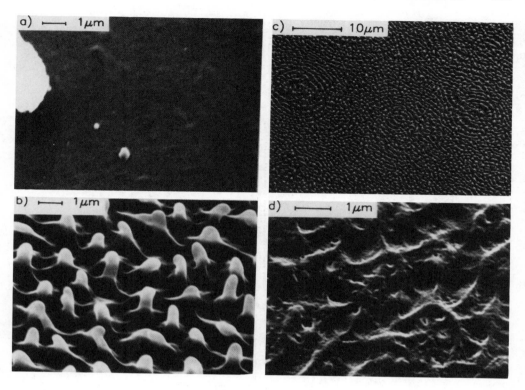

FIGURE 18.   SEM pictures of PET surfaces treated with the laser at 193 nm. (a) 1 pulse; the white spots are debris; (b) 10 pulses; (c) 5 pulses; (d) 10 pulses and subsequent washing in acetone for 10 min. (From Lazare, S. and Srinivasan, R., *J. Phys. Chem.*, 90, 2124, 1986. With permission.)

The fluorescence that the excited product then emitted was detected and analyzed quantitatively. Since the time delay between the excimer laser pulse and the dye laser pulse could be varied, the intensity (= concentration) of the diatomic product in the volume of the plume that was being probed could be determined at various times after the start of the UV pulse that caused the ablation. The time delays that were set were of the order of microseconds so that the width of neither the UV pulse nor the dye-laser pulse introduced any significant error in the timing. From these data, information could be derived on the velocity distribution of the diatomic species in the plume at a particular UV laser fluence. A typical set of curves for the C_2 bandhead signal in the ablation of PMMA at 248 nm is plotted as a function of fluence in Figure 21. The probing wavelength of the dye laser corresponded to 438.2 nm [(2,0) of the $d^3\Pi_g \leftarrow a^3\Pi_u$ transition], and the detection wavelength of 471.5 nm corresponded to (2,1) in the same transition. The most surprising feature of these data is that while the intensities of the fluorescence signal changed by a factor of 200 as the fluence increased from a value (0.15 J/cm^2) which is lower than the threshold for ablation to a value (0.65 J/cm^2) well above it, the maximum velocity changed by only about 30%. It indicated that the production of C_2 and its ejection at a supersonic velocity of 7×10^5 cm/s took place even at a fluence at which etching of the surface of the polymer was too little to be measured by a surface profilometer. An examination of the structure of PMMA shows that the pair of carbons that are likely to be the source of C_2 are a pair situated along the polymer chain (Figure 3) and the number of bonds to be cleaved to form it and eject it at the velocity mentioned above (= 6.1 eV) would require three photons of 248 nm (= 5 eV) wavelength. This is perhaps the most direct evidence that multiphoton photochemistry is important in the ablation of polymers by UV laser radiation.

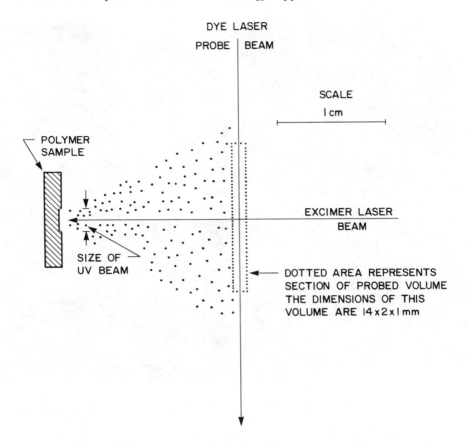

FIGURE 19.   Geometry of UV laser beam and fluorescence pump beam with respect to polymer sample and fluorescence detection system. (From Srinivasan, R., Braren, B., Seeger, D. E., and Dreyfus, R. W., *Macromolecules,* 19, 916, 1986. With permission.)

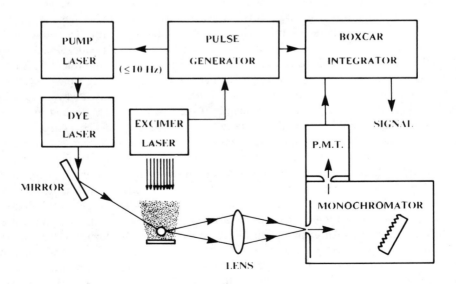

FIGURE 20.   Electronic circuit for triggering UV and fluorescence pump pulses and monitoring fluorescent emission. (From Srinivasan, R., Braren, B., Seeger, D. E., and Dreyfus, R. W., *Macromolecules,* 19, 916, 1986. With permission.)

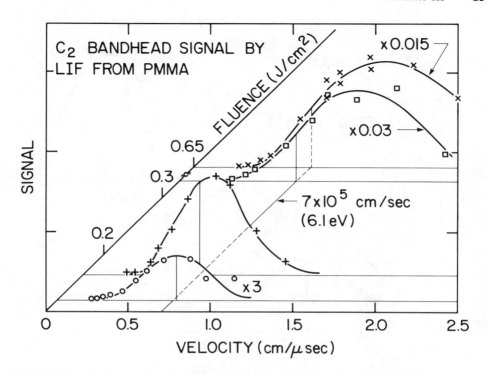

FIGURE 21.   LIF Signal from C_2 radicals indicating the distributions in velocities. Laser irradiation of PMMA at 248 nm. (From Srinivasan, R., Braren, B., Seeger, D. E., and Dreyfus, R. W., *Macromolecules*, 19, 916, 1986. With permission.)

Velocity measurements on C_2 and CN fragments from the ablation of polyimide films with 248-nm laser radiation are shown in Figures 22 and 23. An attempt to fit the velocity distribution of C_2 to a Maxwell-Boltzmann distribution is shown in Figure 24. The measured points are narrowed with respect to the calculated curve. Such a result can be anticipated if the vaporization is not into a freely expanding gas-phase but into a dense colliding medium. This medium is visualized as expanding subsequently into the requisite diffuse gas. One outcome of this picture is that true vaporization occurs not in the laboratory rest frame of reference, but in a moving center-of-gravity frame, which has been discussed elsewhere.[24] The cosine function, usually associated with effusion or vaporization appears as a narrowed distribution because the LIF measurement is made in the fixed laboratory frame. To some degree the velocity/angular dependence in this experiment can be viewed as a supersonic expansion, i.e., the velocity is peaked in the forward direction and therefore the velocity distribution is somewhat narrower than an M-B distribution. The data points can be better fit by a Boltzmann curve plus a constant additive velocity which represents the motion of the center of gravity. These results should be contrasted with velocity measurements obtained by the same technique on the UV ablation of graphite and sapphire.[25]

Velocity measurements on the stable polyatomic product MMA from the ablation of PMMA by 193-nm laser photon have been made by Danielzik et al.[26] by a different detection method. In their work, the ablation products were directed to the entrance slit of a quadrupole mass spectrometer (= QMS) in which they were ionized and mass analyzed. The most common product that was encountered at fluences <0.2 J/cm^2 was MMA, and its velocity distribution at a fluence of 0.08 J/cm^2 is shown in Figure 25. The signal for times less than 50 μs is due to scattered light leaking into the QMS detector. The signal at later times (t > 100 μs), which represents the actual MMA current, corresponds well to a Maxwell-Boltzmann distribution at a temperature of 1200 K. At all fluences <0.15 J/cm^2, the measured

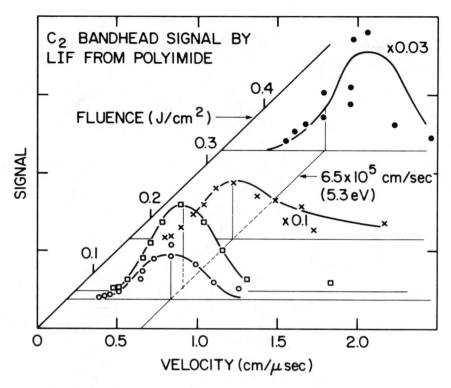

FIGURE 22.   Velocity measurements on C_2 fragments from the ablation of polyimide at 248 nm. (From Srinivasan, R. and Dreyfus, R. W., in *Laser Spectroscopy VII,* Hänsch, T. W. and Shen, Y. R., Eds., Springer-Verlag, Berlin, 1985, 396. With permission.)

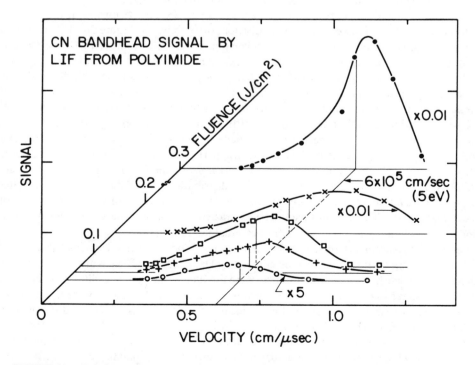

FIGURE 23.   Velocity measurements on CN fragments from the ablation of polyimide at 248 nm. (From Srinivasan, R., Braren, B., and Dreyfus, R. W., *J. Appl. Phys.,* 61, 372, 1987. With permission.)

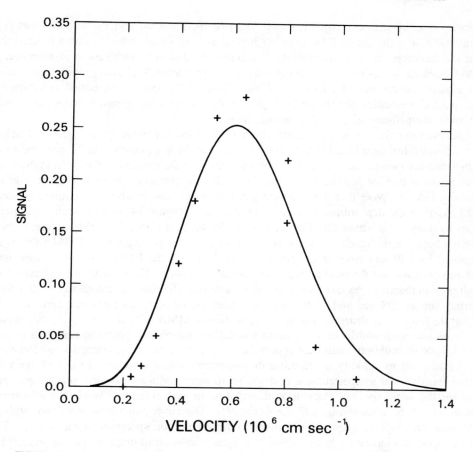

FIGURE 24.   Comparison fit of velocity distribution of C_2 from laser ablation of polyimide at 248 nm to a Maxwell-Boltzmann distribution. (From Srinivasan, R., Braren, B., and Dreyfus, R. W., *J. Appl. Phys.*, 61, 372, 1987. With permission.)

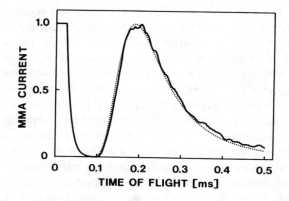

FIGURE 25.   Velocity measurements on MMA (methyl methacrylate) from the ablation of PMMA at 193 nm. The dotted line corresponds to a Maxwell velosity distribution with T = 1200 K. (From Danielzik, B., Fabricius, N., Rowenkamp, M., and Von der Linde, D., *Appl. Phys. Lett.*, 48, 212, 1986. With permission.)

velocity distributions corresponded well to M-B distributions, and the temperatures rose from 800 K at a fluence of 0.06 J/cm² to 3000 K at 0.12 J/cm². Another stable product that was encountered was CO. Its velocity distribution tended to be more complex than that of MMA because it can be produced not only in the primary UV photon:polymer interaction but also in the process of ionization of the MMA. If the latter contribution is subtracted out, the CO velocities also fit an M-B distribution which corresponded to the same characteristic temperature as for MMA at that fluence.

As the laser fluence was increased above 0.20 J/cm², the velocity distribution of MMA was considerably broadened and no longer fitted the M-B equation. There appeared to be both a fast component as well as a component which fit the thermal distribution at the same time in the stream of MMA, the former becoming progressively more important with increasing fluence. Note that the quantum yield of MMA also reaches a maximum at about 0.25 J/cm² in the determination of Srinivasan et al.[22] (Figure 14) and the etch depth/pulse starts to drop off at about this fluence (Figure 5). At first sight, there may appear to be a conflict between the results on velocity distributions of the products from PMMA that were obtained by LIF and from mass spectrometry. Actually, the LIF studies were done with 248-nm photons and the product that was measured, namely, C_2, is undoubtedly formed by multiphoton reactions, as has been pointed out already. The mass spectrometric studies were carried out at 193 nm and at fluences at which one-photon reactions may prevail. The departure from M-B distributions in the velocities of MMA with increasing fluence would be consistent with multiphoton reactions overtaking one-photon photochemistry. It is very desirable to extend both analytical approaches to wider ranges of wavelengths and fluences.

In addition to velocity distribution measurements which provide information on a so-called translational temperature, it is desirable to have a vibrational-rotational temperature also for the same species. Population distribution in each rotational level of a vibrational transition can be obtained by LIF measurements. The entire vibrational spectrum with its rotational envelope can also be obtained from the emission spectrum of the species. The former approach has so far been limited to UV-laser ablation of graphite and sapphire,[25] but the latter has been used quite effectively by Davis et al.[8] for the CH molecule that is observed in the plume from the ablation of PMMA with 193-nm pulses. The (0,0), (1,1), and (2,2) bands of the CH $A^2\Delta \rightarrow X^2\Pi$ transition at 431.4 nm were resolved in a vacuum environment. CH was chosen for detailed study since its large rotational constants make its band structure readily rotationally resolvable and its $A^2\Delta \rightarrow X^2\Pi$ transition is spectroscopically well characterized. The experimental spectrum and the computer fit correspond to vibrational and rotational temperatures being in equilibrium at 3200 ± 200 K. The authors conclude that because of the rapid thermalization rates between rotational and thermal motion, the temperature above the ablation zone is also 3200 K. From the photon energy that is expended in etching and the specific heat of PMMA, it is possible to calculate a maximum temperature that is obtainable when all of the photon energy merely heats the substrate. This temperature is only about 1900 K. They therefore deduce that the higher temperature which exists in the plume is a result of exothermic photochemical reactions which are caused by the photons. In Section II.A, reference was made to the work of the same group on velocity measurements on CH molecules in the same system which yielded a translational temperature of about 11,000 K. This is consistent with the idea that, in supersonic expansions, the translational temperature would be higher than the vibrational-rotational temperatures.

A useful model of the ablation process has been proposed[27] which is consistent with experimental results on the velocity and angular distribution of the ablated material. In this model, the polymer is described by structureless monomer units that are held together by strong attractive interactions. After the laser pulse strikes the sample, a few of the monomer units react photochemically. This process is simulated by allowing each monomer unit to undergo excitation directly from an attractive to a repulsive potential surface. This excitation

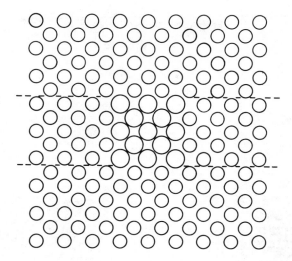

FIGURE 26. Surface layer of the ensemble of monomer units. All odd numbered layers have the same configuration of units. The even-numbered layers have monomer units in the spaces of the first layer. A total of ten layers is included in the calculation. The radii are arbitrary, although the smaller circles correspond to the polymerized form and the larger circles to the reacted monomers. The two horizontal lines show the region of the space that is graphically displayed in Figure 27. (From Garrison, B. J. and Srinivasan, R., *Appl. Phys. Lett.*, 44, 849, 1984. With permission.)

produces a change in the volume occupied by the monomers. For computational simplicity, it is assumed that the polymer is arranged in a face-centered cubic (fcc) crystalline array. The interaction for the bound monomer units is assumed to be pairwise additive with a Morse form. The repulsive interaction is assumed to have the form $1/r^n$.

A polymer with the characteristics of PMMA was chosen as the model system. The surface arrangement of monomer units is shown in Figure 26. It is assumed that the main attractive force holding the monomer units together are the two carbon-carbon bonds along the chain. The strength of a C-C bond is approximately 3.6 eV, which is also taken to be the cohesive energy for one monomer unit in the polymer. The basic repulsive term is $1/r^{12}$. The coefficient of the term for the repulsive interaction is chosen so that the energy of the reacted monomer is 193 nm (6.4 eV) larger than the cohesive energy of the monomer unit in the polymerized form at the same density.

Hamilton's classical equations of motion for the particles are integrated in time. As initial conditions, it is assumed that the monomers in the central region in the top four layers of the crystal interact via the repulsive potential, i.e., they have reacted. In other words, at t = 0, the laser radiation has already been absorbed and the polymer has been fragmented to give monomer units. Initially, the velocities of all of the particles are zero. However, since the reacted particles are at too large a density, i.e., they interact via a repulsive potential, they will gain kinetic energy as time progresses. A sample of the results is shown in Figure 27.

The specific predictions regarding ablation which arise from this calculation are

1. The reacted material will ablate without melting the remainder of the sample.
2. The average perpendicular velocity of the ablated material is about 1000 to 2000 m/s. The precise value depends on the form of the repulsive interaction.
3. The angular distribution of the ablated material is within about 30° of the surface normal.
4. The material ablates layer-by-layer.

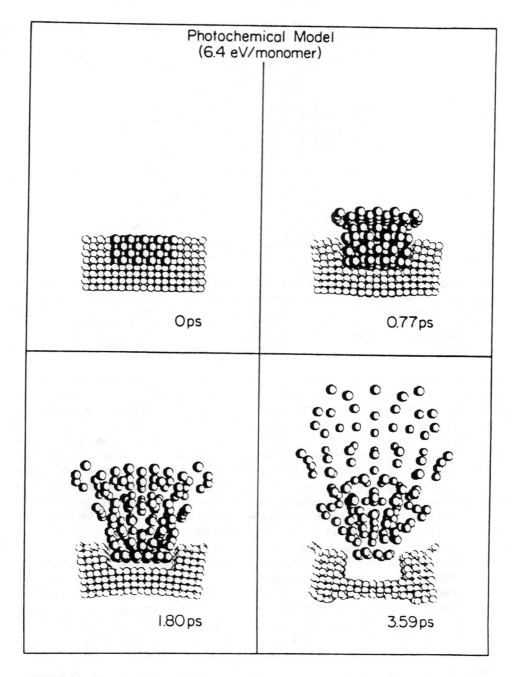

FIGURE 27.   Photochemical model. Movement of the monomers as a function of time. (From Garrison, B. J. and Srinivasan, R., *Appl. Phys. Lett.*, 44, 849, 1984. With permission.)

Experimental measurements of the velocities of the diatomic species that were discussed earlier in this Section are three to four times larger than the values suggested by the model in (1) above. However, the average velocity for the entire mass that is ablated which can be calculated from the photoacoustic measurements that were described earlier indicate a range of 1800 to 2600 m/s, which is consistent with the model. The narrow angular distribution of the ablated material that is predicted by the model has been verified in measurements on the ablation of polyimide by 248-nm pulses.[28] The angular distribution of material ablated from graphite with a 1.06 μm laser beam of fluence about 1000 J/cm² and a pulse width of about 1 ms has been measured and found to spread to over 60° from the surface normal.[29]

The model of ablation dynamics has also been extended to the case of laser pulse in which the wavelength of the photons is not sufficiently energetic to cause an electronic transition in the polymer. In this instance the laser energy will excite vibrational modes and eventually break bonds by thermal decomposition. Several quanta of laser energy may be required to be localized in one bond for that bond to be broken. During the time that the energy is accumulating in one bond, other regions of the sample will become vibrationally excited and possibly melted. The original publication[27] should be consulted to understand the implications of this model but it is sufficient to point out here that there are three principal points on which the thermal model differs from the photochemical model. These are

1. The great excess in the energy input (factor of 2 to 4) that the thermal model requires over the photochemical model for the same amount to be etched. This is the result of the assumption that in the photochemical model, at t = 0, the bonds between the monomers are already broken, whereas, in the thermal model, much more energy has to be supplied in order to cause enough quanta to accumulate in one bond to break it.
2. The thermal model does not give rise to a cleanly etched pit.
3. The ablated material is ejected in a broader cone (60 to 70° of surface normal) in the thermal process than in the photochemical process (25 to 30° of the surface normal).

The data that have been obtained until now on the dynamics of the ablated material are limited to a few products from a few polymers under a limited number of experimental conditions. It is hoped that chemical physicists will turn their attention to more detailed study of a variety of samples in order that a fundamental understanding can evolve in the future.

IV. MECHANISM OF UV LASER ABLATION

The limited scope of the experimental data that exists on the physics and chemistry of the UV-laser ablation of organic solids must be abundantly clear from a reading of Section III. There are hardly two polymers on which the data can be termed comprehensive in details such as product analysis, mass balance, wavelength dependence, pulse width dependence, and other experimental variables. Nevertheless, it is ironic that controversy over the mechanism of this process started almost as soon as the first examples were made known in 1982 and 1983. At that time there was only one firm observation concerning the interaction of UV laser pulses with polymers. It related to the precision of the etch patterns and the lack of thermal damage to the substrate. Since this observation was on the material that did *not* ablate, it was not the most fruitful approach to the understanding of the chemistry of the material that was ablated. This differentiation between the actions of different lasers of wavelengths ranging from the UV to the IR became even more tenuous when it became clear that, even in the UV region, 193-nm pulses gave better results than 248 nm which, in turn, were better than 308-nm pulses. After 4 years of active research by many groups, it is now possible for one to appreciate many of the complexities of UV laser ablation. At the very least one can eliminate certain mechanistic possibilities with confidence. It also seems likely that a simple, general mechanism which is applicable to all organic solids at all UV wavelengths does not exist.

There are two aspects to the mechanism by which UV laser pulses bring about the etching of polymer surfaces with a minimum of thermal damage to the substrate. These are (1) The reaction path in which the polymer bonds are actually broken and (2) a quantitative

ENERGY LEVELS OF A-B

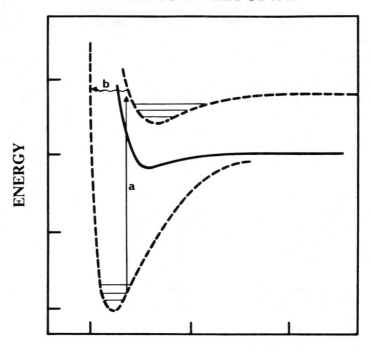

Interatomic Distance

FIGURE 28.   Energy-level diagram for hypothetical bond A-B. (Lower broken line) ground electronic state; (upper broken line and solid line) excited states. (From Srinivasan, R., *Science,* 234, 559, 1986. With permission.)

model of etch behavior as a function of pulse width, wavelength, and fluence. The principal reaction paths that have been proposed can be understood by referring to an energy diagram (Figure 28).[30] It is generally accepted that the absorption of UV photons results in electronic excitation (path a). The excited electronic state can undergo decomposition in that state, which would be a purely photochemical reaction, or, if the excited molecule undergoes internal conversion (path b) to a vibrationally excited ground state, any subsequent decomposition can be considered to be the equivalent of a thermal process. This is the so-called photothermal mechanism in which the photons merely act as a source of thermal energy. Along either pathway the products can be the same (although they do not have to be), and any excess energy over that needed to break the bonds will remain in the products and be dissipated in the ablated fragments. If the time for ablation is of the order of the duration of the laser pulse (~20 ns), the diffusion of the thermal energy to the unirradiated portion of the substrate will extend to 600 Å in a material with a thermal diffusivity of 4×10^{-4} cm/s. Therefore, it would be expected that the region showing thermal damage would be negligibly small.

The expression

$$\ell_f = \frac{1}{\alpha} \log \frac{F}{F_0} \tag{2}$$

which relates the etch depth per pulse to the fluence (F), the fluence threshold (F_0), and the absorptivity (α) at that wavelength, has been derived by more than one group.[31-33] The derivation merely relates to the deposition of the photon energy according to Beer's law and is independent of the assumed mechanism. This equation can be tested by plotting ℓ_f vs. log F when a straight line with a slope of $1/\alpha$ should be obtained. In fact, in many instances there is a linear relationship over a substantial portion of the plot, but the slope is not $1/\alpha$. In other cases, the slope is not linear at all over even a small range of fluence. There is no reported example of a polymer whose etch curve fits Equation 2 exactly.

Early views of this phenomenon implicitly assumed that ablation followed the deposition of all of the photons of the laser pulse in the solid. Note that in Equation 2 there is no time-dependent quantity at all. The photoacoustic experiments mentioned in Section III.A typically showed that, when a fraction of the pulse energy had been deposited in the film, the material started to ablate. The correlation of these data to those obtained by fast UV-scanning spectroscopy of the ablating surface and also to a simulation of UV laser ablation were also pointed out in Sections III.A and C.

A knowledge of the time profile of ablation affects the calculations of the temperature rise in the volume of material that ablated. These calculations, which form the basis for the photothermal mechanism, significantly overestimated the temperature. It can be shown that the actual rate of thermal decomposition would not be adequate to lead to ablation in the time span of the laser pulse. For example, the data on polyimide (Figure 10) show that material begins to ablate in less than 10 ns. At that point, based on the amount of energy absorbed, the temperature rise would be 1200 K in the absorption depth (95%) of 0.07 μm. The half-life for the thermal decomposition of a polyimide at this temperature has been measured to be about 20 s.[34] It can be argued that the rate constants to be used to estimate thermal decomposition half-lives under pulsed laser irradiation conditions are not those obtained from slow pyrolysis studies in static systems. The discordance between the sets of data is so enormous ($\sim 10^9$) that it is difficult to believe that they can be reconciled. An analysis for PMMA that leads to similar conclusions has been reported.[22] Published values for the pyrolysis of PMMA[35] can be combined with the time profile for ablation[11] to show that the rate of pyrolysis will not be fast enough to cause ablation in the duration of a laser pulse.

A. DYNAMIC MODELS
1. Part 1

The first model which can be termed realistic in that it takes into account the changes in the substrate structure even as the laser photons penetrate the material was proposed by Keyes et al.[36] The passage of photons will cause chemical bonds to be broken so that the photon intensity at any depth ℓ will evolve with time:

$$I(\ell,t) = I_0(t)\exp\left[-k' \int_0^\ell d\ell' n(\ell',t) \right] \qquad (3)$$

where I_0 is referred to as the unshielded laser intensity, k is a constant, and n is the density of unbroken polymer bonds. If n is not a function of time, this expression will reduce to Beer's law. The change in n with time is given by

$$n(\ell,t) = -kI(\ell,t)n(\ell,t) \qquad (4)$$

When Equation 3 is substituted into Equation 4 and solved for the initial condition $n(\ell, t = 0) = 0$ when $\ell < 0$; $n(\ell, t = 0) = n_\alpha$ when $\ell > 0$, then

$$\frac{n(\ell,t)}{n_\infty} = [1 + \{\exp(kI_0 t) - 1\}\exp\{-k'n_\infty \ell\}]^{-1} \qquad (5)$$

2. Part 2

The time-dependent passage of the laser pulse through the ablating material that is explained in the theory in Part 1 has been adapted in a more recent approach to a dynamic model.[37]

The ablation process is believed to be a volume explosion which would make the production of small molecules important. It has already been pointed out that the exact chemistry of the decomposition pathways that prevail in the UV-laser ablation process is far from clear. For the purpose of this model, it is sufficient to assume that the absorption of a UV photon from a laser source creates an electronically excited state from which two processes can occur: (1) photofragmentation, which leads to stable photoproducts and (2) relaxation back to the reactant ground state, which merely degrades the energy without causing any photodecomposition.

Either process can be looked upon as being made up of subsidiary pathways which may lead to the same net result. This can be represented in a set of five elementary chemical steps:

$$M + h\nu \rightarrow M^* \qquad [1]$$

$$M^* \rightarrow M \qquad [2]$$

$$M^* \rightarrow Products \qquad [3]$$

Here M represents one absorption center or chromophore in the polymer and M^* represents the same chromophore after excitation by one quantum of UV radiation. Step 2 is a composite of all processes that result in deactivation, whereas step 3 is a composite of all processes that lead to decomposition. Ablation is expected to occur not from the M^* state, but from a higher excited state that is formed from it by the absorption of another photon:

$$M^* + h\nu \rightarrow M^{**} \qquad [4]$$

$$M^{**} \rightarrow Ablation \qquad [5]$$

During the passage of a pulse through the sample, a stationary state in the concentration of M^* will be operative as a balance is created between its formation by step 1 and its disappearance by steps 2 and 4. The basic new idea of this dynamic model is the notion of an absorbed photon flux threshold below which photofragmentation proceeds at a negligible rate. Figure 29 shows, for a typical UV laser pulse, the flux Π defined as the number of photons absorbed per time and volume. Π is proportional to the time-dependent laser intensity (power per unit area) at a given depth ℓ as below:

$$\Pi(t,\ell) = I(t,\ell)(\alpha\lambda/hc) \qquad (6)$$

Here h is Planck's constant and c is the velocity of light. The absorption coefficient α (from Beer's law) of the solid polymer at a wavelength λ is measured from the low intensity absorption spectrum. The photoflux threshold is Π_T and defines the concentration above which absorbed quanta are available for efficient photofragmentation.

The time-dependent effective concentration $\rho(t,x)$ of absorbed photons above threshold at depth ℓ is defined by

$$\rho(t,\ell) = \int_0^t \Theta[\Pi(t',\ell) - \Pi_T]dt' \qquad (7)$$

where

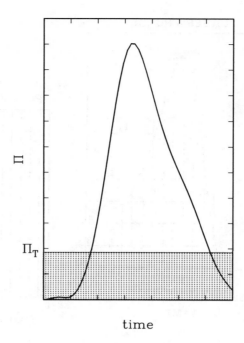

time

FIGURE 29. Schematic picture of the threshold Level Π_T. The flux Π is defined as the number of photons absorbed per time and volume. (From Sutcliffe, E. and Srinivasan, R., *J. Appl. Phys.*, 60, 3315, 1986. With permission.)

$$\Theta(z) = \begin{cases} 0 & \text{if} \quad z < 0 \\ z & \text{if} \quad z \geqslant 0 \end{cases} \tag{8}$$

In order to ablate spontaneously, the irradiated sample must reach a critical density of broken bonds corresponding to a threshold level ρ_T for the effective concentration of the absorbed photons. Once the ablation condition is reached [$\rho(t, \ell) \geqslant \rho_T$], then the internal stress due to the change in volume occupied by the fragments is sufficient to drive them out of the bulk. For practical purposes, the two parameters, Π_T and ρ_T are adjusted to one observable which is the etch curve at one wavelength and are then kept constant for all other experimental conditions. The flux threshold and the ablation condition are, in practice, found to be constant for a given material and independent of the excitation wavelength. However, since the absorption coefficient α is a function of wavelength, the intensity threshold, I_T is related to Π_T by the expression,

$$I_T(\lambda,\alpha) = \Pi_T(hc/\alpha\lambda) \tag{9}$$

It may seem paradoxical that UV ablation will depend only upon the number of absorbed photons per unit volume per unit and not upon the energy contained in each photon. This concept would be untenable if ablation was caused principally by photochemical decomposition from the first excited electronic state. As mentioned earlier, ablation probably depends upon multiphoton excitation of individual chromophores, and the wavelength independence is simply due to the high levels of excitation that are reached before decomposition occurs. It would also be shown that, in spite of this multiphoton excitation mechanism, the reaction is controlled by the rate of excitation to the first excited electronic level, which is the slowest step.

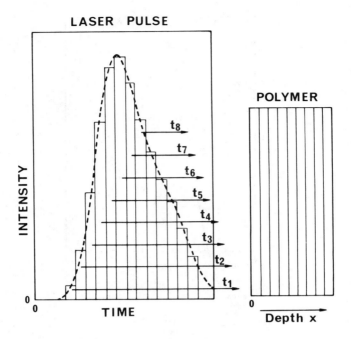

FIGURE 30.   Discretization procedure developed for computing the dynamics. The light pulse (dashed line) is sliced into many short pulses which are successively absorbed in hypothetical layers of the material. A layer ablates when its effective concentration of absorbed quanta reaches the ablation condition. (From Sutcliffe, E. and Srinivasan, R., *J. Appl. Phys.*, 60, 3315, 1986. With permission.)

Since nonradiative relaxation from the excited electronic state is always effective, the fraction of the pulse below the threshold level Π_T will heat the sample. It can be identified with the thermal component of the ablative process. This thermal contribution F_T is given by integrating the intensity below the threshold level over the whole laser pulse of duration $tp[I_0(t) \equiv I(t,0)]$

$$F_t = \int_0^{tp} \min[I_T I_0(t)] dt \qquad (10)$$

$$\min(y,z) = \begin{cases} y & \text{if} \quad y < z \\ z & \text{if} \quad y \geq z \end{cases} \qquad (11)$$

At very low irradiation intensities[8,9] we get $F_T = F$, where F is the total fluence. Thus, all of the absorbed energy is converted to heat. Some complexities which may originate in the thermal component should be pointed out here, although their exact contributions to the ablation process have not been assessed in every instance. The thermal component may itself induce bond scission in the polymer. This is simply the photothermal mechanism which has already been discussed. However, the photodecomposition from the first singlet state may also be thermally activated. This is believed to be (but not proved to be) as slow as the photothermal process. Strongly exothermic decompositions such as the case of nitrocellulose may be an exception to this idea.

B. METHOD AND INTERPRETATION: CASE I AND CASE II

A schematic presentation of the numerical procedure that was developed to compute the dynamics of ablation is given in Figure 30. The sample is split into many successive layers (typically 10^3 to 10^4), the thickness of which defines the resolution of the computation. Similarly a discretization of the laser pulse is achieved by replacing it with a sequential train of short rectangular pulses (typically 20 ps) of constant intensity. At each time step t_i a pulse of corresponding intensity $I_0(t_i)$ and duration Δ_i is sent through the sample. Based on Beer's law, the number of photons absorbed in each layer at depth ℓ_j is readily evaluated. From the volume of a sample layer and the duration of a pulse element, the time evolution of the number of quanta absorbed per unit time and unit volume $\Pi(t_i\ell_j)$ is fully defined as a function of the sample depth. Replacing the integral in Equation 7 by a sum leads to the time and depth-dependent effective concentration of absorbed photons according to

$$\rho(t_k,\ell_j) = \sum_{i=1}^{k} \Theta[\Pi(t_i,\ell_j) - \Pi_T]\Delta t_i \qquad (12)$$

The indices in the sum might run over more than one single laser pulse. The consequences of pulse repetition are naturally imbedded in the model. At each time step the ablation condition is tested throughout the sample $[\rho(t_k\ell_j) \geq \rho_T?]$. When a layer within the sample reaches the ablation condition, two further alternatives may be considered:

1. Case 1, in which the fragments are almost transparent at the excitation wavelength which is simulated by an instantaneous disappearance of the affected layer.
2. Case 2, where the outgoing fragments filter the remaining part of the incoming pulse with the same efficiency as the nonirradiated sample. The model accommodates this shielding quantitatively by waiting until the entire pulse has gone through the sample to permit the layer to ablate.

Thus, the dynamic model provides a basis for the classification of the sensitivities of different polymers to UV laser radiation. Those that can be considered to be weak absorbers ($\alpha < 5000$ cm^{-1}) fall into case 1, whereas the strong absorbers fall under case 2. The fundamental implications of this separation will be discussed in a later section.

C. RESULTS: PMMA — A WEAK ABSORBER

The etch curve of PMMA at 193 nm (Figure 5) was fitted to the dynamic model on the basis that this polymer is a weak absorber. The resulting graph is compared to the experimental data (which were used in the fitting process) in Figure 31. The half width of the laser pulse was taken to be 13 ns. The model is seen to give a quantitative description of the data from the threshold fluence to about 0.3 J/cm^2. Near the fluence threshold, the curve is very sensitive to Π_T, whereas ρ_T prevails at higher fluences. These two parameters are thus precisely and unambiguously defined; the set of values leading to the etch curve in Figure 31 is unique. At fluences greater than 0.3 J/cm^2, the experimental points start to level off and ultimately reach a plateau. It is possible that this is due to a secondary effect brought about by the etch products in the plume, somehow interfering with the flux of incoming photons. It has been pointed out in Section III that the character of the products changes at about this fluence as determined by chemical and mass spectrometric analysis. The dynamic model has not yet been developed to deal with this secondary effect.

Figure 32 compares case 1 to case 2 in PMMA. Case 2 was computed with the same parameters as case 1. Up to about 0.08 J/cm^2, both cases are practically indistinguishable. Beyond that point, filtering by the ejected fragments (case 2) tends to hinder the etching process. This effect is nevertheless weak in PMMA because the absorption coefficient is small and $1/\alpha >>$ etch depth. Since adjusting the parameters to case 1 or case 2 gives almost the same final result, PMMA is an ideal system for modeling.

A key premise in this model is that Π_T and ρ_T are constants characterizing a polymer

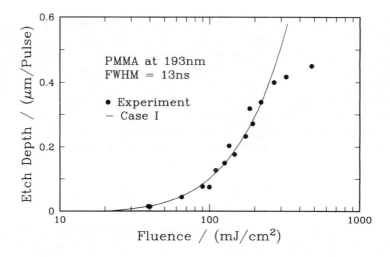

FIGURE 31.   Etch depth per pulse as a function of fluence for PMMA at 193 nm. (Solid line) best fit assuming Case I; (•) experimental values. (From Sutcliffe, E. and Srinivasan, R., *J. Appl. Phys.*, 60, 3315, 1986. With permission.)

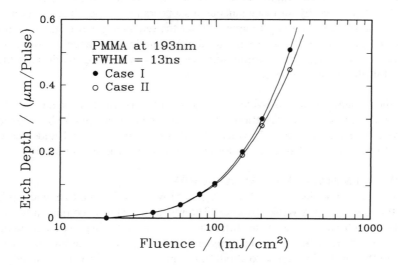

FIGURE 32.   Comparison of case 1 to case II for PMMA at 193 nm. (•) case I; (O) case II. (From Sutcliffe, E. and Srinivasan, R., *J. Appl. Phys.*, 60, 3315, 1986. With permission.)

and do not depend upon the irradiation conditions such as wavelength. It is possible to test this by applying the fit to the etch curve at 248 nm (Figure 33). There are no adjustable parameters at this point, since the absorption coefficient at 248 nm was determined by experiment. The fit of the curve to experiment is seen to be very satisfactory up to a fluence of 2.0 J/cm^2. Once again, the leveling in the etch curve at high fluence is not included in the model. Case 2 gives the same curve as case 1 at this wavelength, since the absorption coefficient is an order of magnitude smaller at 248 nm than at 193 nm.

In this model the possibility of accumulating fragments from laser pulse to laser pulse is a very important consequence of the basic assumptions underlying the model. Practically, if the energy delivered to the sample by the first pulse is not sufficient to reach the ablation condition, several pulses are then required in order to accumulate enough fragments below

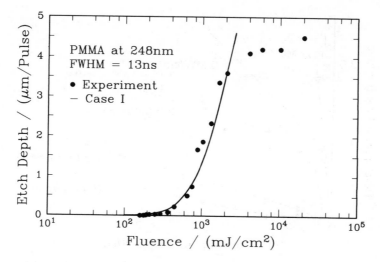

FIGURE 33. Etch curve for PMMA at 248 nm. (Solid line) case I; (•) experimental values. (From Sutcliffe, E. and Srinivasan, R., *J. Appl. Phys.*, 60, 3315, 1986. With permission.)

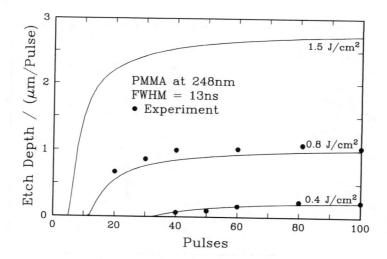

FIGURE 34. Etch depth as a function of the number of laser pulses for PMMA. Several "incubation" pulses are required to trigger the etching process. (Solid line) Case I; (•) experimental values. (From Sutcliffe, E. and Srinivasan, R., *J. Appl. Phys.*, 60, 3315, 1986. With permission.)

the interface. Figure 34 shows the average etch depth for various numbers of incident pulses. It is seen that there are a certain number of "incubation" pulses necessary at each fluence before etching is realized. The prediction agrees well with experimental results. It can be readily demonstrated that the incubation pulses actually decompose the polymer chain without ablating them. In the early experiments of Kawamura et al.[38] exposure to the UV laser pulses were followed by wet development of the sample. At 248 nm, their data (Figure 35) show that, at fluences where ablation would not occur, the polymer can be dissolved out by wet development. These results are more dramatically illustrated in the data shown in Figure 36. In this instance, following exposure to each pulse of UV-laser radiation, the polymer surface was developed in a solvent. It is seen that even when there is spontaneous etching

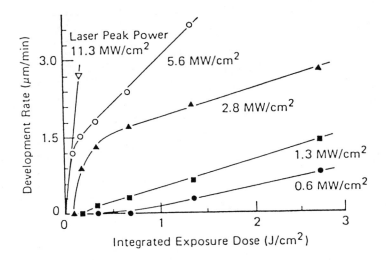

FIGURE 35.   UV Photoetching characteristics of PMMA at 248 nm followed by wet development of the sample. (From Kawamura, Y., Toyoda, K., and Namba, S., *Appl. Phys. Lett.*, 40, 374, 1982. With permission.)

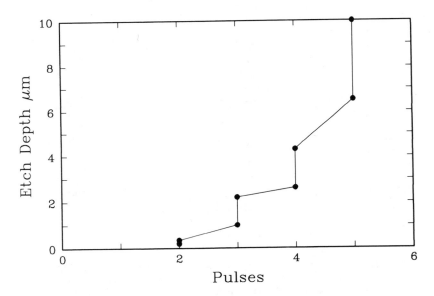

FIGURE 36.   Laser ablation of PMMA at 248 nm. After irradiation of a number of pulses the sample was wet developed. Each vertical line corresponds to a wet development.

by ablation, wet development causes a further amount of polymer fragments to be dissolved out. This clearly shows that the first pulse is unique. Subsequent pulses always interact with partially photolyzed material.

From a knowledge of the flux threshold that is derived from experiment, it is possible to calculate the relationship that exists between the absorbed intensity at a given wavelength that is useful in causing ablation to the part that is degraded to thermal energy. Such a relationship is plotted for PMMA at three wavelengths in Figure 37. At 193 nm, since the flux threshold is small (Table 4), most of the absorbed energy is used in etching which also means that the heating effect is a minimum. At 248 nm, the threshold increases, and the

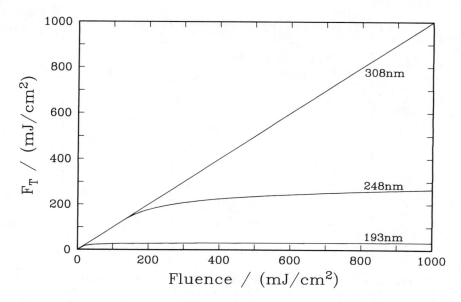

FIGURE 37.   Thermal component of PMMA as a function of the laser fluence at 193, 248, and 308 nm. The diagonal line at 308 nm corresponds to a full conversion of the absorbed energy into heat. (From Sutcliffe, E. and Srinivasan, R., *J. Appl. Phys.*, 60, 3315, 1986. With permission.)

TABLE 4
Model Parameters and Derived Quantities for PMMA and Polyimide

	PMMA	Polyimide	Units
Π_T	4.3×10^{27}	3.1×10^{29}	photons/s/cm^3
ρ_T	4.1×10^{21}	4.1×10^{21}	photons/cm^3
ρ_T	0.62	1.9	photons/monomer
I_T(193 nm)	0.97	0.77	MW/cm^2
I_T(248 nm)	7.5	0.98	MW/cm^2
I_T(308 nm)	60	1.7	MW/cm^2

thermal effect increase correspondingly. For example, at an incident fluence of 0.4 J/cm^2, at 193 nm, the thermal contribution is only 10%, whereas, at 248 nm, at the same incident fluence, the thermal contribution will be 50%. At 308 nm, the polymer has no significant absorption, which means that essentially 100% of the incident energy will go to heat the sample. These conclusions are consistent with observations made on the thermal damage to the substrate at these wavelengths (Section III).

D. POLYIMIDE: A STRONG ABSORBER

The intense absorption spectrum of polyimide (PMDA-ODA) is illustrated in Figure 3. Unlike PMMA, in this instance, the differences between case 1 and case 2 are no longer negligible, that is, whether the ejected fragments are transparent are not transparent at the laser wavelength has major consequences on the shielding effect. The etch curve can be fitted (the values of Π_T and ρ_T can be determined) only at a wavelength at which $\alpha^{-1} >>$ etch depth. This was achieved using 351-nm radiation and restricting the fit to fluences just above the threshold. The temporal shape of the pulse was also carefully determined in order to optimize the fitting. After the parameters were determined from the etch curve at 351 nm, the reliability of the fit was tested on the experimental etch curve at 308 nm (Figure

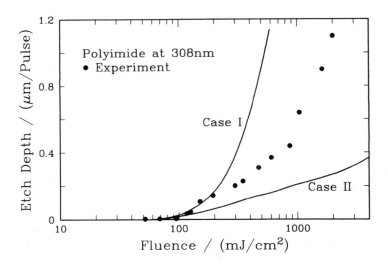

FIGURE 38. Etch curve at 308 nm for polyimide. Comparision between experimental (•) values and the model predictions from Cases I and II. (From Sutcliffe, E. and Srinivasan, R., *J. Appl. Phys.*, 60, 3315, 1986. With permission.)

38). There is a dramatic difference between cases 1 and 2. At the threshold, the predicted curve seems to fit the data well. With increasing fluence, as more and more material is ejected, the shielding effect sharply decreases the etch depth per pulse. However, the correction introduced by case 2 seems excessive, meaning that the ejected material may not absorb as intensely as the original polymer at this wavelength.

The parameters used in fitting the data for polyimide and polystyrene, both of which are intense absorbers of UV radiation, are compared to the parameters for the weak absorbers in Table 4. It is noteworthy that ρ_T, the ablation condition, varies very little among these widely differing polymers. It suggests that the deposition of a certain amount of absorbed photons in unit volume of the polymer in unit time is sufficient to bring about ablation irrespective of the chemical structure. This concept which seems to contradict fundamental ideas in photochemistry bears further examination.

From the elementary steps 1, 2, 4, and 5, it can be derived that

$$\text{Rate of ablation } (dp/dt) = \frac{I^2\alpha^*\alpha^{**}[M]}{k_2 + I\alpha^{**}}$$

where α^* is the absorptivity for the first excitation step, while α^{**} is the absorptivity for the second excitation step. It is known that invariably $\alpha^{**} >> \alpha^*$ and $I\alpha^{**} >> k_2$. This expression will therefore reduce to

$$dP/dt = I\alpha^*[M]$$

which explains why the application of Beer's law is so successful in simulating the reaction dynamics. This approach will certainly fail if the photon energy is insufficient to promote the molecule to the first electronic excited state.

V. PRACTICAL CONSIDERATIONS

The practice of UV-laser ablation of polymers is extremely simple. The laser pulse as it is emitted from a commercial excimer laser can be directed to the surface of a polymer

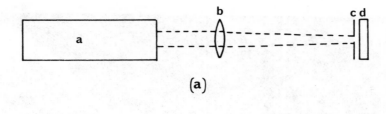

(a)

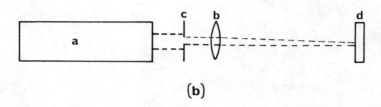

(b)

FIGURE 39.   Schematic arrangement for achieving an etch pattern. (a) metal mask and (b) projection mask: (a) laser; (b) lens; (c) mask; and (d) polymer.

with a focusing lens, and the etching can be conducted in an air atmosphere. In the schematic drawing (Figure 39), the definition of the etch pattern is achieved with a metal mask that is in contact with the work. Alternatively, the pattern can be defined by a transparent mask that is placed between the lens and the laser. It is obvious that this mask should be made of a UV-transparent material such as fused quartz. A contact mask cannot be made of quartz because the ablated material has to be ejected into the atmosphere. Only metal masks can be used, and the usual limit for the smallest dimension that can be etched in this manner is 30 to 50 μm. By the use of projection optics, features as small as 1 to 3 μm have been etched by ablation.[39]

The geometry of the hole that is etched has been studied in detail by Braren and Srinivasan.[40] The hole is always formed with overcut walls, the minimum angle (which is obtainable at the highest fluences) being 1 to 2° off the vertical. With decreasing fluence, the wall angle becomes progressively less steep. The aspect ratio that is obtainable is also affected by the fluence since the wall angle has to be as small as possible to obtain a high aspect ratio. It is relatively easy to achieve an aspect ratio of five or more.

If the etched hole does not penetrate the organic film completely, some interesting features of the bottom surface of the hole become apparent. There are ridges and valleys on the bottom surface which originate near the wall. The first ridge is the steepest, and the first valley is the deepest (Figure 40). These features probably are caused by diffraction fringes originating at the edge of the hole. The reflection of the incident beam off the side wall of the hole is a contributing factor.

It was pointed out in Section III that a large fraction of the ablated polymer leaves the surface as finely divided solid material (Figure 16). This gives rise to several practical problems. Although the solid particles leave the surface with considerable velocity, a sizeable fraction tends to return to the surface. It has been shown[41] that these particles acquire an electrostatic charge which may contribute to their tendency to settle on the surface of the work. The problem is particularly severe if the polymer surface is kept horizontal, and the laser beam impinges on it vertically.

At fluences which are just above the threshold for ablation, the particulate debris tends to fall back on the etched surface and often sticks to it. During continued ablation and etching of such a surface, the debris creates a shadowed area under it (Figure 41) which results in unevenness which is undesirable.[42] It is advantageous to operate at a fluence that

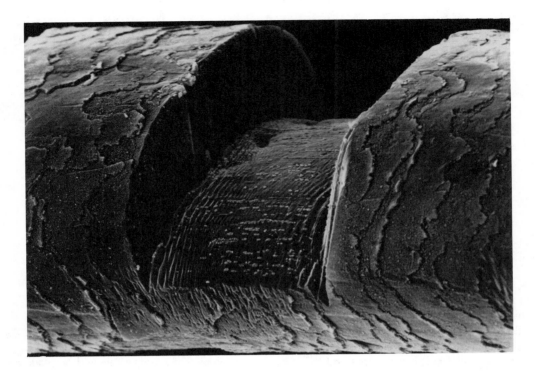

FIGURE 40. Human hair etching by 193-nm laser pulses shows ridges and valleys on the bottom surface.

FIGURE 41. Cones created by debris from laser ablation of polyimide at 308 nm.

is well above the threshold value, since the debris is then blown away from the irradiated region. Whenever patterning of a surface is attempted in a step-and-repeat fashion, the avoidance of debris in the irradiated area becomes a serious issue. Powerful jets of nitrogen blown perpendicular to the laser beam and across the surface can help to sweep the debris as it is formed. Debris which clings to the work at the end of the laser exposure can be cleaned with an aqueous solution of an electrolyte such as sodium chloride.[43]

REFERENCES

1. **Srinivasan, R. and Leigh, W. J.,** Ablative photodecomposition: action of far-ultraviolet (193 nm) laser radiation on poly(ethylene terephthalate) films, *J. Am. Chem. Soc.,* 104, 6784, 1982.
2. **Srinivasan, R. and Mayne-Banton, V.,** Self-developing photoetching of poly(ethylene terephthalate) films by far-ultraviolet excimer laser radiation, *Appl. Phys. Lett.,* 41, 576, 1982.
3. **Linsker, R., Srinivasan, R., Wynne, J. J., and Alonso, D. R.,** Far-ultraviolet laser ablation of atherosclerotic lesions, *Lasers Surg. Med.,* 4, 201, 1984.
4. **Trokel, S. T., Srinivasan, R., and Braren, B.,** Excimer laser surgery of the cornea, *Am. J. Ophthalmol.,* 96, 710, 1983.
5. **Pummer, H.,** The excimer laser: 10 years of fast growth, *Photonics Spectra,* May, 1985.
6. **Koren, G. and Yeh, J. T. C.,** Emission spectra, surface quality, and mechanism of excimer laser etching of polyimide films, *Appl. Phys. Lett.,* 44, 1112, 1984.
7. **Koren, G. and Yeh, J. T. C.,** Emission spectra and etching of polymers and graphite irradiated by excimer lasers, *J. Appl. Phys.,* 56, 2120, 1984.
8. **Davis, G. M., Gower, M. C., Fotakis, C., Efthimiopoulos, T., and Argyrakis, P.,** Spectroscopic studies of ArF laser photoablation of PMMA, *Appl. Phys.* A36, 27, 1985.
9. **Dyer, P. E. and Sidhu, J.,** Excimer laser ablation and thermal coupling efficiency to polymer films, *J. Appl. Phys.,* 57, 1420, 1985.
10. **Gorodetsky, G., Kazyaka, T. G., Melcher, R. L., and Srinivasan, R.,** Calorimetric and acoustic study of ultraviolet laser ablation of polymers, *Appl. Phys. Lett.,* 46, 828, 1985.
11. **Dyer, P. E. and Srinivasan, R.,** Nanosecond photoacoustic studies on ultraviolet laser ablation of organic polymers, *Appl. Phys. Lett.,* 48, 445, 1986.
12. **Srinivasan, R., Dyer, P. E., and Braren, B.,** Far-ultraviolet laser ablation of the cornea: photoacoustic studies, *Lasers Surg. Med.,* 6, 514, 1987.
13. **von Gutfeld, R. J., McDonald, F. A., and Dreyfus, R. W.,** Surface deformation measurements following excimer laser irradiation of insulators, *Appl. Phys. Lett.,* 49, 1059, 1986.
14. **Srinivasan, R., Braren, B., and Dreyfus, R. W.,** Ultraviolet laser ablation of polyimide films, *J. Appl. Phys.,* 61, 372, 1987.
15. **Estler, R. C. and Nogar, N. S.,** Mass spectroscopic identification of wavelength dependent UV laser photoablation fragments from poly(methyl methacrylate), *Appl. Phys. Lett.,* 49, 1175, 1986.
16. **Gupta, A., Llang, R., Tsay, F. D., and Moacanin, J.,** Characterization of a dissociative excited state in the solid state: photochemistry of poly(methyl methacrylate). Photochemical processes in polymeric systems, *Macromolecules,* 13, 1696, 1980.
17. **Zandee, L. and Bernstein, R. B.,** Laser ionization mass spectrometry: extensive fragmentation via resonance-enhanced multiphoton ionization of a molecular benzene beam, *J. Chem. Phys.,* 70, 2574, 1979.
18. **Zandee, L. and Bernstein, R. B.,** Resonance-enhanced multiphoton ionization and fragmentation of molecular beams: NO, I_2, benzene, and butadiene, *J. Chem. Phys.,* 71, 1359, 1979.
19. **Koplitz, B. D. and McVey, J.,** Fragment power dependence in the laser-induced ionization of benzene-d_6, *J. Phys. Chem.,* 89, 4196, 1985.
20. **Srinivasan, R. and Lazare, S.,** Modification of polymer surfaces by far-ultraviolet radiation of low and high (laser) intensities, *Polymer,* 26, 1297, 1985.
21. **Srinivasan, R., Braren, B., Dreyfus, R. W., Hadel, L., and Seeger, D. E.,** Mechanism of the ultraviolet laser ablation of poly(methyl methacrylate) at 193 nm and 248 nm: laser-induced fluorescence analysis, chemical analysis, and doping studies, *J. Opt. Soc.,* A3, 785, 1986.
22. **Srinivasan, R., Braren, B., Seeger, D. E., and Dreyfus, R. W.,** Photochemical cleavage of a polymeric solid: details of the ultraviolet laser ablation of poly(methyl methacrylate) at 193 and 248 nm, *Macromolecules,* 19, 916, 1986.

23. **Srinivasan, R. and Dreyfus, R. W.,** Laser induced fluorescence studies on ultraviolet laser ablation of polymers, in *Laser Spectroscopy VII,* Hänsch, T. W., and Shen, Y. R., Eds., Springer-Verlag, Berlin, 1985, 396.

24. **Utterback, N. G., Tang, S. P., and Friichtenicht, J. F.,** Atomic and ionic beam source utilizing pulsed laser blow off, *Phys. Fluids,* 19, 900, 1976.

25. **Dreyfus, R. W., Kelly, R., Walkup, R. E., and Srinivasan, R.,** Studies of excimer laser etching mechanism using laser induced fluorescence measurements, *SPIE Excimer Lasers Opt.,* 710, 46, 1987.

26. **Danielzik, B., Fabricius, N., Rowekamp, M., and von der Linde, D.,** Velocity distribution of molecular fragments from poly(methyl methacrylate) irradiated with UV laser pulses, *Appl. Phys. Lett.,* 48, 212, 1986.

27. **Garrison, B. J. and Srinivasan, R.,** Microscopic model for the ablative photodecomposition of polymers by far-ultraviolet radiation (193 nm), *Appl. Phys. Lett.,* 44, 849, 1984.

28. **Srinivasan, R. and Liu, S.-H.,** unpublished data, 1984.

29. **Covington, M. A., Liu, G. N., and Lincoln, K. A.,** Free-jet expansions from laser-vaporized planar surfaces, *AIAA J.,* 15, 1174, 1977.

30. **Srinivasan, R.,** Ablation of polymers and biological tissue by ultraviolet lasers, *Science,* 234, 559, 1986.

31. **Deutsch, T. F. and Geis, M. W.,** Self-developing UV photoresist using excimer laser exposure, *J. Appl. Phys.,* 54, 7201, 1983.

32. **Andrew, J. E., Dyer, P. E., Forster, D., and Key, P. H.,** Direct etching of polymeric materials using a XeCl laser, *Appl. Phys. Lett.,* 43, 717, 1983.

33. **Jellinek, H. H. G. and Srinivasan, R.,** Theory of etching of polymers by far-ultraviolet, high-intensity pulsed laser and long-term irradiation, *J. Phys. Chem.,* 88, 3048, 1984.

34. **Sroog, C. E.,** Polyimides, *J. Polym. Sci. Macromol. Rev.,* 11, 186, 1976.

35. **Jellinek, H. H. G. and Luh, M. D.,** Thermal degradation of polymethylmethacrylate, *Macromol. Chem.,* 115, 89, 1968.

36. **Keyes, T., Clarke, R. H., and Isner, J. M.,** Theory of photoablation and its implications for phototherapy, *J. Phys. Chem.,* 89, 4194, 1985.

37. **Sutcliffe, E. and Srinivasan, R.,** Dynamics of UV laser ablation of organic polymer surfaces, *J. Appl. Phys.,* 60, 3315, 1986.

38. **Kawamura, Y., Toyoda, K., and Namba, S.,** Effective deep ultraviolet photoetching of polymethyl methacrylate by an excimer laser, *Appl. Phys. Lett.,* 40, 374, 1982.

39. **Jain, K. and Kerth, R. T.,** Excimer laser projection lithography, *Appl. Opt.,* 23, 648, 1984.

40. **Braren, B. and Srinivasan, R.,** Optical and photochemical factors which influence etching of polymers by ablative photodecomposition, *J. Vac. Sci. Technol.,* B3, 913, 1985.

41. **von Gutfeld, R. J. and Srinivasan, R.,** Electrostatic collection of debris resulting from 193 nm laser etching of polyimide, *Appl. Phys. Lett.,* 51, 15, 1987.

42. **Dyer, P.E., Jenkins, S. D., and Sidhu, J.,** Development and origin of conical structures on XeCl laser ablated polyimide, *Appl. Phys. Lett.,* 49, 453, 1986.

43. **Braren, B. and Srinivasan, R.,** unpublished data, 1985.

44. **Braren, B. and Seeger, D. E.,** Low-temperature UV laser etching of PMMA: on the mechanism of ablative photodecomposition (APD), *J. Polym. Sci. Polym. Lett.,* 24, 371, 1986.

45. **Davis, G. M. and Gower, M. C.,** Time resolved transmission studies of poly(methyl methacrylate) films during ultraviolet laser ablative photodecomposition, *J. Appl. Phys.,* 61, 2090, 1987.

46. **Schafer, B. and Hess, P.,** Measurement of time-of-flight distributions for wavelength-dependent IR laser stimulated desorption, *Chem. Phys. Lett.,* 105, 563, 1984.

47. **Srinivasan, R. and Braren, B.,** Ablative photodecomposition of polymer films by pulsed far-ultraviolet (193 nm) laser radiation: dependence of etch depth on experimental conditions, *J. Polym. Sci. Polym. Chem.,* 22, 2601, 1984.

48. **Hargis, P. J.,** Photochemical and thermal effects in the ultraviolet laser ablation of low density materials, *Am. Inst. Phys. Conf. Proc.,* in press.

49. **White, R. M.,** Generation of elastic waves by transient surface heating, *J. Appl. Phys.,* 34, 3559, 1963.

50. **Geis, M. J., Randall, J. N., Deutsch, T. F., Efremow, N. N., Donnelly, J. P., and Woodhouse, J. D.,** Nitrocellulose as a self-developing resist with submicrometer resolution and processing stability, *J. Vac. Sci. Technol.,* B1, 1178, 1983.

51. **Golombok, M., Gower, M. C., Kirby, S. J., and Rumsby, P. T.,** Photoablation of plasma polymerized polyacetylene films, *J. Appl. Phys.,* 61, 1222, 1987.

52. **Brannon, J. H., Lankard, J. R., Baise, A. I., Burns, F., and Kaufman, J.,** Excimer laser etching of polyimide, *J. Appl. Phys.,* 58, 2036, 1985.

53. **Koren, G.,** CO_2 laser-assisted UV ablative photoetching of Kapton films, *Appl. Phys. Lett.,* 45, 10, 1984.

54. **Lazare, S., Hoh, P. D., Baker, J. M., and Srinivasan, R.,** Controlled modification of organic polymer surfaces by continuous wave far-ultraviolet (185 nm) and pulsed-laser (193 nm) radiation: XPS studies, *J. Am. Chem. Soc.,* 106, 4288, 1984.

55. **Yeh, J. T. C.,** Laser ablation of polymers, *J. Vac. Sci. Technol.,* A4, 653, 1986.
56. **Lazare, S. and Srinivasan, R.,** Surface properties of poly (ethylene terephthalate) films modified by far-ultraviolet radiation at 193 nm (laser) and 185 nm (low intensity), *J. Phys. Chem.,* 90, 2124, 1986.
57. **Lazare, S. and Srinivasan, R.,** Modification of surface properties of polyimide by far-ultraviolet radiation at high and low intensity: XPS study, in *Proc. 2nd Int. Conf. Polyimides,* Ellenville, New York, 1985, 119.
58. **Marinero, E. E.,** Laser multiphoton processes in thin films: nonlinear photochemistry of organosilane polymers, *Chem. Phys. Lett.,* 115, 501, 1985.
59. **Küper, S. and Stuke, M.,** Femtosecond UV excimer laser ablation, *Appl. Phys. A,* in press.

Chapter 6

HOLOGRAPHIC SPECTROSCOPY AND HOLOGRAPHIC INFORMATION RECORDING IN POLYMER MATRICES

C. Bräuchle and H. Anneser

TABLE OF CONTENTS

I. INTRODUCTION

Since its invention by Gabor[1] and the development of lasers as coherent light sources, holography has found a great variety of different applications. It is not only used for the production of three-dimensional images[2] but also, for example, for holographic interferometry,[3,4] a nondestructive method to detect very small deformations of objects under mechanical stress, and for medical and biomedical purposes.[5] Holographic techniques allow the production of different kinds of optical elements for focusing and deflecting light.[6,7] These holographic optical elements (HOEs) can be used as passive devices in integrated optics and optical communication technologies.[8-11] One of the most interesting applications of holographic techniques lies in the field of optical data storage and data processing.[8,11-14]

In this chapter we will show the usefulness of holographic techniques for the investigation of photochemical and photophysical processes of molecules. Continuous wave holographic gratings, as well as transient holographic grating experiments, are used to obtain kinetic parameters of different photochemical reactions. We will consider a newly developed holographic grating technique, called phase-modulated holography (PMH). For the first time this technique allows a separate and simultaneous monitoring of amplitude and phase gratings, including the absolute signs of these contributions to the overall hologram efficiency. From the absorption grating, the change in the absorption coefficient and the photochemical quantum yield can be determined. From the phase grating the total change of the molar refraction arising in the photopolymer can be evaluated, allowing one to directly draw conclusions on the magnitude and the sign of photoinduced matrix effects. Additionally, PMH offers an up to eight orders of magnitude improvement of the detection sensitivity with respect to phase-insensitive holographic techniques (PIH), and in principle a quantum-limited detection sensitivity. This extremely high sensitivity opens totally new applications for this technique. As will be shown here by several examples, these advantages render PMH an ideal tool for the investigation of new materials for holographic optical recording processes, with possible applications in optical data storage and processing as well as in the production of holographic optical elements.

II. THE PRINCIPLE OF THE HOLOGRAPHIC GRATING TECHNIQUE FOR THE INVESTIGATION OF PHOTOCHEMICAL AND PHOTOPHYSICAL PROCESSES

A. CONTINUOUS WAVE HOLOGRAPHIC GRATINGS
1. The Production of a Holographic Grating

For the investigation of photochemical and photophysical processes by holography it is reasonable to use the simplest of holograms, i.e., a plane wave hologram.[14-17] It is produced by the interference of two coherent waves (reference and object wave) with the intensities I_R and I_0 and the same wavelength on the sample, as shown in Figure 1.

The interference pattern of both plane waves is a simple cosine modulation of intensity $I(x)$ in the x-direction across the sample:

$$I(x) = (I_R + I_0)\left[1 + V\cos\left(\frac{2\pi x}{\Lambda}\right)\right] \quad (1)$$

The fringe spacing Λ in this equation depends on the wavelength λ and the angle of incidence θ:

$$\Lambda = \frac{\lambda}{2\sin\theta} \quad (2)$$

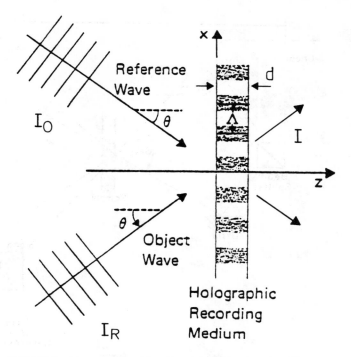

FIGURE 1. The production of a plane wave hologram. For meanings of symbols see test.

The contrast or fringe visibility V is given by

$$V = \frac{\sqrt{2I_R I_0}}{I_R + I_0} \tag{3}$$

With both beams having the same intensity ($I_R = I_0$), V equals unity, and totally dark and maximally bright stripes occur in the interference pattern.

If the holographic sample contains molecules that are photochemically or photophysically sensitive to the incident light, they will change their chemical or physical properties, i.e., changes of the index of refraction n or the absorption coefficient a of the molecules will occur as a spatial modulation:

$$n(x) = n_0 + n_1 \cos\left(\frac{2\pi x}{\Lambda}\right) \tag{4a}$$

$$a(x) = a_0 + a_1 \cos\left(\frac{2\pi x}{\Lambda}\right) \tag{4b}$$

with the average values of the index of refraction and the absorption coefficient n_0 and a_0 and the corresponding amplitudes of modulation n_1 and a_1. This modulation shows the same periodic behavior as the intensity pattern I(x) in Equation 1. Holograms due to a modulation of the index of refraction in Equation 4a are called phase holograms, whereas those caused by a variation of the absorption coefficient in Equation 4b are called amplitude holograms.

When a reading beam strikes such a holographic grating along the direction of the reference wave, some of the light from this beam will be diffracted by the holographic grating, and the absent object wave will be reconstructed. One possible experimental setup for the recording of hologram growth curves is depicted in Figure 2.

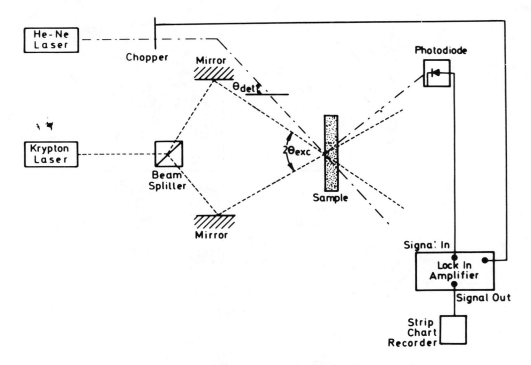

FIGURE 2.   Experimental setup for the recording of hologram growth curves.

The ratio of the intensity I of the diffracted beam divided by the intensity I_R of the reading beam is called the diffraction efficiency η and characterizes the hologram. For thick holographic gratings (thickness d >> fringe distance Λ) Kogelnik obtained the important expression:[18]

$$\eta = \frac{I}{I_R} = D^2[P^2 + A^2] \tag{5}$$

with the attenuation

$$D = \exp(-a_0 d/\cos\theta) \tag{6}$$

which accounts for the absorption of the reading beam in the sample, and with the phase contribution

$$P = \sin\left(\frac{\pi n_1 d}{\lambda\cos\theta}\right) \tag{7}$$

and the amplitude contribution

$$A = \sinh\left(\frac{a_1 d}{2\cos\theta}\right) \tag{8}$$

For applications of this holographic technique, where the total diffraction efficiency is small ($\eta \leq 0{,}01$) Equation 5 can be approximated by Equation 9:

$$\eta = \left[\left(\frac{\pi n_1 d}{\lambda\cos\theta}\right)^2 + \left(\frac{a_1 d}{2\cos\theta}\right)^2\right]\exp\left(-\frac{2a_0 d}{\cos\theta}\right) \qquad (9)$$

In these cases η is directly proportional to the square of the amplitudes n_1 and a_1.

2. Holography and its Relation to Reaction Kinetics

To relate the growth of the hologram to the underlying photochemical reaction,[19-21] the changes in index of refraction and absorption coefficient must be related to the variation in concentration of reactants and products. Considering a pure phase hologram and a simple one-step reaction $A_1 \xrightarrow{k} A_2$ with the overall reaction rate k, the changes of the concentrations of A_1 and A_2 with time are given by the kinetic rate equations:[22]

$$C_1(t) = C_1(0)e^{-kt} \qquad (10a)$$

$$C_2(t) = C_1(0) - C_1(0)e^{-kt} \qquad (10b)$$

where t equals time and $C_1(0)$ is the initial concentration of A_1.

The spatial variation of photochemical product formation under holographic conditions leads to a change in the index of refraction which can be obtained from the photochemical changes by using the Lorentz-Lorenz relation.[23] After some manipulations,[19] the modulation amplitude of the index of refraction n_1 can be expressed as

$$n_1 = \frac{(n_0^2 + 2)^2}{6000\,n_0}\sum_i R_i \Delta C_i(t) \qquad (11)$$

where R_i is the molar refraction, $\Delta C_i(t)$ gives the time-dependent changes of the concentration at the maxima of the interference pattern, and the subscript i runs over all components of the reaction. For very small diffraction efficiencies in the early stages of hologram formation, a relationship for the efficiency as a function of time is obtained by putting Equations 10 in 11 and 11 in 9.[19,20,24-26]

$$\eta = (F \cdot C \cdot k)^2 t^2 \qquad (12)$$

In Equation 12 t equals time and F and C are constants representing geometrical and material-specific parameters:

$$F = \frac{\pi d}{\lambda\cos\theta} \qquad (13)$$

$$C = \frac{(n_0^2 + 2)^2}{6000\,n_0}\Delta R C_1(0) = Q\Delta R C_1(0) \qquad (14)$$

ΔR accounts for the change of the molar refraction and $C_1(0)$ is the starting concentration of A_1. As the holographic growth curves $\eta(t)$ are quadratic in time,

$$\eta(t) = at^2 \qquad (15a)$$

$$\eta^{1/2}(t) = a^{1/2}t \qquad (15b)$$

the overall rate constant can easily be obtained by recording η as a function of time.

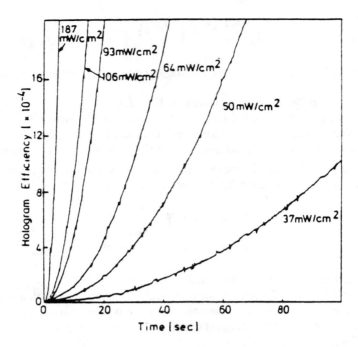

FIGURE 3.   Hologram growth curves at different laser beam intensities.

The time dependence of hologram growth has also been calculated for more complex reaction schemes[15,19,21] such as parallel and consecutive reactions, which shall not be discussed in this paper.

Furthermore, by analysis of hologram growth curves it is possible to evaluate quantum yields and the number of mechanistically necessary photons for a given energy level scheme of a reaction. By recording hologram growth curves with different intensities of the hologram producing beams (see Figure 3), an overall rate constant k can be determined for each intensity.

This dependence of k on the intensity can be explained by referring to the energy level schemes in Figure 4.

On the left side of Figure 4 a one-photon two-level (1P2L) process is shown. Here the absorption of one photon excites the molecule into the reactive state A_1^*. The right side of Figure 4 illustrates a two-photon four-level (2P4L) scheme, indicating that two photons are necessary to reach the reactive state A_2^*. Since the rate constant k obtained above describes only the overall reaction process ($A_1 \rightarrow A_2$ for the left side or $A_1 \rightarrow A_3$ for the right side of Figure 4), it involves the intensity-dependent "absorption rates" I_a. In the steady-state approximation it can be easily shown that k is related to the intensity I according to

$$k = \xi I^r \tag{16}$$

In Equation 16 r gives the number of mechanistically necessary photons, and ξ contains the photochemical quantum yield $\emptyset_p$. For the one-photon process (r = 1) one gets

$$\xi = 2303\epsilon_1\Phi_p \tag{17a}$$

and for the two-photon process (r = 2) ξ is given by

$$\xi = (2303)^2\epsilon_1\epsilon_2\tau_2\Phi_2\Phi_p \tag{17b}$$

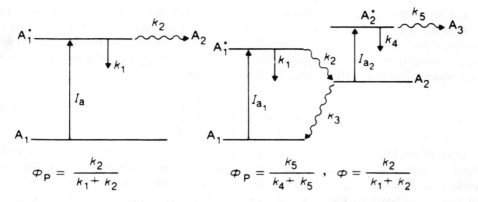

FIGURE 4. Energy level diagrams for a 1P2L (left side) and a 2P4L (right side) photoreaction. The rate constants and the quantum yields for the kinetic processes are indicated, as are expressions for the quantum yields.

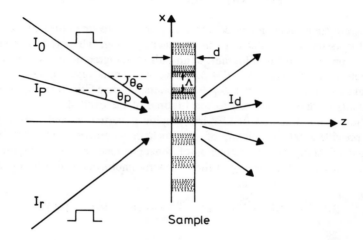

FIGURE 5. Schematic illustration of the transient grating experiment. The crossed excitation pulses I_0 and I_r generate an interference pattern in the sample. A CW beam I_p is Bragg-diffracted and yields the grating signal I_d.

In Equations 17, ϵ_1 and ϵ_2 are the molar extinction coefficients of the absorption steps, τ_2 is the lifetime and ϕ_2 the quantum yield for population of state A_2. By plotting log k as a function of log I, r is obtained as the slope and logξ as the intercept of a straight line; with known extinction coefficients the photochemical quantum yield ϕ_p can be evaluated. By this way hydrogen abstraction by benzophenone in a polymethylmethacrylate matrix[27,29] and the dissociation of dimethyl-*s*-tetrazine[24,25,28] have been studied. These reactions have been shown to follow a one-photon photochemistry. 1P2L-systems and 2P4L-systems will be further discussed in Section IV.

B. TRANSIENT HOLOGRAPHIC GRATINGS

Pulsed holographic experiments are widely used for the investigation of molecular and solid-state dynamics. Here we will concentrate on the transient grating technique as a tool for the investigation of excited state dynamics,[15,21] and of irreversible photochemical reactions by performing single-shot experiments.[30,31]

A schematic illustration of a transient grating experiment is given in Figure 5.

Two mutually coherent laser pulses with a pulse width of about a few nanoseconds are crossed inside the holographic sample and form a holographic grating with changes in the index of refraction and absorbance due to excitation of molecules to higher electronic or vibrational states. In the absence of any other processes besides the relaxation of the excited states to the ground state, the time dependence of the intensity $I_d(t)$ of a diffracted probe beam will be determined by the excited state lifetime τ:

$$I_d(t) = A \exp\left(-\frac{2}{\tau}t\right) \qquad (18)$$

In Equation 18 A is a constant containing all time-independent holographic parameters. Therefore, with Equation 18 it is possible to obtain lifetimes of excited states performing transient grating experiments.[30-34] However, if a photoreaction occurs from excited states of molecules it can be shown that for a 1P2L reaction scheme (see Figure 4), the time dependence of the diffraction efficiency should be given by[15,21]

$$\eta(t) = (K_1 + K_2 \exp[-(k_1 + k_2)t])^2 \qquad (19)$$

K_1 and K_2 are time-independent constants, k_1 and k_2 are rate constants as indicated in Figure 4, and k_1 describes the decay of the transient grating, whereas k_2 is responsible for the growth of a photochemical hologram. With pulsed holographic measurements it is only possible to determine the sum $(k_1 + k_2)$. However, if this result is combined with the photochemical quantum yield $\phi_p = k_2/k_1 + k_2$ obtained from a CW-holographic experiment, the rate constants k_1 and k_2, as well as the lifetime of the excited state $\tau = 1/k_1$, can be evaluated. Thus, a combination of CW-holographic experiments and pulsed holographic methods can provide a detailed picture of a photochemical reaction. This will be shown by an investigation of the hydrogen abstraction of benzophenone (BP) in a polymethylmethacrylate (PMMA) matrix[30,31] according to the following photochemical reaction scheme:[35-38]

$$^1BP \xrightarrow{h\nu} {}^1BP^* \xrightarrow{PS} {}^3BP^* \xrightarrow{k_H} (BPH\cdot + R\cdot) \rightarrow LAT$$

$$\downarrow slow$$

$$products$$

After excitation to the state S_1, BP relaxes within picoseconds and virtually to unit quantum yield to the lowest triplet state T_1.[39] From there it either deactivates to the ground state or, if possible, abstracts a hydrogen from the solvent or host to form a radical pair. This radical pair can react further to form the so-called LAT (light-absorbing transient), and in a second photoinduced step finally stable photoproducts occur. With a CW holographic technique we have shown that the overall quantum yield for the reaction BP → LAT is $\phi = 0.2$.[29]

The obtained grating signal $S(t)$ (see Figures 6 and 7) is a superposition of contributions from the triplet state T_1 of BP, the formed radical pair, and the thermal grating induced by the relaxation processes:[30]

$$S(t) \sim [\Delta n_i(t) + \Delta n_j(t) + \ldots + \Delta n_{th}(t)]^2 + [\Delta K_i(t) + \Delta K_j(t) + \ldots]^2 \qquad (20)$$

In Equation 20 Δn and ΔK are the modulation amplitudes, i and j denote the various excited states and species present in the sample, which can give rise to phase as well as amplitude gratings, depending on the wavelength of the probe beam, and the index th stands

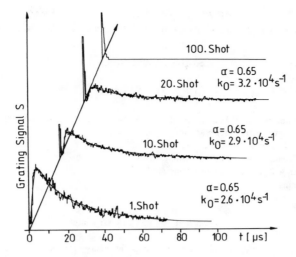

FIGURE 6.   Transient grating data for BP/PMMA excited with $\lambda_e = 355$ nm and $\theta_e = 4.90°$ and probed with a HeNe laser at $\lambda_p = 633$ nm. 1. shot indicates the signal of a fresh sample. nth.shot indicates the single shot signal after n-1 shots have excited the sample. The solid lines correspond to a theoretical fit.

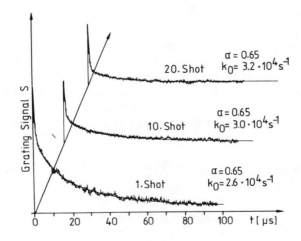

FIGURE 7.   Transient grating data for BP/PMMA excited with $\lambda_e = 355$ nm and $\theta_e = 4.90°$ and probed with an Ar^+ laser at $\lambda_P = 514$ nm. (1.shot) the signal of a fresh sample; (nth.shot) the single-shot signal after $n - 1$ shots have excited the sample. The solid lines correspond to a theoretical fit.

for the thermal grating, which is a pure phase grating. It is independent from the wavelength and shows a fast decay.

To distinguish among the underlying processes, one can change two parameters: the probe wavelength λ_p and the fringe spacing Λ. Variation of Λ allows the separation of diffusive and nondiffusive processes. Variation of λ_p gives access to the various species and states involved in the photoprocess due to their different absorption and refraction spectra. After an extensive fit procedure[30,31] assuming that the triplet state T_1 of BP decays according to a Williams-Watts law or stretched exponential,

TABLE 1
Parameters and Processes Observed by Holographic Grating Methods [15-17,32]

cw-Holographic Gratings

Overall reaction rates k:A $\xrightarrow{\text{k}}$ B
Quantum yields Φ
Information about the photophysical/chemical level scheme: number of mechanistically necessary photons
Complex reactions: consecutive, parallel reactions
Computer simulation of hologram growth curves
Photoaction spectra (3-beam experiment) → reactive states
Separation of photochemical steps; coherent/incoherent light
Growth of polymer chains
Photomechanical effects (polymer matrices)

Transient Holographic Gratings

Dynamics of excited states → lifetime, photochemical constant, energy transfer
Thermal diffusion
Translational diffusion (polymer self-diffusion)
Rotational diffusion
Light-induced phonon scattering
Photoinduced charge transport

$$[T_1] \sim \exp[-(k_0 t)^\alpha] \tag{21}$$

the mean rate constant k_0 and the dispersion parameter α have been determined.

The dispersion parameter α accounts for a distribution of inequivalent sites in the amorphous host and, accordingly, for a distribution of first-order quenching or deactivation constants. The mean rate constant k_0 is the sum of the rate constant k_{de} for the deactivation to the ground state and k_H for hydrogen abstraction from the host. With $k_0 = 2.6 \times 10^4 s^{-1}$ from the fit and the quantum yield $\emptyset = 0,2$ from CW experiments, we obtain the pure rate constant for hydrogen abstraction $k_H = \emptyset k_0 = 5 \times 10^3 s^{-1}$.

C. PARAMETERS AND PROCESSES OBSERVED BY HOLOGRAPHIC GRATING METHODS

In the previous sections we have discussed the use of transient and CW-holographic grating experiments for the investigation of photochemistry and photophysics of molecules. In Table 1 we present a short survey of some of the parameters and processes which can be observed by holographic grating methods.[15-17,32]

III. PHASE-MODULATED HOLOGRAPHIC GRATING TECHNIQUE (PMH)

A. ONE-BEAM VERSION OF PMH

All of the holographic experiments described in Section II have been performed by PIH. With these techniques it is only possible to obtain the sum of the phase and amplitude hologram and the square of P and A (see Equations 5 to 8).

PMH has been developed[40-43] to allow a separate and simultaneous investigation of the P and A contributions to the hologram formation and to obtain the true signs of the changes of the refraction index n and of the absorption coefficient a. Further, the detection sensitivity is considerably improved when P and A (both quantities are much smaller than one) are observed instead of P^2 and A^2. The experimental setup for PMH in the one-beam version is shown in Figure 8.

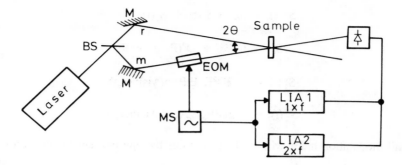

FIGURE 8.   Schematics of the basic PMH setup. Using a beamsplitter BS, the laser beam is split into a reference (r) and a modulated (m) beam, which are subsequently recombined with two mirrors (M) under an angle 2 θ onto the sample. The m beam is phase-modulated with an electrooptic modulator EOM which is driven by a sinusoidally varying voltage generated with some modulation source MS, e.g., a function generator. Homodyne detection is accomplished with two lock-in amplifiers (LIA 1 and LIA 2), which are synchronized to a reference signal (ref) derived from the MS. The fundamental and second harmonic contributions in the photodetector (PD) output are recorded separately by using the two LIAs in fundamental (1 × f) and second harmonic (2 × f) mode.

The photochemically induced cosine modulation of the index of refraction and the absorption coefficient (see Equation 4) and the correspondingly produced phase and absorption gratings cause a diffraction of the r-beam into the m-beam direction. The diffracted beam electric field amplitude in m-beam direction is given by

$$E_d = -D(iP + A)E_r \tag{22}$$

with the r-beam electric field amplitude E_r before the sample and the loss factor D due to Equation 6. P and A are the phase and absorption hologram efficiencies, which in case of small diffraction amplitudes, can be written as

$$P = \pi n_1 d/\lambda \cos\theta \tag{23a}$$

$$A = a_1 d/2\cos\theta \tag{23b}$$

The $\pi/2$ phase difference between dispersive (P) and absorptive (A) response of a medium (the i in Equation 22) allows the separation and simultaneous recording of P and A. Therefore, one beam is modulated (m-beam) with an electrooptic modulator (EOM) (see Figure 8) with a modulation frequency ω_m and a modulation index M. The m-beam electric field amplitude after the sample is given by

$$E_m' = DE_m \exp[iM\sin(\omega_m t)] \tag{24}$$

The m-beam and the diffracted r-beam generate a homodyne beat signal $S^m(t)$ on a square law photodetector:

$$S^m(t) = |E_m' + E_d|^2 \tag{25}$$

Taking into account the $\pi/2$ phase shift between the diffraction contributions due to P and A, $S^m(t)$ is obtained as a sum of three terms by inserting Equations 22 and 24 into Equation 25:

$$S^m(t) = S_0 + S_1(t) + S_2(t) \tag{26a}$$

$$S_0^m = D^2E_m^2 + D^2(P^2 + A^2)E_r^2 \tag{26b}$$

$$S_1^m(t) = -2D^2E_mE_rP\sin(M\sin\omega_m t) \tag{26c}$$

$$S_2^m(t) = -2D^2E_mE_rA\cos(M\sin\omega_m t) \tag{26d}$$

For a small modulation index $M \ll 1$ one obtains the approximated contributions:

$$S_0^{m\prime} = D^2E_m^2 + D^2(P^2 + A^2)E_r^2 - (1 - M^2/4)D^2AE_rE_m \tag{27a}$$

$$S_1^{m\prime}(t) = -2D^2E_rE_mPM\sin\omega_m t \tag{27b}$$

$$S_2^{m\prime}(t) = -(D^2/2)E_rE_mAM^2\cos^2(2\omega_m t) \tag{27c}$$

In Equation 27 the dc term stemming from the expansion of cos ($M \sin \omega_m t$) is added to S_0^m to yield $S_0^{m\prime}$.

The phase hologram efficiency P, including its sign, can be determined from the homodyne term $S_1^{m\prime}(t)$, which is synchronous with the fundamental of the applied phase modulation. The absorption hologram efficiency A, including its sign, is obtained from the second harmonic homodyne term $S_2^{m\prime}(t)$. With blocked m-beam ($E_m = 0$) the situation corresponds to the conventional PIH detection technique, and the signal reduces to the well-known formula

$$S^m(t) = D^2[P^2(t) + A^2(t)]E_r^2 \tag{28}$$

For a simple one-step photoreaction $A_1 \xrightarrow{k} A_2$ with an underlying 1P2L mechanism, the diffraction amplitudes of the phase grating P and of the amplitude grating A are linear in time in the small reaction yield limit:[24,40-43]

$$P(t) = c_0(2303)\epsilon_E Q\Delta R_{eff}\left[\frac{\pi d}{\lambda\cos\theta}\right]I\emptyset t \tag{29a}$$

$$A(t) = c_0(2303)\epsilon_E\Delta\epsilon(2,303/4)\left[\frac{d}{\cos\theta}\right]I\emptyset t \tag{29b}$$

In Equation 29 $\Delta\epsilon = \epsilon_p - \epsilon_E$ is the difference between the extinction coefficients of product (ϵ_p) and educt (ϵ_E), c_0 is the initial concentration of the educt, I is the recording intensity, $\emptyset$ is the quantum yield for the product formation, and ΔR_{eff} is the total change of molar refraction. The factor Q arises from the Lorentz-Lorenz relation and is given by:

$$Q = \frac{(n_0^2 + 2)^2}{6000\, n_0} \tag{30}$$

From A(t) the photochemical quantum yield of the underlying photoreaction can be determined, whereas the ratio P/A gives the total change of molar refraction ΔR_{eff}. Comparing Equations 27 and 28 to 29 shows the quadratic time dependence of $S^{m\prime}(t)$ and the linear time dependence of $S_1^{m\prime}(t)$ and $S_2^{m\prime}(t)$, as well as the impossibility to separate A and P by PIH measurements.

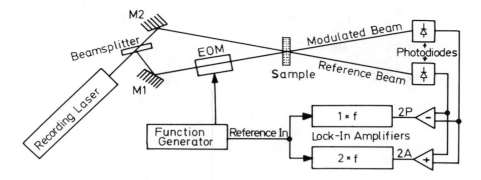

FIGURE 9. Experimental setup for PMH in the dual beam version.

B. DUAL-BEAM VERSION OF PMH

A dual-beam PMH setup as shown in Figure 9 considerably improves the detection sensitivity with a signal enhancement by a factor of two and with a five to six orders of magnitude improvement of the signal-to-noise ratios with respect to single-beam PMH.[40] The latter is caused by an effective suppression of most of the technical excess noise. In order to demonstrate these advantages of the dual-beam modification, PMH growth curves $U_1(t)$ (proportional to $P(t)$) and $U_2(t)$ (proportional to $A(t)$) for the system 2-methyl-2-nitroso-propane (MNP) in poly(ethyl-α-cyanoacrylate) (p-EtCAc) (see Section IV.A) have been recorded alternately with single and dual beam detection. The corresponding curves are shown in Figure 10.

Due to the interference of the diffracted and the transmitted beam, homodyne beat signals $S^m(t)$ in the direction of the modulated beam and $S^r(t)$ in the direction of the reference beam are obtained. In changing from m-detection (see Section III.B) to r-detection, the fundamental contribution to the homodyne signal changes sign,

$$S_1^r(t) = -S_1^m(t) \tag{31}$$

whereas the dc and second order contributions are not affected

$$S_0^r(t) = S_0^m(t) \tag{32}$$

$$S_2^r(t) = S_2^m(t) \tag{33}$$

$S^m(t)$ and $S^r(t)$ are evaluated and further processed with adding and subtracting circuity (see Figure 9). The fundamental homodyne signal is recorded with a lock-in amplifier operating in the fundamental mode and is proportional to twice the dispersive diffraction amplitude P:

$$S_1^m(t) - S_1^r(t) = -2D^2 E_r E_m 2PM\sin\omega_m t \tag{34}$$

The corresponding difference of the m- and r-beam detector outputs is given by

$$U_1(t) = -2P(t)\sigma RD^2 (F_m F_r)^{1/2} M\sin(\omega_m t) \tag{35}$$

The second harmonic contribution is obtained as the sum of $S_2^m(t)$ and $S_2^r(t)$ and is proportional to $2A$:

$$S_2^m(t) + S_2^r(t) = -(D^2/2)E_r E_m 2AM^2\cos 2\omega_m t \tag{36}$$

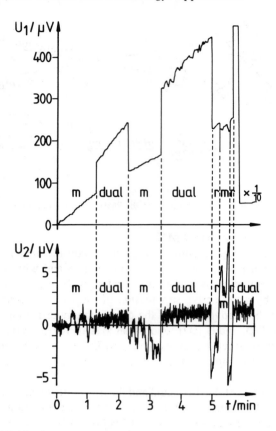

FIGURE 10.   Dual- vs. single-beam PMH growth curves of MNP/p-EtCAc. The beams actually used for detection are indicated in the figure.

The sum of the m- and r-beam detector outputs is recorded with a lock-in amplifier operating at the second harmonic mode and is given by

$$U_2(t) = -2A(t)\sigma RD^2(F_m F_r)^{1/2}(M^2/4)\cos(2\omega_m t) \qquad (37)$$

In Equations 35 and 37 σ is the photodetector sensitivity, R is the photodetector load resistance, and $F_{m,r}$ are the m- and r-beam powers. As will be shown later (see Section III.C.), dual-beam PMH offers an up to eight orders of magnitude improvement of detection sensitivity as compared to PIH and, in principle, a quantum-limited sensitivity for the detection of P(t).

C. DETECTION SENSITIVITY LIMITS OF PMH

The homodyne detection used in PMH is considerably more sensitive than the direct detection in conventional PIH. Here we will briefly discuss the several noise contributions to PMH signal detection and the detection sensitivity limitations of dual-beam PMH in comparison to PIH-detection.[40] The different noise contributions and detection sensitivity limits of dual-beam PMH are shown in Table 2, those of PIH measurements are depicted in Table 3.

For an examination of noise contributions we have to evaluate the observed photocurrent in the m-beam direction given by

$$i^m(t) = \sigma \int_0^\infty I_{PD}^m(r,t)2\pi r dr \qquad (38)$$

TABLE 2
Detection Sensitivity Limitations of Dual-Beam PMH

	Thermal noise	Shot noise	Source noise	Mechanical noise
Noise power	$4k_BT\Delta f$	$4e\sigma D^2FR\Delta f$	$\mu N_{sc}(\omega)(\sigma D^2F)^2R\Delta f$ $4N_{sc}(\omega)(\sigma D^2F)^2R\Delta f$	$\mu(A/P)^2N_{mech}(\omega)R\langle i_1^2\rangle\Delta f$ $\mu(P/A)^2N_{mech}(\omega)R\langle i_2^2\rangle\Delta f$
P_{min}^2	$\dfrac{2k_BT\Delta f}{R\sigma^2D^4M^2F^2}$ 0.4×10^{-16}	$\dfrac{2e\Delta f}{\sigma D^2M^2F}$ 1.6×10^{-16}	$\dfrac{\mu N_{sc}(\omega)\Delta f}{4M^2}$ 2.5×10^{-18}	
A_{min}^2	$\dfrac{32k_BT\Delta f}{R\sigma^2D^4M^4F^2}$ 0.7×10^{-15}	$\dfrac{16e\Delta f}{\sigma D^2M^4F}$ 1.2×10^{-15}	$\dfrac{16N_{sc}(\omega)\Delta f}{M^4}$ 1.6×10^{-10}	$\mu P^2N_{mech}(\omega)\Delta f$
$(A/P)_{min}^2$	$\dfrac{128k_BT\Delta f}{R\sigma^2D^4M^4F^2}$ 0.7×10^{-15}	$\dfrac{64e\Delta f}{\sigma D^2M^4F}$ 1.2×10^{-15}	$\dfrac{16N_{sc}(\omega)\Delta f}{M^4}$ 1.6×10^{-10}	$\mu N_{mech}(\omega)\Delta f$ 10^{-9}

Note: The technical noise powers in dual beam PMH are different for the fundamental and the second harmonic photocurrents and are given separately in the upper and lower part of the row "noise power", respectively. The numerical examples have been calculated using the following data (which correspond to usual experimental conditions and are based on measurements or conservative estimates of technical noise powers and mismatch factors): $T = 300$ K, $F_m = F_r = F$, $\sigma D^2F = 1$ mA, $R = 50$ Ω, $\Delta f = 1$ Hz, $N_{sc}(\omega) = 10^{-11}/$Hz, $N_{mech}(\omega) = 10^{-3}/$Hz, $\mu = 10^{-6}$, $M = 1$.

TABLE 3
Detection Sensitivity Limitations of Conventional PIH

	Thermal noise	Shot noise	Source noise
Noise power $(P^2 + A^2)_{min}$	$4k_BT\Delta f$ $\dfrac{3}{\sigma D^2F_r}\sqrt{\dfrac{2k_BT\Delta f}{R}}$ 1.7×10^{-8}	$2e\sigma D^2(P^2 + A^2)F_r/3R\Delta f$ $\dfrac{6e\Delta f}{\sigma D^2F_r}$ 0.5×10^{-15}	$N_{sc}(\omega)\sigma^2D^4(P^2 + A^2)^4(F_r/3)^2R\Delta f$ Unlimited (0) 0

Note: The numerical examples are based on the same data as in Table 2.

with the photodetector sensitivity σ and the time-dependent radial intensity distribution $I_{PD}^m(r,t)$ on the photodetector. After integration and using the Bessel function approximation, we obtain the total photocurrent as a sum of three terms:

$$i_0^m(t) = \sigma D^2F_m + \sigma D^2[P(t)^2 + A(t)^2](F_r/3) - 2\sigma D^2J_0(M)[\sqrt{F_mF_r}/2]A(t) \quad (39a)$$

$$i_1^m(t) = -2\sigma D^2J_1(M)\sqrt{F_mF_r}P(t)\sin(\omega_m t) \quad (39b)$$

$$i_2^m(t) = -2\sigma D^2J_2(M)\sqrt{F_mF_r}A(t)\cos(2\omega_m t) \quad (39c)$$

Fundamental noise is generated in the detection process and is independent of frequency. It comprises thermal and shot noise. As their contributions at the m- and r-photodetector outputs are statistically independent, fundamental noise cannot be suppressed by the dual beam PMH post-detection arithmetics.

Thermal or Johnson noise power is given by the Nyquist formula

$$R\langle i_{th}^2\rangle = 4k_BT\Delta f \quad (40)$$

whereas shot or quantum noise power is given by

$$R\langle i_{SN}^2 \rangle = 2e\langle i \rangle R \Delta f \tag{41}$$

In Equations 40 and 41 the brackets $\langle \ \rangle$ denote a time average, k_B is the Boltzmann constant, T is the absolute temperature of the detector load resistance R, Δf is the detection bandwidth, e is the elementary charge, and $\langle i \rangle$ is the average photocurrent. Further mathematical treatment of Equations 40 and 41 in combination with Equations 39 yield the fundamental noise contributions and detection limits P_{min}^2 and A_{min}^2 shown in Table 2. P_{min}^2 and A_{min}^2 are defined such that the corresponding signal-to-noise ratios equal unity.

Technical noise is due to intensity fluctuations of the laser (source noise) and mechanical instabilities of the optical setup (mechanical noise). It is frequency dependent, and its noise power rapidly falls off at higher frequencies. The technical noise contributions at the m- and r-beam detector outputs are correlated and can be suppressed by the PMH postdetection arithmetics.

The source noise power is determined by a specific frequency dependent power density $N_{SC}(\omega)$ and is given by

$$R\langle i_{SC}^2 \rangle = R N_{SC}(\omega) \langle i \rangle^2 \Delta f \tag{42}$$

Mechanical noise is due to mechanical instabilities which cause an additional fluctuating phase shift $\Delta\phi(t)$ between the m- and r-beams. A frequency-dependent power density $N_{mech}(\omega)$ of the phase difference fluctuations is defined by

$$\langle \Delta\phi(t)^2 \rangle = N_{mech}(\omega) \Delta f \tag{43}$$

The leak between absorptive and dispersive signal contributions induced by mechanical noise only affects PMH detection, whereas for hologram formation the only consequence is a slight reduction in grating amplitude. The derived expressions for the noise power, P_{min}^2, A_{min}^2, and $(A/P)_{min}^2$ are shown in Table 2.

The PIH noise contributions are obtained in an analogous way from the formulas mentioned before using the PIH photocurrent power (calculated from Equation 39a with $F_m = 0$).

$$R\langle i_0^2 \rangle = R[\sigma D^2(P^2 + A^2)F_r/3]^2 \tag{44}$$

The corresponding formulas and numerical values are shown in Table 3.

With dual-beam PMH, theoretically, the fundamental sensitivity limit in detection of the dispersive signal P can be reached, which is calculated to be $P^2 \gtrsim 10^{-16}$. PIH is thermal noise limited, and the corresponding detection limit is $(P^2 + A^2) \gtrsim 10^{-8}$. Therefore, in principle an eight orders of magnitude improvement of detection sensitivity can be reached with dual-beam PMH experiments.

Experimentally, a nearly quantum-limited detection sensitivity was demonstrated[44] by PMH growth curves obtained with systems of MNP dissolved in poly-(α-cyanoacrylate) (c = $6{,}2 \times 10^{-7}$ g/g). The dispersive growth curve P(t) is shown in Figure 11.

The smallest detectable value of P(t) is given by

$$P_{min} = \left(\frac{2e\Delta f}{\sigma W} \right)^{1/2} \cdot \frac{1}{D2J_1(M)} \tag{45}$$

with the elementary charge e, the detector sensitivity σ, the beam power W, the detection bandwidth Δf, the attenuation D, and the first order Bessel function $J_1(M)$. For a small index of modulation M, i.e., $M \ll 1$, $2J_1(M)$ equals M. The calculated value P_{min} for the

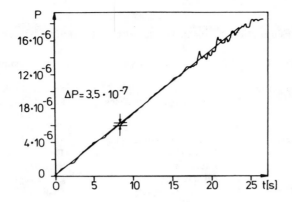

FIGURE 11. Dispersive PMH growth curve P(t) for MNP/p-(α-CAc) (c = 6.2 $\times$ 10^{-7} g/g). Experimental conditions: (σ) 0.30 A/W;(W) 3.0 mW; (D) 1.0;(f) 13.0 kHz.

experimentally used parameters was P_{min} = 1.1 $\times$ 10^{-7}. From Figure 11 P'_{min} is experimentally determined to P'_{min} = 3,5 $\times$ 10^{-7}. Therefore, the quantum-limited sensitivity was nearly achieved by the experiment. The rest of the noise results from the electronic instruments, which can be improved additionally.

IV. HOLOGRAPHIC RECORDING MECHANISMS WITH PHOTOPOLYMERIC SYSTEMS

A. CONVENTIONAL HOLOGRAPHIC RECORDING MATERIALS

To be appropriate for holographic recording a material has to show a variety of different properties. It should have a reasonable high hologram efficiency, and the time to reach the maximum efficiency should be as short as possible, with low or medium light intensity. The formed hologram should not be erased during reading. This normally affords a fixing process for the recorded interference pattern. Furthermore, the recording material should have a high sensitivity in wavelength regions accessible to commercially available lasers. Of special interest is the use of semiconductor lasers emitting in the near IR-wavelength region from 680 to 1100 nm. These lasers are very small, cheap, and easy to handle. For these reasons the search of holographic recording materials which are sensitive in this wavelength region is of great actuality. In the next chapters (see Sections IV.B and C.) we will show in two examples how this sensitivity can be obtained in two different ways. Table 4 shows the properties of some holographic recording materials[45,46] and illustrates the point that no material satisfies all of the desired requirements.

B. 1P2L SYSTEMS FOR NEAR-IR RECORDING

From liquid solution and gas-phase spectroscopy it is known[47,48] that tertiary nitrosoalkanes in their monomer form show a low intensity absorption maximum in the wavelength region from 600 to 790 nm, i.e., up to the near IR. This results from a singlet-singlet transition of the lone electron pair on the nitrogen. A light-induced homolysis of the C-N bond produces alkyl and nitric oxide radicals. In order to ensure that this photochemistry also holds true for nitrosoalkanes (RNO) dissolved in solid polymer matrices, we investigated[49] the photochemistry of MNP, 2-chloro-2,3-dimethyl-3-nitrosobutane (TENC), and 4-nitroso-4-chloropentanoic acid (NCPA) (see Figure 12) in a matrix of poly(methyl-α-cyanoacrylate) [p-(MeCAc)] and p-(EtCAc).

Furthermore, the holographic recording mechanism had to be studied with respect to possible applications of these near-IR-sensitive photopolymer systems, e.g., in integrated

TABLE 4
Properties of Some Holographic Recording Materials

Material	Advantage	Disadvantage
Dichromated gelatin	Large n_1	Spectral sensitivity
	High hologram efficiency	Recording speed
	Thin coatings	Passivation
		Shelf-life
Photopolymer	Light-induced developing	Small n_1
	No shrinkage	Thick coatings
	Inert	Spectral sensitivity
	Coating Ease	Low hologram efficiency
Photographic film	Spectral sensitivity	Medium efficiency
	Chemical amplification	Substrate choices
	Reproducibility	Resolution
Photoresist	High hologram efficiency	Spectral sensitivity
	Coating ease	Recording speed
	Stability	Linearity
Thermoplastic	Self-developing	Spectral sensitivity
	High sensitivity	Low hologram efficiency
	Erasibility	Resolution

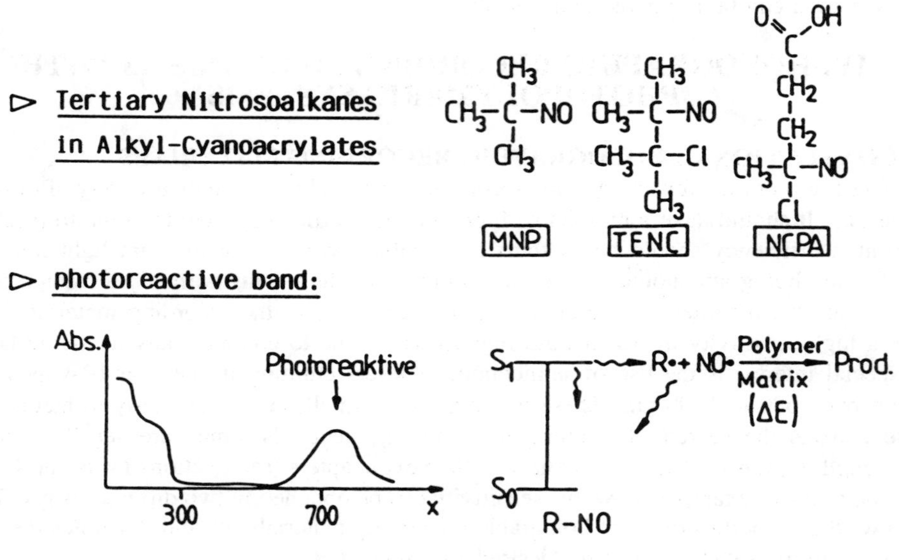

▷ **Tertiary Nitrosoalkanes**

in Alkyl-Cyanoacrylates

▷ **photoreactive band:**

FIGURE 12.   Reaction scheme of the photoreaction of RNO in p-(alkyl-CAc) and the structures of the three RNO molecules investigated.

optics, holographic optical elements, and as materials for holographic recording with semi-conductor lasers. As shown in Figure 12, all three RNO undergo photolysis according to a 1P2L scheme (see Section II.A.2, Figure 4) when illuminated with red or near-IR light up to 799.3 nm. After photolysis the long-wavelength absorption maximum vanishes. The quantum yields of the reaction in the solid polymer matrices could be determined from transmission spectroscopy,[49] as well as from measurements of the growth curve of the absorption grating [A(t)] with PMH. The values obtained show good agreement (see Table 5).

The quantum yields found in p-(EtCAc) are higher than in p-(MeCAc) due to the higher activation energy associated with the higher glass transition temperature of p-(MeCAc) compared to p-(EtCAc).

TABLE 5
Central Wavelength of the Absorption Maximum λ_0

	λ_0[nm]	$E_A\left[\dfrac{\text{kcal}}{\text{mol}}\right]$	Φ_{TS}	Φ_{PMH}	ϕ^*
MNP/p-(EtCAc)	678	0.77	$(1.12 \pm 0.12) \times 10^{-2}$	$(1.03 \pm 0.26) \times 10^{-2}$	
TENC/p-(EtCAc)	688	0.70	$(7.00 \pm 2.30) \times 10^{-2}$	$(8.08 \pm 2.26) \times 10^{-2}$	1.94[a]
NCPA/p-(EtCAc)	648	0.83	$(8.91 \pm 3.52) \times 10^{-3}$	$(7.94 \pm 1.35) \times 10^{-3}$	1.05[b]
MNP/p-(MeCAc)	678	1.55	$(4.67 \pm 0.70) \times 10^{-3}$	$(5.30 \pm 1.13) \times 10^{-3}$	
TENC/p-(MeCAc)	688	1.42	$(2.64 \pm 0.32) \times 10^{-2}$	$(2.20 \pm 0.45) \times 10^{-2}$	
NCPA/p-(MeCAc)	648	1.38	$(2.83 \pm 0.78) \times 10^{-3}$	$(3.35 \pm 0.71) \times 10^{-3}$	

Note: Activation energy E_A, and quantum yield of the photodissociations, Φ_{TS} determined from transmission spectroscopy; Φ_{PMH} determined from PMH measurements for RNO/p-(alkyl-CAc).

[a] Quantum yield of the photolysis of TENC in methanol.[49]
[b] Quantum yield of the photolysis of NCPA in methanol.[50]

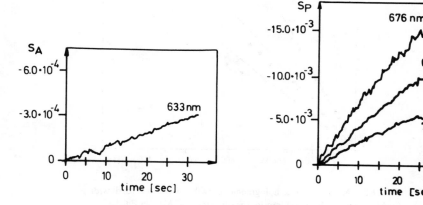

FIGURE 13.   Signals S_P and S_A proportional to the phase (P) and amplitude (A) contribution to the holographic grating monitored simultaneously for a sample MNP/p-(EtCAc) with the apparatus in Figure 9.

With the dual-beam setup of PMH, as shown in Figure 9, the signals proportional to the phase (P) and amplitude (A) contribution of the holographic grating depicted in Figure 13 were monitored simultaneously. The effective change of the molar refraction ΔR_{eff} can be evaluated from the ratio $P(t)/A(t)$ and is shown in Table 6 for the different systems investigated.

A comparison of the values of ΔR_{eff} with those for the change of molar refraction ΔR_{PC}, calculated with the help of the Kramers-Kronig relation[46] from the change of the absorption spectra, shows that ΔR_{eff} is by two orders of magnitude larger than ΔR_{PC}, ΔR_{eff} is determined by

$$\Delta R_{eff} = \Delta R_{PC} + K \tag{46}$$

where K is the contribution from a photomechanical effect, i.e., a density change of the photopolymer system induced by photochemistry. K is related to the relative density change $\Delta\rho/\rho_p c_0$

$$K = (n^2 - 1)(10^3 \Delta\rho/[(n^2 + 2)c_0\rho_P]) \tag{47}$$

TABLE 6
Calculated Change of Molar Refraction (ΔR_{PC})

	$\Delta R_{PC}\left[\dfrac{cm^3}{mol}\right]$	P/A	$\Delta R_{eff}\left[\dfrac{cm^3}{mol}\right]$	$\|\Delta R_{eff}\|\left[\dfrac{cm^3}{mol}\right]$	$K\left[\dfrac{cm^3}{mol}\right]$
MNP/p-(EtCAc)	−0.33	$(5.71 \pm 1.08) \cdot 10^2$	-26.38 ± 5.01	28.39 ± 6.53	-28.06 ± 6.45
TENC/p-(EtCAc)	−0.39	$(9.52 \pm 1.99) \cdot 10^2$	-46.20 ± 9.70	36.22 ± 5.43	-35.83 ± 5.37
NCPA/p-(EtCAc)	0.45	$(4.20 \pm 1.09) \cdot 10^2$	-27.50 ± 7.15	27.83 ± 7.19	-28.28 ± 7.31
MNP/p-(MeCAc)	−0.33	$(2.41 \pm 0.53) \cdot 10^2$	-11.13 ± 2.45	12.70 ± 2.29	-12.37 ± 2.23
TENC/p-(MeCAc)	−0.39	$(3.06 \pm 0.70) \cdot 10^2$	-14.85 ± 3.44	15.65 ± 2.19	-15.26 ± 2.14
NCPA/p-(MeCAc)	0.45	$(1.86 \pm 0.45) \cdot 10^2$	-12.15 ± 2.92	9.42 ± 2.26	-9.87 ± 2.37

Note: Ratio of phase to amplitude hologram efficiency (P/A). Change of molar refraction measured by PMH (ΔR_{eff}) and K^a.

[a]   All values are calculated for and measured at $\lambda = 632.8$ nm.

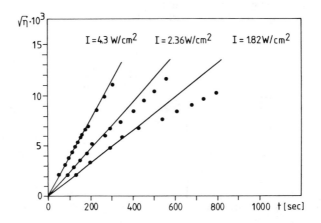

FIGURE 14. Linearized holographic growth curves recorded with a GaAlAs laser at $\lambda = 788$ nm in a fresh sample of MNP/p-EtCAc.

In Equation 46 ρ_p is the density of the material after the photoreaction is completed. The negative values of K indicate an expansion of the matrix which is mainly responsible for the hologram formation. For the RNO molecules in p(EtCAc) this expansion is about −7% (obtained from the relative density change).

Two possible mechanisms can account for the observed facts. First a rearrangement of the polymer chains can produce an expansion of the polymer framework. This rearrangement can be induced by the incorporation of photochemically generated NO or alkyl radicals in the polymer chains. Second, radical-induced chain scission could produce low molecular weight chain fragments and thus a lower density.

A photopolymer for holographic recording can be characterized by its photosensitivity S:[15,50]

$$S = \eta^{1/2}/It \qquad (48)$$

S is obtained from linearized holographic growth curves, as shown in Figure 14:

The sensitivities in the near IR are of the same order of magnitude as those of the best photopolymer systems known to date in the UV-visible region.[14] The values for the various systems RNO/p-(alkyl-CAc) are compared in Table 7 for three different near-IR wavelengths.

TABLE 7
Holographic Sensitivity S(cm²/J) for the Photopolymer Systems RNO/p (alkyl-CAc) Measured at Three Different Near-IR Wavelengths

$SE^2 \left[\dfrac{kcal}{mol}\right]$	MNP/p-EtCAc	MNP/p-MeCAc	TENC/p-EtCAc	TENC/p-MeCAC	NCPA/p-EtCAc	NCPA/p-MeCAc
Freshly Prepared Samples						
$\lambda = 752$ nm	$(4.30 \pm 0.98) \times 10^{-3}$	$(8.37 \pm 1.50) \times 10^{-4}$	$(2.61 \pm 0.40) \times 10^{-2}$	$(4.95 \pm 0.69) \times 10^{-3}$	$(1.45 \pm 0.25) \times 10^{-4}$	$(2.57 \pm 0.61) \times 10^{-5}$
$\lambda = 788$ nm	$(6.07 \pm 1.21) \times 10^{-5}$		$(1.28 \pm 0.30) \times 10^{-4}$			
$\lambda = 799$ nm	$(1.86 \pm 0.69) \times 10^{-5}$		$(3.00 \pm 1.10) \times 10^{-5}$			
Old Sample (10 days)						
$\lambda = 752$ nm	$(3.45 \pm 0.79) \times 10^{-4}$	$(1.03 \pm 0.18) \times 10^{-4}$	$(2.53 \pm 0.39) \times 10^{-3}$	$(4.58 \pm 0.64) \times 10^{-4}$	$(1.20 \pm 0.17) \times 10^{-5}$	$(2.14 \pm 0.24) \times 10^{-6}$

For the first time these sensitivities in the near IR make it possible to use commercially available semiconductor lasers for hologram formation, which are much smaller, less expensive, and easier to handle than ion lasers emitting in the visible wavelength region.

C. 2P4L SYSTEMS FOR GATED HOLOGRAPHIC RECORDING

Holographic recording systems following a 1P2L reaction scheme show three main disadvantages. For many applications the wavelengths of reading and writing beams are the same and lie within the absorption region of the recording medium. Therefore, holograms resulting from changes in the absorption coefficient with a maximum theoretical diffraction efficiency[51] of the 3.7% for a pure amplitude hologram have to be taken into account. In these cases a pure phase hologram with a theoretical maximum hologram efficiency of 100% is impossible, and, unless a fixing process has been applied, the hologram will be erased by the reading process due to absorption and photochemistry of the reading beam. Furthermore, the spectral sensitivity range of 1P2L recording materials is limited. Only few systems undergo efficient photochemistry in the red or IR, and these materials tend to be thermally unstable.

However, the spectral sensitivity of recording materials in the near-IR wavelength region is of great interest, as mentioned before (see Sections IV.A and B). Possible difficulties arising for 1P2L systems can be overcome if a 2P4L photoreaction is used for holographic recording.[52-56] The 2P4L scheme and an arrangement for hologram formation are shown in Figure 15.

With the first photon of wavelength λ_1 a molecule is excited from its ground state A_1 to its excited (singlet) state A_1^*. This state relaxes to an intermediate state or chemical species A_2, which then can be excited again with a second photon of wavelength λ_2 to A_2^*. Here it is important that only A_2^* and no other excited state can form photoproducts. Furthermore, the wavelength λ_1 for the excitation of A_1 has to be shorter than λ_2. This means that photochemical reactions and hologram formation can only take place if the sample is simultaneously illuminated with λ_1 and λ_2. For holographic recording λ_1 is an incoherent uniform light beam, and the hologram is written with two coherent laserbeams with the near-IR wavelength λ_2. By turning on and off the incoherent light source, it is possible to gate the hologram formation temporally.

This advantage is followed by three others. Therefore, 2P4L-systems are self-developing, the reading process with λ_2 can take place immediately after turning off the λ_1-source. Since no absorption process occurs if only light of wavelength λ_2 hits the sample, pure phase holograms are created with a maximum theoretical efficiency of 100%. In addition, λ_2 can be in the near-IR wavelength region and easily produce photochemistry because the molecule is already excited to metastable state, and now photons of small energy are able to break bonds.

2P4L hologram recording has been performed with poly-(alkyl-α-cyanoacrylates),[56] carbazole dissolved in a poly-methyl-methacrylate matrix,[52] and with α-diketones like benzil, camphorquinone, and biacetyl in a poly-α-cyanoacrylate (p-α-CAc) host.[15,53,55]

In order to illustrate the characteristics and advantages of 2P4L hologram recording, we will discuss the investigations on biacetyl (BA) dissolved in a p-(α-CAc) matrix. The experimental setup for these experiments is shown in Figure 16.

The high energetic λ_1-radiation was supplied by a 200-W Hg lamp, and therefore was spatially and temporally incoherent. The wavelength region was restricted to 300 to 400 nm. The λ_2-radiation was provided by a Kr-ion laser at 752,5 nm and was used for writing and detecting the holograms. Hologram formation was observed up to $\lambda_2 = 1{,}06$ μm with a pulsed Nd:YAG laser. This shows that 2P4L reaction schemes allow an extension of the spectral sensitivity of recording materials in the IR region.

Another advantage of these systems is the possibility of gating the hologram formation in space and time. The temporal gating process is illustrated in Figure 17.

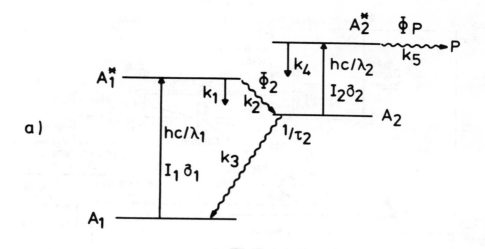

a)

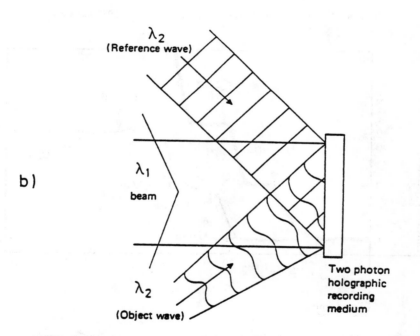

b)

FIGURE 15. (a) Level scheme for a 2P4L photochemistry; and (b) the arrangement for the production of holograms with 2P4L photochemically active materials.

The UV lamp is alternately turned on and off. Hologram growth stops when the UV light is shut off. In fresh samples, however, a self-enhancement reaction in the dark is detected,[54] which is explained by a residual polymerization. The maximum hologram efficiency obtained in these 2P4L systems was 70%.[57] Holograms can be written by semiconductor lasers and read without erasure immediately after turning off the UV light. The photosensitivity of the materials can be considerably increased by choosing polymer matrices that allow self-enhancement reactions, e.g., by residual polymerization, crosslinking, or photodegradation.

2P4L materials may be applied in the production of HOEs in the wavelength region of semiconductor lasers and perhaps as recording media for optical data storage.

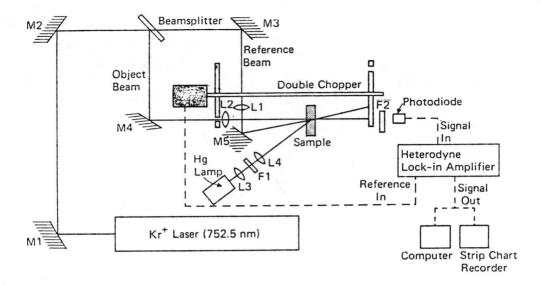

FIGURE 16.   Experimental setup for the investigation of 2P4L materials.

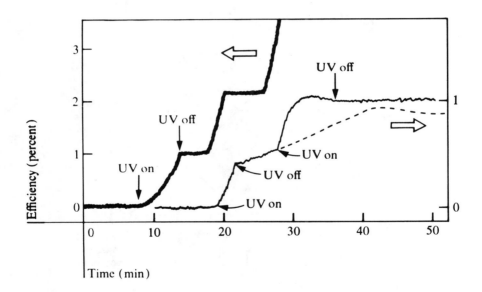

FIGURE 17.   (Bold-line curve) Gating of a hologram in a 200-μm 15% BA/CAc sample used 12 h after preparation; (Lighter-line curve) A hologram grown in 5% BA/CAc under the conditions specified for the 15% BA/CAc sample, except that this sample was used only 30 min after preparation; (Dotted line) the course of the self-enhancement reaction in the dark.

V. APPLICATIONS OF HOLOGRAPHY AND HOLOGRAPHIC RECORDING MATERIALS

The development of cheap and new photopolymeric holographic recording materials is of great interest with respect to their possible application in the production of HOEs,[6,7] which can play an important part as passive devices in the fast-growing area of integrated optics[9] and optical communication technologies. The production and properties of such HOEs shall be illustrated with a few examples.

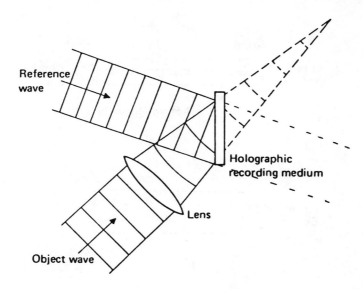

FIGURE 18. The arrangement of the coherent light waves necessary to record the holographic image of a lens.

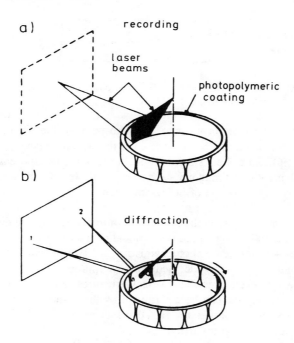

FIGURE 19. Holographic optical scanner. (a) Writing of the holograms; (b) scanned illumination of an object.

The simplest of the HOEs, the holographic grating, has already been described (see Section II.A.1.). Further, a holographic lens can be generated by writing the holographic image of a real lens into a recording material (see Figure 18). If such a hologram is illuminated with a reference wave, the object wave is reconstructed, and the holographic lens focuses the light like a real lens.

Another example of a HOE is a holographic optical scanner,[58] as shown in Figure 19. A series of holograms are written in the light-sensitive layer of a glass or polymer ring.

(a)

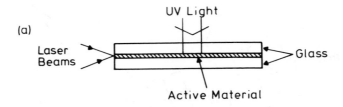

(b)

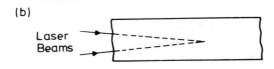

(c)

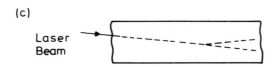

FIGURE 20.   Formation of a thin-film waveguide directional coupler (a, b) and its function (c).

Each hologram directs the laser beam onto a certain point of the object, and thus, during rotation of the ring, the beam can be scanned over the whole object.

Advantages of HOEs are their small weight, large aperture, and easy reproducibility. Furthermore, one element can simultaneously contain different functions (e.g., lens, beam-splitter, etc.), and by superposition of holograms many such elements can be stored in the same sample.

The usefulness of holography in the fabrication of integrated optical devices is illustrated by the formation of an optical waveguide directional coupler as shown in Figure 20, whose active material is suitable for 2P4L photochemistry. The two interfering IR beams intersect inside the narrow dielectric waveguide, and with an appropriate positioning of the planes of the UV and IR beams, even three-dimensional structures can be produced. The hologram efficiency inside the waveguide can be determined by the duration of the UV illumination. Thus, an optical directional coupler for IR beams is produced which splits an IR beam according to the obtained hologram efficiency.

A further and very important application of holographic techniques and holographic recording materials lies in the area of optical data storage. Figure 21 gives a short survey of different conventional materials and processes for data storage with the corresponding storage density. Today optical information storage is already possible by optical disc files. Whereas the maximum storage density of conventional magnetic memories is about 10^6 bit per square centimeter, with optical systems the upper limit of storage density is about 10^8 bit per square centimeter due to the limitation of beam focusing. Permanent optical memories, i.e., compact discs which form a read-only memory (CD-ROM), as well as magnetooptic memories and optically induced phase transition memories (both characterized by the possibility of erasure and directly read after write [E-DRAW]) might be surpassed by several

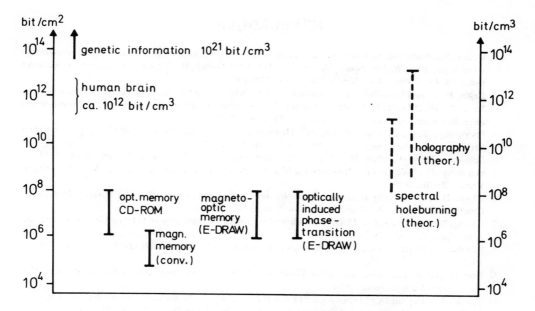

FIGURE 21.   Storage densities of different information recording systems.

orders of magnitude in storage density with spectral holeburning and holographic information storage.[59] In spectral hole during the third dimension is given by the wavelength, whereas in holographic techniques the volume of the storage medium can be used to reach theoretical storage densities up to 10^{12} to 10^{13} bit per cubic centimeter. These are comparable to the human brain. Furthermore, data rates up to 10^{12} bit per second should be possible.

VI. CONCLUSIONS

In this article we have shown on the one hand how holographic grating experiments can be used to obtain information about parameters of photochemical reactions and reaction mechanisms, as well as details about hologram formation mechanisms. On the other hand, we realized that a better knowledge of these processes enables one to find new classes of holographic recording materials.

Further a phase-modulated holographic technique, PMH, was presented, which allows a separate and simultaneous recording of the amplitude [A(t)] and phase [P(t)] contributions to the holographic diffraction efficiency with the possibility to determine quantum yields, total changes of molar refraction, and, in comparison with absorption spectroscopy, photoinduced density changes (matrix effects) in polymer matrices. In addition PMH offers an up to eight orders of magnitude improvement in detection sensitivity compared with PIH experiments. Thus, with PMH reaction mechanisms can be measured which are far beyond the detection limit of PIH.

PMH was applied to the investigation of tertiary nitrosoalkanes in a polymer alkyl-cyano-acrylate matrix following a 1P2L reaction scheme. The advantages of a 2P4L reaction scheme have been demonstrated with the system of biacetyl dissolved in p-(α-CAc). Both photopolymeric systems are sensitive in the near-IR wavelength region and allow hologram formation with readily available and less expensive semiconductor lasers, i.e., GaA1As lasers emitting in the near-IR (680 to 1000 nm).

The applications of photopolymeric holographic recording materials and holographic techniques have been shown to range from simple three-dimensional image formation over the production of HOEs and the use in integrated optics to the very important field of optical data storage.

REFERENCES

1. **Gabor, D.**, A new microscopic principle. *Nature,* 161, 777, 1948. **Gabor, D.**, Microscopy by reconstructed wave-fronts, *Proc. R. Soc. London Ser. A,* 197, 454, 1949; **Gabor, D.**, Microscopy by reconstructed wave fronts. II, *Proc. Phys. Soc. London Ser. B,* 64, 449, 1951.
2. **Leith, E. N. and Upatnieks, J.**, Reconstructed wavefronts and communication theory, *J. Opt. Soc. Am.,* 52, 1123, 1962; **Leith, E. N. and Upatnieks, J.**, Wavefront reconstruction with diffused illumination and three-dimensional objects, *J. Opt. Soc. Am.,* 54, 1295, 1964.
3. **Ostrovsky, Y. J., Butusov, M. M., and Ostrovskaya, G. V.**, *Interferometry by Holography,* Springer Series in Optical Sciences, Vol. 20, Springer-Verlag, Berlin, 1980.
4. **Schuman, W. and Dubas, M.**, *Holographic Interferometry,* Vol. 16, Springer-Verlag, Berlin, 1979.
5. **von Bally, G.**, *Holography in Medicine and Biology,* Springer Series in Optical Sciences, Vol. 18, Springer-Verlag, Berlin, 1979.
6. **Close, D. H.**, Optically recorded holographic optical elements, in *Handbook of Optical Holography,* Coulfield, H. J., Ed., Academic Press, New York, 1979, 573.
7. **Chang, B. J. and Leonard, C. D.**, Dichromated gelatin for the fabrication of holographic optical elements, *Appl. Opt.,* 18, 2407, 1979.
8. **Streibl, N.**, Parallel optical interconnections and computer architecture, *Laser and Optoelektron.,* 20, 53, 1988.
9. **Hunsperger, R. G.**, *Integrated Optics: Theory and Technology,* Springer Series in Optical Sciences, Vol. 33, Springer-Verlag, Berlin, 1984.
10. **Tamir, T.**, Integrated Optics, in *Topics in Applied Physics,* Vol. 7, Springer-Verlag, Berlin, 1979.
11. **Ebeling, K. J.**, Hardware for digital optical data processing, *Laser Optoelektr.* 20, 35, 1988.
12. **Casasent, D.**, Optical Data Processing, in *Topics in Applied Physics,* Vol. 23, Springer-Verlag, Berlin, 1978.
13. **Lee, S. H.**, Optical Information Processing, in *Topics in Applied Phys.,* Vol. 48, Springer-Verlag, Berlin, 1981.
14. **Harikaran, P.**, *Optical Holography,* Cambridge University Press, London, 1984.
15. **Bräuchle, C. and Burland, D. M.**, Holographische Methoden zur Untersuchung photochemischer und photophysikalischer Eigenschaften von Molekülen, *Angew. Chem.,* 95, 612, 1983: Holographic methods for the investigation of photochemical and photophysical properties of molecules, *Angew. Chem. Int. Ed. Engl.,* 22, 582, 1983 and references cited therein.
16. **Eichler, H. J., Günter, P., and Pohl, D. W.**, *Laser-Induced Dynamic Gratings,* Springer Series in Optical Sciences, Vol. 50, Springer-Verlag, Berlin, 1986.
17. **Eichler, H. J., Ed.**, *IEEE J. Quant. Electron.,* 22, (8), 1986.
18. **Kogelnik, H.**, Coupled wave theory for thick hologram gratings, *Bell Syst. Tech. J.,* 48, 2909, 1969.
19. **Burland, D. M. and Bräuchle, C.**, The use of holography to investigate complex photochemical reactions, *J. Chem. Phys.,* 76(9), 4502, 1982.
20. **Deeg, F. W., Pinsl, J., Bräuchle, C., and Voitländer, J.**, The evaluation of photochemical quantum yields by holography, *J. Chem. Phys.,* 79(3), 1229, 1983.
21. **Bräuchle, C.**, Holographische Methoden zur Untersuchung photoreaktiver Festkörper, in *Photoreaktive Festkörper,* Sixl, H. and Friedrich, J., Eds., Wahl-Verlage, Karlsruhe, West Germany, 1984, 415.
22. **Szabo, Z. G.**, in *Theory of Kinetics,* Bamford, C. H. and Tipper, C. F. H., Eds., Elsevier, Amsterdam, 1969.
23. **Jackson, J. D.**, *Classical Electrodynamics,* John Wiley & Sons, New York, 1962, 155.
24. **Bjorklund, G. C., Burland, D. M., and Alvarez, D. C.**, A holographic technique for investigating photochemical reactions, *J. Chem. Phys.,* 73, 4321, 1980.
25. **Burland, D. M., Bjorklund, G. C., and Alvarez, D. C.**, Use of holography to investigate photochemical reactions, *J. Am. Chem. Soc.,* 102, 7117, 1980.
26. **Bräuchle, C.**, Holography as a new tool for investigating photochemical reactions in the solid state, *Mol. Cryst. Liq. Cryst.,* 96, 83, 1983.
27. **Bräuchle, C., Burland, D. M., and Bjorklund, G. C.**, Hydrogen abstraction by benzophenone studied by holographic photochemistry, *J. Phys. Chem.,* 85, 123, 618, 1981.
28. **Bräuchle, C., Burland, D. M., and Bjorklund, G. C.**, Study of the photolysis of dimethyl-s-tetrazine using a holographic technique, *J. Am. Chem. Soc.,* 103, 2515, 1981.
29. **Deeg, F. W., Pinsl, J., and Bräuchle, C.**, Hydrogen abstraction of benzophenone from polymer matrices: evaluation of quantum yields and photomechanical effects, *J. Phys. Chem.,* 90, 5715, 1986.
30. **Deeg, F. W., Pinsl, J., and Bräuchle, C.**, Investigation of irreversible photochemical reactions by transient grating techniques, *J. Phys. Chem.,* 90, 5710, 1986.
31. **Deeg, F. W., Pinsl, J., and Bräuchle, C.**, New grating experiments in the study of irreversible photochemical reactions, *IEEE J. Quantum Electron.* QE-22(8), 1476, 1986.
32. **Fayer, M. D.**, Dynamics of molecules in condensed phases: picosecond holographic grating experiments, *Annu. Rev. Phys. Chem.,* 33, 63, 1982, and references cited therein.

33. **Eichler, H. J.**, Laser-induced grating phenomena, *Opt. Acta,* 24, 631, 1977, and references cited therein.
34. **Eichler, H. J., Glotzbach, P., and Kluzowski, B.**, Investigation of spatial "hole-burning" in a ruby laser by diffraction of light, *Z. Angew. Phys.,* 28, 303, 1970.
35. **Wagner, P. J.**, Chemistry of excited triplet organic carbonyl compounds, *Top. Curr. Chem.* 66, 1, 1976.
36. **Schenck, G. O., Cziesla, M., Eppinger, K., Matthias, G., and Pape, M.**, Isobenzpinakol als Ursache des spektralen Effekts bei der Photoreduktion des Benzophenons, *Tet. Lett.,* 3, 193, 1967.
37. **Filipescu, N. and Minn, F. L.**, On the photoreduction of benzophenone in isopropyl alcohol, *J. Am. Chem. Soc.,* 90, 1544, 1968.
38. **Chilton, J., Giering, L., and Steel, C.**, The effect of transient photoproducts in benzophenone-hydrogen donor systems, *J. Am. Chem. Soc.,* 98, 1865, 1976.
39. **Hochstrasser, R. M., Lutz, H., and Scott, G. W.**, The dynamics of populating the lowest triplet state of benzophenone following singlet excitation, *Chem. Phys. Lett.,* 24, 162, 1974.
40. **Gehrtz, M., Pinsl, J., and Bräuchle, C.**, Sensitive detection of phase and absorption gratings: phase-modulated, homodyne-detected holography, *Appl. Phys.,* B43, 61, 1987.
41. **Pinsl, J., Gehrtz, M., and Bräuchle, C.**, Phase-modulated holography: a new technique for investigation of solid-state photochemistry and hologram formation mechanisms, *J. Phys. Chem.,* 90, 6754, 1986.
42. **Bräuchle, C.**, Optical solid state spectroscopy by holographic grating techniques with quantum-limited sensitivity, *Ber. Bunsenges. Phys. Chem.,* 91, 1247, 1987.
43. **Bräuchle, Chr.**, Laserphotochemistry by holographic grating experiments with quantum-limited detection sensitivity and holographic optical information storage by photopolymer systems, *Makromol. Chem.,* Makromol. Symp., 18, 175, 1988.
44. **Hrebabetzky, F. and Bräuchle, C.**, in preparation.
45. **Forshaw, M. R. B.**, Thick holograms: a survey *Opt. Laser Technol.,* 2, 28, 1974; **Kurtz, R. L. and Owen, R. B.**, Holographic recording materials — a review, *Opt. Eng.,* 14, 393, 1975; **Hariharan, P.**, Holographic recording materials: recent developments, *Opt. Eng.,* 19, 636, 1980; **Smith, H. M.,** Ed., Holographic recording materials, in *Topics in Applied Physics,* Vol. 20, Springer-Verlag, Berlin, 1977.
46. **Tomlinson, W. J. and Chandross, E. A.**, Organic photochemical refractive — index image recording systems, *Adv. Photochem.,* 12, 201, 1980.
47. **Patai, S.,** Ed., *The Chemistry of Amino, Nitroso and Nitro Compounds and their Derivatives,* John Wiley & Sons, Chichester, England, 1982.
48. **Feuer, H.,** Ed., *The Chemistry of Nitro and Nitroso Groups,* John Wiley & Sons, New York, 1969, 137.
49. **Pinsl, J., Gehrtz, M., Reggel, A., and Bräuchle C.**, Photochemistry of tertiary nitrosoalkanes in solid polymer matrices: a promising new class of organic materials for holographic recording with seminconductor lasers, *J. Am. Chem. Soc.,* 109, 6479, 1987.
50. **Bartolini, R. A., Bloom, A., and Weakliem, H. A.**, Volume holographic recording characteristics of an organic medium, *Appl. Opt.,* 15, 1261, 1976.
51. **Collier, R. J., Burckhardt, C. B., and Lin, L. H.**, *Optical Holography,* Academic Press, New York, 1971.
52. **Bjorklund, G. C., Bräuchle, C., Burland, D. M., and Alvarez, D. C.**, Two-photon holography with continuous-wave lasers, *Opt. Lett.,* 6(4), 159, 1981.
53. **Bräuchle, C., Wild, U. P., Burland, D. M., Bjorklund, G. C., and Alvarez, D. C.**, Two photon holographic recording with continuous-wave lasers in the 750 to 1100-nm range, *Opt. Lett.,* 7(4), 177, 1982.
54. **Bräuchle, Chr., Wild, U. P., Burland, D. M., Bjorklund, G. C., and Alvarez, D. C.**, A new class of materials for holography in the infrared, *IBM J. Res. Dev.* 26(2), 217, 1982.
55. **Bräuchle, C.**, Photochemical laser studies in polymer films using holography, *Polymer Photochem.* 5, 121, 1984.
56. **Pinsl, J., Deeg, F. W., and Bräuchle, C.**, Two-photon four-level hologram recording in poly-(alkyl-α-cyanoacrylates), *Appl. Phys.,* B40, 77, 1986.
57. **Gerbig, V., Grygier, R. K., Burland, D. M., and Sincerbox, G.**, Near-infrared holography by two-photon photochemistry, *Opt. Lett.,* 8(7), 404, 1983.
58. **Pole, R. V., Werlich, H. W., and Krusche, R. J.**, Holographic light deflection, *Appl. Opt.,* 17, 3294, 1978.
59. **Kämpf, G.**, Polymere als Träger und Speicher von Informationen, *Ber. Bunsenges. Phys. Chem.,* 89, 1179, 1985.

Chapter 7

APPLICATION OF HOLOGRAPHIC GRATING TECHNIQUES TO THE STUDY OF DIFFUSION PROCESSES IN POLYMERS

H. Sillescu and D. Ehlich

TABLE OF CONTENTS

I. INTRODUCTION

The diffusive decay of optical gratings was first investigated by forced Rayleigh scattering (FRS) in heat conduction experiments of Eichler and collaborators.[1,2] The same principle provides a technique for measuring diffusion coefficients of photoreactive molecules. Here, a grating of photoexcited or photochemically altered molecules is produced by interference of two coherent laser beams, and the time evolution of this grating is monitored by light scattering. First applications to liquid crystals,[3] polymer solutions,[4] and polymer melts[5] were followed by an increasing number of diffusion studies in polymer systems. Rather than attempting a comprehensive review of these results, we discuss in Section IV some illustrative examples demonstrating the potential of the technique in polymer science. Sections II and III provide a self-contained introduction to the physical principles and experimental requirements of FRS as applied to diffusion in polymers. There may be some overlap with the corresponding sections in Volume IV, Chapter 6; however, we concentrate on the particular aspects of diffusion, including experimental details of sample preparation and probe design. Further applications of FRS to thermal energy excitations, laser-induced ultrasonics, flow studies, charge transport in semiconductors and electrooptic materials, exciton diffusion in solids, photochemistry, and optical damage can be found in the book by Eichler et al.[2]

II. PHYSICAL PRINCIPLES

A. HOLOGRAM FORMATION

The simplest possible hologram is the one used in diffusion studies by FRS. It is formed by recording in a light-sensitive material the interference pattern of two coherent plane wave fields of equal intensity. If the light beams intersect at an angle θ as shown in Figure 1, the light intensity along the x-axis is given by

$$I = I_0[1 + \cos(2\pi x/d)] \tag{1}$$

$$d = \frac{\lambda}{2\sin(\theta/2)} \tag{2}$$

Each of the two coherent beams has the initial intensity $I_0/2$ and wave length λ. The approximation of plane waves is applicable to Gaussian beams if the beam diameter is larger than about 100 d, where d is the distance between intensity maxima.[2] It should be noted that inside a sample with an index of refraction n the wave length is $\lambda_i = \lambda/n$, and the intersecting angle θ_i is subject to Snellius' refraction law: $n \sin(\theta_i/2) = \sin(\theta/2)$. Thus, Equation 2 is independent of n. The situation is, however, different for the formation of a "reflection hologram"[2] where the two light beams come from above and below the sample and cross at an angle of $\alpha_i = \pi - \theta_i$ (see Figure 1). In this case the grating is parallel to the sample surface and

$$d_r = \frac{\lambda_i}{2\sin(\alpha_i/2)} = \frac{\lambda}{2[n^2 - \sin^2(\theta/2)]^{1/2}} \tag{3}$$

since $n \cos(\alpha_i/2) = \sin(\theta/2)$. The minimum grating distance $\lambda/2n$ is obtained in the limit of antiparallel incidence ($\alpha_i \to \pi; \theta \to 0$). For a simple one-photon process the concentration c of the photoreactive dye is related with the light intensity I by

$$-\frac{dc}{dt} = kIc \tag{4}$$

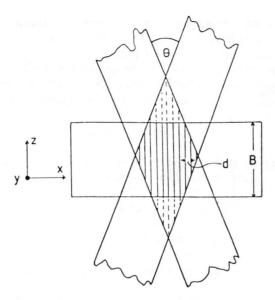

FIGURE 1. Hologram formation by interference of two coherent light beams crossing at angle θ in a sample of thickness B (see Equation 2 for grating distance d).

where k depends upon the extinction and the quantum yield. The concentration profile after short reaction times is obtained from Equations 1 and 4 as

$$c(x) = c_{min} + 1/2(c_{max} - c_{min})[1 - \cos(2\pi x/d)] \tag{5}$$

where $c_{max} = c_0$ is the original concentration at $t = 0$, and c_{min} corresponds to the intensity maxima. For longer reaction times c(x) deviates from Equation 5 and becomes a series of δ-functions at $x = id$ ($i = 0, 1, 2, \ldots$) in the limit of $t \to \infty$. However, only the Fourier component of c(x) which is proportional to Equation 5 is monitored when "reading" the hologram at the Bragg angle (see below). In addition to the grating c(x), a grating of the photoproduct concentration $c_p(x) = c_0 - c(x)$ is formed in the one-photon process. $c_p(x)$ is shifted in phase by d/2 with respect to c(x). The different contributions of the dye and the photoproduct to the absorption and refraction of the sample result in a spatially dependent complex index of refraction, which is analyzed in the following section.

B. HOLOGRAM READING AND EFFICIENCY

The theory of light scattering from holographic gratings is well described in the literature, e.g., in Reference 2. The gratings used in diffusion studies are called "thick holograms", since the sample thickness B is much larger than the grating distance d. In this case, most scattered light is detected at the Bragg angle given by Equation 2. If the wavelength λ of the reading beam is identical with that of the writing beam, the geometry of Figure 1 applies also to detection of the scattered light (see Section III). Otherwise, the detector must be adjusted according to Equation 2. The efficiency η defined by the ratio of diffracted over incident light intensity is given by[2],[6]

$$\eta = \frac{I}{I_0} = \exp\left[\frac{-aB}{\cos(\theta/2)}\right]\left\{\sin^2\left[\frac{\pi \Delta nB}{\lambda\cos(\theta/2)}\right] + \sinh^2\left[\frac{\Delta aB}{4\cos(\theta/2)}\right]\right\} \tag{6}$$

where the grating is characterized by the maximum deviations Δa and Δn from the average absorption constant a and index of refection n, respectively. For pure phase holograms (a = Δ a = 0) the maximum efficiency is η_{max} = 1. Pure absorption or amplitude holograms (Δn = 0) have a maximum efficiency of only η_{max} = 0.037. Therefore, the reading wavelength should be chosen in a region of the spectrum where Δn is as large as possible and a very small in order to avoid further photoreactions in the sample.

For small values of ΔnB and ΔaB an expansion of Equation 6 yields the efficiency

$$\eta = k_1 \Delta n^2 + k_2 \Delta a^2 \tag{7}$$

Since Δn and Δa are proportional to the concentration change $\Delta c = c_{max} - c_{min}$ defined in Equation 5, which is an exponential decay function in simple cases,

$$\Delta c(t) = const \cdot exp(-t/\tau) \tag{8}$$

the FRS intensity is also an exponential

$$I(t) = [A exp(-t/\tau) + B]^2 + C \tag{9}$$

The parameters B and C originate from impurities inside and spurious light sources outside the sample contributing to the coherent and incoherent background, respectively.

C. ANALYSIS OF DIFFUSION PROCESSES

Let us assume that after a short hologram writing pulse a spatially dependent concentration c(x) given by Equation 5 exists in the sample which provides the initial condition (t = 0) of a diffusion process following Fick's law;

$$\frac{\partial}{\partial t} c(x,t) = D \frac{\partial^2}{\partial x^2} c(x,t) \tag{10}$$

The solution is then easily obtained as

$$c(x,t) = \frac{1}{2}(c_{max} - c_{min})cos(2\pi x/d)exp(-t/\tau) + \frac{1}{2}(c_{max} + c_{min}) \tag{11}$$

$$\tau^{-1} = 4\pi^2 D/d^2 \tag{12}$$

In order to measure the diffusion coefficient D by FRS, one has to determine τ as a fit parameter of Equation 9, and the grating distance d from the intersecting angle of the writing beams via the Bragg equation, Equations 2 or 3.

The concentration c(x,t) may change by chemical reactions in addition to diffusion. For simple cases the FRS decay is still described by Equation 9; however, with a decay constant given by

$$\tau^{-1} = 4\pi^2 D/d^2 + \tau_{chem}^{-1} \tag{13}$$

D and τ_{chem} can eventually be determined from the slope and intercept of a plot of τ^{-1} vs. d^{-2}, provided the FRS decay is recorded as a function of θ.

The photoreactive dye and its photoproduct may diffuse with different diffusion coefficients D_1 and D_2, resulting in a more complex FRS decay:[7]

$$I(t) = [A_1 exp(-t/\tau_1) + A_2 exp(-t/\tau_2) + B]^2 + C \tag{14}$$

where τ_1 and τ_2 are related with D_1 and D_2 by Equations 12 or 13. A_1 and A_2 can have different signs, since the contributions Δn_1, Δn_2, Δa_1, and Δa_2 to the total index of refraction and absorption changes may have different signs, and the product grating is shifted by $d/2$ with respect to the educt grating. This has the interesting consequence that $I(t)$ may decay to the background value $B^2 + C$ at some time t_0, where

$$A_1 \exp(-t_0/\tau_1) + A_2 \exp(-t_0/\tau_2) = 0 \tag{15}$$

subsequently rise again and come back to $B^2 + C$ in the limit $t \gg \tau_1, \tau_2$. In this case, D_1 and D_2 can be determined with high accuracy from a fit of Equation 14.[7]

We should mention that higher order diffraction peaks at integer multiples of the Bragg angle can be detected if the concentration grating differs from the simple cosine function of Equation 5.[2,8] A Fourier expansion of $c(x)$ will then contain higher order terms in addition to Equation 5 which correspond to periodicities $d/2$, $d/3$, $d/4$, . . . , and contribute to the higher order diffraction peaks.

D. FLUORESCENCE REDISTRIBUTION AFTER PATTERN PHOTOBLEACHING (FRAPP)

Originally, the method was devised for the fluorescence microscope, where a small spot was bleached, and the diffusion of unbleached dye molecules into this area was monitored by recording the growing fluorescence intensity.[9] Later on, the spot was replaced by a pattern generated through a shadow mask and finally a holographic grating.[10,11] Different from the FRS technique, the system must now be stable within a fraction of the grid distance over the whole detection period. The problem of long time instability can, however, be circumvented by a lock-in technique keeping the detection in optimum condition and further improving the high sensitivity of the FRAPP method.[12] A similar sensitivity improvement of the FRS technique is possible through phase modulation[2] (see Volume IV, Chapter 6).

III. EXPERIMENTAL ASPECTS AND SAMPLE PREPARATION

A. EXPERIMENTAL SETUP

The setups presently used in FRS diffusion studies differ somewhat in detail; however, they are rather similar in the basic requirements.[2,13] In the following, we describe the apparatus used in our laboratory in Mainz for diffusion in bulk polymer systems.[14,15] The essential components are shown in Figure 2. The laser beam is generated by a 4-W argon ion laser operating in single mode at 458 or 488 nm. The laser beam is focused in the sample to a spot size of 0.2 to 1.0 mm. A crystal beam splitter (BS) (Malvern) continously adjustable for the intersecting angle θ between 0 and 10° is used as shown in Figure 2 for the small angle regime. For large angles (60 to 150°) semireflecting mirrors are used for beam splitting, and the two coherent beams are guided by appropriately adjusted mirrors to the sample. It should be noted that the two optical path lengths between BS and sample must be equal within a few millimeters if the laser is in multimode operation, but can be rather different for single mode operation, where the coherence length is large (>1 m). The vibrational stability is most critical for all components affecting the path length between BS and sample. We apply the usual vibration damping of the table and optical components used, e.g., in quasielastic light scattering experiments. Furthermore, we shield the whole setup from external light sources by a dark curtain. Electronic shutters are used for generating the writing and reading pulses (S1) and shutting off (S2) one of the two coherent beams during the reading period. In our experiments, the reading wavelength was identical with that of the writing beam. Thus, the intensity of the reading beam was reduced (A1) by $\sim 10^{-4}$ in order to avoid further photoreactions. If a different wavelength is used for detection, provisions

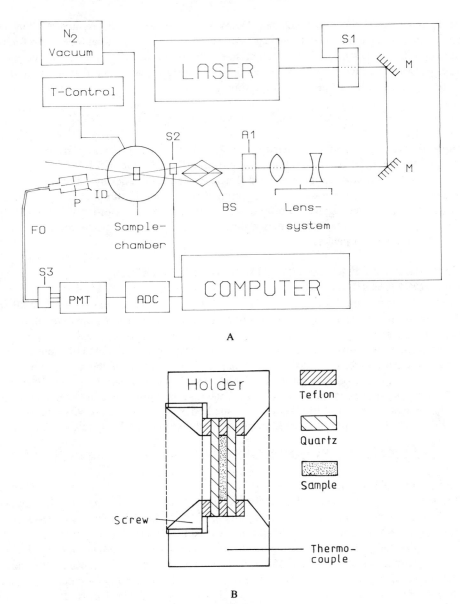

FIGURE 2. (A) Experimental setup.(S1, S2, S3) Shutters; (M) mirror; (A1); Attenuators; (BS) beam splitter; (ID) iris diaphragm; (P) pinhole; (FO) fiber optics; (PMT) photomultiplier; and (ADC) analogue digital converter; (B) sample holder (in sample chamber of Figure A).

for precision adjustment of the reading angle are necessary. The Bragg-diffracted reading beam hits the freely adjustable entrance of an optical cable (F O) connected with a photo-multiplier which is protected against the high-intensity writing beam by an automatic shutter. The system is operated by a computer which also analyzes the digitized FRS intensity (see Equations 9 and 14).

The sample is placed between quartz plates, as shown in Figure 2B (see C,1) in a brass sample holder which can be set to temperatures of 20 to 240°C by a regulated electrical heating device. The temperature difference (<2°C) between the position of the thermocouple, and the hologram in the sample was determined by observing the melting clearing temperature in a set of test samples with known melting points. The sample holder is mounted on an x-y positioning device which allows for selecting up to $\sim$30 different spots on the pellet for subsequent diffusion experiments. The sample chamber can be evacuated and flushed with N_2 for protection of O_2-sensitive samples; it is enclosed by a glass cylinder which allows for irradiation from any horizontal direction.

B. PHOTOLABELS

The selection of appropriate photolabels is less difficult in the FRAPP technique, since a considerable number of fluorescent labels, even functionalized for labeling biological or synthetic macromolecules under mild conditions, are commercially available.[39] Good labels for FRS studies are less abundant, since they have to fulfill a number of requirements which sometimes must be met simultaneously:

1. They must be light sensitive at one of the laser lines which are between 360 and 514 nm for an Ar^+ laser. The sensitivity should be sufficiently high in order to avoid sample heating by large irradiation power. However, too high sensitivity may require to prepare the samples in the dark.
2. The photoreaction should provide a large change of the index of refraction or the absorption constant at the reading wave length. The latter must be in a part of the spectrum where no further photobleaching occurs during detection, or it must be possible to choose a sufficiently low reading beam intensity.
3. For polymer diffusion, the label must be chemically bound to the polymer molecule and should perturb the polymer diffusion as little as possible.
4. Secondary reactions after bleaching should not alter the dynamics of the polymer system (e.g., by cross-linking) which is investigated. For reversible photochromic labels the thermal back reaction should not be faster than the diffusion process to be studied (see Equation 13).

The most important photolabels used in past and present FRS studies are listed in Table 1 along with the temperature range and some other properties useful for FRS studies. More details can be found in the literature references given in the last column.

C. SAMPLE PREPARATION
1. Monomer Dyes in Polymers

For polymers that are solid at room temperature the best method seems to be freeze drying of a solution, e.g., in benzene or dioxane. If diffusion in plasticized polymers is of interest the plasticizer can be dissolved in the solution if its vapor pressure at the freeze drying temperature is sufficiently low. The homogeneous solution should be filtered through a μ-pore-filter before freeze drying. The powder is then pressed to a pellet in a press that should be heated slightly above T_g if the polymer is very brittle. The pellet is placed into the sample holder shown in Figure 2B and annealed at the diffusion temperature until no flow can be detected in the FRS decay curves. Other possibilities are to dissolve the label in a nonsolvent of the finely divided polymer and dry the mixture before pressing a pellet from the powder or to prepare a film by evaporating a polymer solution containing the dye. Dissolution of the dye in the monomer and subsequent bulk polymerization may also be useful in special cases.

2. PHOTOLABELED POLYMERS

Unambiguous information on the molecular weight (mol wt) dependence of polymer diffusion coefficients D can only be obtained if the mol wt distribution of the labeled polymers is narrow. Polydispersity corrections can be applied if the mol wt dependence of D is known or can be obtained from the results in a self-consistent manner. However, this is not possible for broad distributions, since D can depend upon the diffusants and the environmental matrix mol wt. The best way for getting narrowly distributed labeled polymers is to combine the termination of anionic polymerization with the labeling reaction. In most of our PS diffusion studies labeling with the ONS dye (see Table 1) was done following the scheme:

Since the living polymers terminated by impurities are not end functionalized, the mol wt distribution of the labeled polymer is always narrower than that of the corresponding unlabeled one. If end labeling is not possible through functionalized chain ends, one has to use random labeling along the chain by polymer analogous reactions (e.g., chloromethylation[40]) or copolymerization with functionalized monomers. In principle, labeling in the center of a chain is possible by terminating the anionic polymerization with a bifunctional dye or a terminating molecule to which the dye can be bound in a subsequent reaction.

IV. APPLICATIONS

A. DIFFUSION OF POLYMERS IN SOLUTIONS

The first FRS application to polymer diffusion[4] had the purpose of testing the reptation model as well as some scaling laws proposed by de Gennes for semidilute solutions.[16] In good agreement with the theoretical prediction, Léger et al.[4,17] reported that the diffusion coefficient D or photolabeled polystyrene (PS) is proportional to M^{-2} and $C^{-1.75}$, where M and C are the molecular weight and the concentration, respectively, of the polymer, Subsequent experiments have revealed that the situation is more complex. In Figure 3 we show an example which exhibits some of the yet unsolved problems.[18]

Diffusion coefficients of photolabeled PS were measured in toluene solutions containing up to 20% of unlabeled PS. Since the D values of labeled PS were found to depend upon the molecular weight of the matrix PS, the latter was chosen sufficiently large in order to obtain the asymptotic limit value D_{tr}^{∞} which is drawn in Figure 3. It should be noted that the observation[18] $D_{tr} > D_{tr}^{\infty}$ for equal tracer, and matrix molecular weight is already at variance with the prediction that "tube renewal" has no influence upon D_{tr} unless the matrix chains are sizeably shorter than the tracer chains.[16] Furthermore, the molecular weight dependence does not follow the scaling. $D \alpha M^{-2}$, as predicted by the reptation model. Though the deviations from the reptation model cannot yet be explained in any quantitative way, one might speculate about relating them with nonpropagative concentration fluctuations due to "pre gel" aggregation observed in quasielastic light scattering experiments.[19] This might result in additional tube fluctuations not contained in the model of tube renewal via reptation of matrix chains.[16]

B. POLYMER-POLYMER DIFFUSION

In bulk polymers, the reptation model was tested by introducing chemical cross-links in the matrix, thus suppressing any lateral diffusion of the tracer chain and leaving only the

TABLE 1
Properties of Dye Labels

Type of compound (R = usual position of polymerchain)	Photoreaction	Grating and Bleaching λ/nm	UV-VIS spectra: – dye, ···· product (E in arbitrary units)	Remarks	References
Spiropyrane	Ring opening reversible at room temperature 35s	Phase/amplitude 351		Ring opened merocyanine-dye is stabilized in water (irreversible)	17,31,32
Azobenzene	*cis-trans*-isomerisation Reversible	Phase/amplitude 458,488		Differently substituted derivatives have been used up to 78°C maximum (spectrum of 4-Dimethylaminoazobenzene shown)	18,31
Tetrahydrothiopheneindigo	*cis-trans*-isomerization Reversible at 160°C	Phase/amplitude 458—514		High thermal stability exceeding 200°C Tetramethyl derivative in use (DTBT)	15,24,33

TABLE 1 (continued)
Properties of Dye Labels

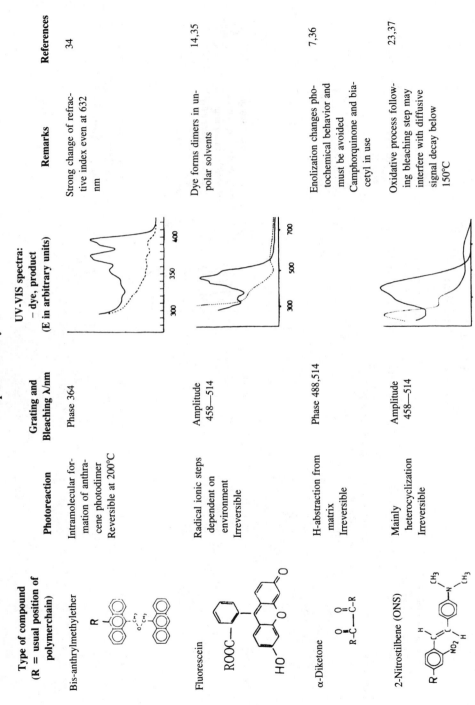

Type of compound (R = usual position of polymerchain)	Photoreaction	Grating and Bleaching λ/nm	UV-VIS spectra: – dye, ⋯ product (E in arbitrary units)	Remarks	References
Bis-anthrylmethylether	Intramolecular formation of anthracene photodimer Reversible at 200°C	Phase 364		Strong change of refractive index even at 632 nm	34
Fluorescein	Radical ionic steps dependent on environment Irreversible	Amplitude 458—514		Dye forms dimers in unpolar solvents	14,35
α-Diketone	H-abstraction from matrix Irreversible	Phase 488,514		Enolization changes photochemical behavior and must be avoided Camphorquinone and biacetyl in use	7,36
2-Nitrostilbene (ONS)	Mainly heterocyclization Irreversible	Amplitude 458—514		Oxidative process following bleaching step may interfere with diffusive signal decay below 150°C	23,37

2-Nitrotolane

Heterocyclization
irreversible

Amplitude
458—514

Light sensitivity requires
handling in the dark; in
absence of oxidants ther-
mally stable >200°C

38

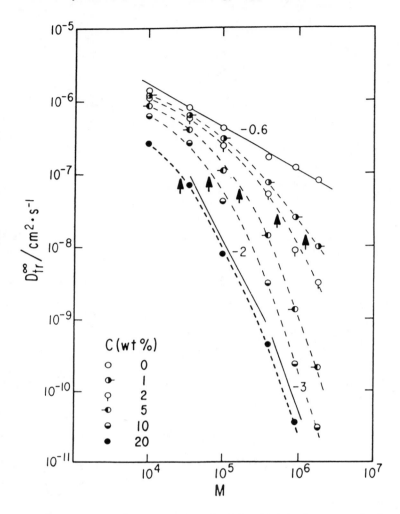

FIGURE 3.   Tracer diffusion coefficients D_{tr}^{∞} of photolabeled polystyrene molecules of molecular weight M in toluene solutions of unlabeled very long polystyrene chains having the concentration C. (Tracer concentration: 0.05 wt%.)

"reptative" curvilinear motion along the chain contour. Our finding that the diffusion coefficient of photolabeled PS in highly cross-linked PS is almost equal to that in linear PS of very large molecular weight confirms the curvilinear transport mechanism in the noncross-linked melt.[20] In Figure 4, we compare the diffusion coefficients of photolabeled linear, star-branched, and intramolecular cross-linked PS in a network of unlabeled PS. A three-arm star molecule can only diffuse by reptation if one arm end is retracted in its tube to the center and threaded into a neighboring mesh of the network. For this mechanism de Gennes has predicted an exponential molecular weight dependence[16] which is confirmed by our experiments (see Figure 4).[21] However, our absolute D values are surprisingly large; only for high molecular weights are they much smaller than those of the linear diffusant molecules. Similarly, the diffusion of labeled intramolecular cross-linked PS is slowed down by a relatively small amount, except for the largest molecular weights, although the curvilinear motion should be suppressed in the whole range where the reptation model applies to linear chains. Thus, we should conclude that cooperative processes play a more important role for the diffusion of medium size nonlinear macromolecules than is expected from the reptation

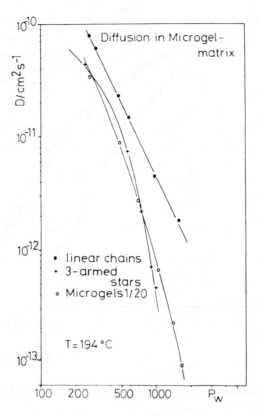

FIGURE 4. Tracer diffusion coefficients D of photolabeled PS molecules in microgel matrix[21] of intramolecular cross-linked PS. (Microgels 1/20 have an average number of 20 monomers between cross-links.) (P_w)Weight average degree of polymerization of labeled PS.

C. DIFFUSION OF MONOMER DYES IN POLYMERS

Diffusion studies of small probe molecules in polymers have used various techniques which are, however, all limited to the regime of relatively fast diffusion, $D \gtrsim 10^{-10}$ cm^2 s^{-1}.[22] The FRS technique has extended this range down to $\sim 10^{-16}$ cm^2 s^{-1} and has thus provided a valuable tool for studying polymer dynamics in the vicinity of the glass transition.[7,15,23,24] In Figure 5, we show an example of dye diffusion in PS ($\bar{P}_n = 2300$) above and below the glass transition, which documents the applicability range of the technique and the type of information available. The drawn curves have been obtained as best fits of Vogel-Fulcher type equations to the experimental data points above T_g.

The glass transition temperature T_g of PS was reduced from 100 to 69°C by adding about 10% of the plasticizer tricresylphosphate (TCP) and to 66°C by reducing the average degree of polymerization to $\bar{P}_n = 20$ in another sample. The diffusion coefficient D of the dye DTBT (see Table 1) dissolved in trace amounts shows qualitatively the same temperature dependence if referenced to the respective T_g value. Small differences can be determined by fitting with parameters of the free volume theory.[7] The absolute D value measured at T_g depends very much upon the size and flexibility of the dye molecule and the type of polymer studied. Thus, the fraction of free volume sampled by the diffusing probe molecule, as well as the internal flexibility of polymers at $T \lesssim T_g$ (β-process), can be investigated quantitatively by performing diffusion studies of dyes used as probe molecules.[15,24] Finally, we note that annealing the samples at $T < T_g$ results in reduced D values drawn in Figure 5 in columns at some temperatures below T_g.

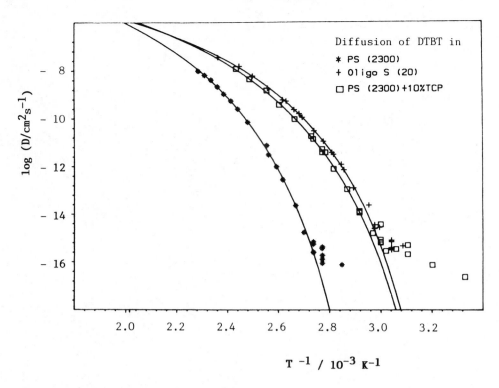

FIGURE 5.   Tracer diffusion coefficients of the photoreactive dye DTBT (see Table 1) in different polystyrene matrices (see text).

D. COMPARISON WITH OTHER TECHNIQUES

A considerable number of other techniques applicable to polymer diffusion studies has been developed or improved in recent years.[25,26] The IR-scanning [27] and the NMR pulsed field gradient technique[28] have been applied very successfully to diffusion of polymers in solutions and melts in a range of $D \gtrsim 10^{-10}$ cm^2 s^{-1}. The application of small-angle neutron scattering requires preparing a nonequilibrium distribution of hydrogenated and deuterated polymers, mostly a multilayer sandwich.[29] The diffusion coefficient is obtained by analyzing the elastic scattering function after annealing the sample. Relatively few diffusion coefficients have been determined by this technique because of beam time limitations. Deuterated polymers are also required for application of forward recoil spectrometry (FRES),[30] where a thin layer (<30 nm) of D-polymer is placed upon the H-polymer. After annealing at the diffusion temperature the concentration profile of the D-polymer is determined by irradiating the sample with α-particles ($\sim$3 MeV) and analyzing the energy loss spectrum of the scattered ^{2}H$^+$ ions freed from the D-polymer. The FRES method is applicable in a D-range of 10^{-12} to 10^{-15} cm^2 s^{-1} and can also be used for investigating interdiffusion of compatible polymers.[31] It has been applied by Kramer and co-workers to a large number of diffusion problems and is also increasingly used by other groups. The FRES and FRS techniques are complementary in many respects, since they use different kinds of tracers and sample preparation, thus covering a wide field of polymer diffusion problems. Further techniques for investigating polymer diffusion can be found in recent reviews.[25,26]

REFERENCES

1. **Eichler, H. J., Salje, G., and Stahl, H.**, Thermal diffusion measurements using spatially periodic temperature distributions induced by laser light, *J. Appl. Phys.*, 44, 5383, 1973.
2. **Eichler, H. J., Günther, P., and Pohl, D. W.**, *Laser-Induced Dynamic Gratings*, Springer-Verlag, Berlin, 1986.
3. **Hervet, H., Urbach, W., and Rondelez, F.**, Mass diffusion measurements in liquid crystals by a novel optical method, *J. Chem. Phys.*, 68, 2725, 1978.
4. **Hervet, H., Léger, L., and Rondelez, F.**, Self-diffusion in polymer solutions, *Phys. Rev. Lett.*, 42, 1681, 1979.
5. **Coutandin, J., Sillescu, H., and Voelkel, R.**, Application of holography to measuring slow diffusion of labeled molecules, *Makromol. Chem., Rapid Commun.*, 3, 649, 1982.
6. **Kogelnik, H.**, Coupled wave theory for thick hologram gratings, *Bell Syst. Tech. J.*, 48, 2909, 1969.
7. **Zhang, J., Wang, C. H., and Ehlich, D.**, Investigation of the mass diffusion of camphorquinone in amorphous poly (methylmethacrylate) and poly (tert-butylmethacrylate) hosts by the induced holographic grating relaxation technique, *Macromolecules*, 19, 1390, 1986.
8. **Coutandin, J.**, Dissertation, Universität Mainz, 1984.
9. **Peters, R., Peters, J., Tews, K. H., and Bähr, W.**, A microfluorimetric study of translational diffusion in erythrocyte membranes, *Biochem. Biophys. Acta*, 367, 282, 1974.
10. **Davoust, J., Devaux, Ph. F., and Léger, L.**, Fringe pattern photobleaching, a new method for the measurement of transport coefficients of biological macromolecules, *EMBO J.*, 1, 1233, 1982.
11. **Smith, B. A., Samulski, E. T., and Winnik, M. A.**, Tube renewal versus reptation: polymer diffusion in molten poly (propyleneoxide), *Phys. Rev. Lett.*, 52, 45, 1984.
12. **Léger, L.**, private communication.
13. **Fayer, M. D.**, Dynamics of molecules in condensed phases: picosecond holographic grating experiments, *Annu. Rev. Phys. Chem.*, 33, 63, 1982.
14. **Antonietti, M., Coutandin, J., Grütter, R., and Sillescu, H.**, Diffusion of labeled macromolecules in molten polystyrenes studied by a holographic grating technique, *Macromolecules*, 17, 798, 1984.
15. **Ehlich, D.**, Dissertation, Universität Mainz, 1989.
16. **De Gennes, P. -G.** *Scaling Concepts in Polymer Physics*, Cornell University Press, Ithaca, NY, 1979.
17. **Léger, L., Hervet, H., and Rondelez, F.**, Reptation in entangled polymer solutions by forced rayleigh light scattering, *Macromolecules*, 14, 1732, 1981.
18. **Kim, H., Chang, T., Yohanan, J. M., Wang, L., and Yu, H.**, Polymer diffusion in linear matrices: polystyrene in toluene, *Macromolecules*, 19, 2742, 1986.
19. **Koberstein, J. T., Picot, C., and Benoit, H.**, Light and neutron scattering studies of excess low-angle scattering in moderately concentrated polystyrene solutions, *Polymer*, 26, 673, 1985.
20. **Antonietti, M. and Sillescu, H.**, Self-diffusion of polystyrene chains in networks, *Macromolecules*, 18, 1162, 1985.
21. **Antonietti, M. and Sillescu, H.**, Diffusion of intramolecular cross-linked and three-arm-star branched polystyrene molecules in different matrices, *Macromolecules*, 19, 798, 1986.
22. **Crank, J. and Park, G. S.**, Eds., *Diffusion in Polymers*, Academic Press, London, 1968.
23. **Countandin, J., Ehlich, D., Sillescu, H., and Wang, C. -H.**, Diffusion of dye molecules in polymers above and below the glass transition temperature studied by the holographic grating technique, *Macromolecules*, 18, 587, 1985.
24. **Ehlich, D. and Sillescu, H.**, *Macromolecules*, in press.
25. **Tirrell, M.**, polymer self-diffusion in entangled systems, *Rubber Chem. Technol.*, 57, 523, 1984.
26. **Binder, K. and Sillescu, H.**, Diffusion polymer-polymer, in *Encyclopedia of Polymer Science and Engineering*, 2nd ed., John Wiley & Sons, New York, in press.
27. **Klein, J.**, Evidence for reptation in an entangled polymer melt, *Nature*, 271, 143, 1978.
28. **Von Meerwall, E.**, Self-diffusion in polymer systems, measured with field-gradient spin echo NMR methods, *Adv. Polym. Sci.*, 54, 1, 1984.
29. **Bartels, C. R., Graessley, W. W., and Crist, B.**, Measurement of self-diffusion coefficient in polymer melts by a small angle neutron scattering method, *J. Polym. Sci. Lett.*, 21, 495, 1983.
30. **Green, P. F., Mills, P. J., Palmstrom, Ch. J., Mayer, J. W., and Kramer, E. J.**, Limits of reptation in polymer melts, *Phys. Rev. Lett.*, 53, 2145, 1984.
31. **Fischer, E.**, Photochromie and reversible photoisomerisierung, *Fortschr. Chem. Forsch.*, 7, 605, 1967.
32. **Heiligman-Rim, R., Hirshberg, Y., and Fischer, E.**, Photochromism in spiropyrans. V. On the mechanism of phototransformation, *J. Phys. Chem.*, 66, 2470, 1962 and references therein.
33. **Hermann, H. and Lüttke, W.**, Die Eigenschaften der 3.3′-Dioxo-4,4,4′,4′ - tetramethyl -2,2′-bithiolan-ylidens, einer Verbindung mit dem Grundchromophorsystem der Thioindigo-Farbstoffe, *Chem. Ber.*, 101, 1715, 1968.

34. **Tran-Long, Q., Chang, T., Han, C. C., and Nishijima, Y.,** Application of a photodimerizable probe to the forced rayleigh scattering technique for measurement of self-diffusion of polymer chains, *Polymer,* 27, 1705, 1986.
35. **Linquist, L.,** A Flash photolysis study of fluorescein, *Ark. Kemi,* 16(8), 79, 1961.
36. **Rubin, M. B.,** Recent photochemistry of α-diketones, *Topics Curr. Chem.,* 129, 1-56, Springer-Verlag, Berlin, 1985.
37. **Splitter, J. S. and Calvin, M.,** The photochemical behavior of some *o*-Nitrostilbenes, *J. Org. Chem.,* 20, 1086, 1955.
38. **Heyne, J.,** Diploma thesis, Universität Mainz, 1987.
39. **Haughland, R. P.,** *Handbook of Fluorescent Probes and Research Chemicals,* Molecular Probes, Junction City, OR, 1985.
40. **Feinberg, R. S. and Merrifield, R. B.,** Zinc chloride-catalyzed chloromethylation of resins for solid phase peptide synthesis, *Tetrahedron,* 30, 3209, 1974.
41. **Composto, R. J., Mayer, J. W., Kramer, E. J., and White, D. M.,** Fast mutual diffusion in polymer blends, *Phys. Rev. Lett.,* 57, 1312, 1986.

Chapter 8

LASER PHOTOCHEMICAL SPECTRAL HOLE BURNING: APPLICATIONS IN POLYMER SCIENCE AND OPTICAL INFORMATION STORAGE

Mark A. Iannone and Gary W. Scott

TABLE OF CONTENTS

I. INTRODUCTION

When the chromophores which absorb in one portion of an inhomogeneously broadened absorption band are depleted by a photochemical or photophysical process, a hole in the absorption spectrum is formed. A number of spectroscopic techniques are based on such selective processes. Some of these techniques, in which a transient hole in the absorption spectrum is formed by simple saturation of the absorption, have been in use for years. "Permanent" spectral hole-burning, in which the hole is due to a photochemical or environmental change affecting the absorbers, is possible at low temperatures and was discovered in 1974.

Gorokhovskii et al.[1] found that a hole was formed in the 0-0 absorption band of phthalocyanine in n-octane at a temperature of 5 K after irradiation at 694.3 nm with the fundamental of a ruby laser. This particular hole was observed in both absorption and fluorescence excitation spectroscopy. Since n-octane is a Shpol'skii matrix[2] for phthalocyanine, it gives greatly narrowed spectral bands in comparison to a liquid solvent. Nevertheless, the hole-burning experiment shows that the observed bandwidths were not entirely due to the homogeneous linewidth of the solute. In this case the absorption bands are still inhomogeneously broadened by perturbations caused by the matrix, and thus, hole burning can be carried out in the 0-0 band.

In the same year, Kharlamov et al.[3] observed stable "gaps", subsequently named nonphotochemical holes, in the absorption spectra of perylene and 9-aminoacridine in an ethanol glass at 4.2 K. A HeCd laser was used to burn out the hole, which was subsequently detected by a broadband light source/monochromator in combination with a Fabry Perot etalon and detector, thereby allowing accurate measurement of the hole width, from which the homogeneous linewidth was calculated.

It is interesting to note that these initial studies of the spectral hole-burning phenomenon included both photochemical and nonphotochemical processes, detection using both absorption and fluorescence excitation spectroscopy, and a discussion of both homogeneous vs. inhomogeneous broadening contributions to the observed linewidths. All of these issues have received much subsequent study for a wide variety of systems which are discussed below. There have been several recent reviews of the literature on spectral hole burning.[4-8] In the present chapter we emphasize the applications of spectral hole burning to polymer science and to optical information storage in polymeric materials.

In the discussion which follows we shall use the terms host and matrix interchangeably to denote the solid solvent, usually in the form of a glassy polymer. Guest molecule, absorber, or solute will all refer to the guest chromophore, present in low concentration in the host, which is responsible for the absorption spectrum in which the hole burning occurs.

II. BACKGROUND

A. HOMOGENEOUS LINEWIDTHS

Homogeneous broadening of spectral lines in an optical absorption spectrum is caused by any process that affects the line centers and widths of all the transition for molecules in the sample in exactly the same way. Thus, in a spectrum showing only homogeneous broadening, all molecules in the sample which contribute absorption to a given spectral line give a contribution with exactly the same peak frequency and width. An obvious example of homogeneous broadening is that due to the finite lifetimes of the states involved in the transition. In fact, the minimum observable lower limit on a spectral linewidth of a molecule is set by the finite lifetime (τ_1) of the excited state. This is a consequence of the uncertainty principle:

$$\Delta E \tau_1 \geqslant \hbar$$

$$\Delta \omega \geqslant (\tau_1)^{-1} \tag{1}$$

In Equation 1, $\Delta \omega$ is equal to the full width at half maximum (FWHM) in units of angular frequency of a spectral line for which the only source of broadening is due to the finite lifetime. The line shape may be derived by assuming an exponential decay with time constant τ_1, of the excited state dipole moment time-correlation function; the real part of the Fourier transform of this quantity is the absorption line shape.[9,10] The result is a Lorentzian lineshape given by

$$g(\omega) = C[(\omega_0 - \omega)^2 + (\Gamma/2)^2]^{-1} \tag{2}$$

with FWHM $\Gamma = 1/\tau_1$, in which ω_0 is the center of the line, and C is a constant proportional to $|\langle \mu \rangle|^2$ and μ is the transition moment for the molecule.

In addition to the intrinsic finite lifetime of the molecule, other processes that may be due to the environment can affect the homogeneous linewidth. In the case of a guest molecule in a solid host, random thermal motion of the host and perturbations of the host or in the configuration of the guest molecule within the host caused by excitation of the molecule generally have little effect on the lifetime of the excited state. These processes do, however, affect the coherence time of the excited state. That is, the phase relationship between one excited molecule and other molecules excited at the same frequency is lost in a finite time due to such processes. Since dephasing is a random process, it may be characterized by a time constant τ_2. When dephasing is included in the optical Bloch equations,[11] the result for the absorption lineshape is a Lorentzian with a linewidth given by

$$\Gamma = 1/\tau_1 + 2/\tau_2 \qquad (3)$$

Thus, dephasing may also contribute to the linewidth. Dephasing processes, and thus the observed linewidth, depend on the nature of the matrix as well as on the temperature; in particular, Γ approaches $1/\tau_1$ as the temperature goes to 0 K. Several theories have been proposed to describe the temperature dependence of Γ in polymers (e.g., see References 12 to 16). We describe below experimental studies of Γ vs. temperature in polymeric hosts and discuss conclusions derived about the relation of the solute to the glassy polymer at low temperatures.

As described in Section VI.C, under appropriate experimental conditions the homogeneous linewidth, Γ, may be found from PHB experiments. The observed hole width is equal to 2Γ if other sources of broadening are carefully excluded. (We have used the symbols τ_1 for the lifetime and τ_2 for the pure dephasing time instead of the more standard T_1 and T_2 (or T_2^*, or T_2'), respectively, to avoid confusion with other symbols. However, in keeping with common practice, we will use $T_2 \equiv 1/\Gamma$. The symbol Γ is reserved for homogeneous linewidths, not hole widths.)

B. INHOMOGENEOUS BROADENING AND SITE SELECTION

When an absorbing molecule is present in an amorphous host (solid or liquid), or even in an ordered host such as a strained crystal, inhomogeneous broadening of the absorption bands of the guest molecule may be observed because of the perturbing effect of the many dissimilar environments in which these molecules are found (Figure 1). While for organic molecules the homogeneous linewidth is usually <1 cm^{-1} for the lowest energy absorption band, the inhomogeneous linewidths may be as much as several hundred cm^{-1} in glassy hosts. At higher temperatures, thermal motion causes the environment of any molecule to fluctuate rapidly, and the absorption at any single wavelength can have contributions from most or even all of the molecules in the sample during a very short time period. At low temperatures, typically below 4 K, the molecular environment changes much more slowly. Thus, monochromatic light absorbed by guest molecules in the sample affects only a fraction of the total population of guest molecules. The affected molecules are those which are in sites for which the gas-to-solid state shift causes the energy difference between the ground and excited state to be equal to the energy of an absorbed photon. This process of selecting a subset of the guest molecules according to their transition energies is referred to as site selection. It should be noted that the selected guest sites are not necessarily physically identical. The absorbance of the sample at a given wavelength is appropriately characterized by a molecular absorption cross-section and a convolution of the molecular absorbance lineshape, weighted by the site population, with the bandwidth of the interrogating light.

Shpol'skii matrices,[2] referred to above, are often used along with several other techniques to provide a more nearly homogeneous environment for a guest molecule in a solvent than does a glass host. In these matrices, the guest molecule "fits" into the crystal structure of the host, often an n-alkane, substituting for one or more solvent molecules in such a way that each guest substitutes into the matrix the same way. Hole-burning experiments in such hosts have shown that, although guest spectral lines may be narrowed, the site environments are still not entirely homogeneous.

Several processes can wash out the effects of site selection, including strong coupling of the electronic transition to host vibrations and photons as well as lifetime broadening due to rapid radiationless processes such as energy transfer, vibrational relaxation, and internal conversion. These processes are discussed in more detail below.

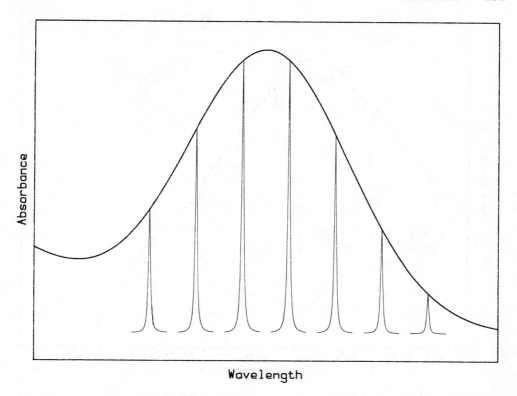

FIGURE 1. An inhomogeneously broadened absorption band showing several contributions from individual molecules in the form of homogeneously broadened Lorentzian line shapes. The inhomogeneous band is the sum of one such contribution from each molecule in the sample, and the total absorbance at a given wavelength depends on the number of molecules absorbing at that wavelength.

At low temperatures, photochemical reactions initiated by absorption of monochromatic light can affect only the site-selected subset of the population of molecules. If the absorption spectrum of the product does not overlap that of the reactant, or if this absorption is not narrow-band, the depletion of molecules absorbing at the irradiation wavelength will produce a hole in the absorption spectrum (Figure 2). This hole will be ''permanent'' in the sense that spectral diffusion processes that can cause hole filling only occur at any appreciable extent after a long time, providing the sample is maintained at a low temperature. This type of hole burning may be contrasted with transient hole burning experiments at higher temperatures, such as the Lamb dip experiment and transient, saturated absorption spectroscopy.[17,18]

The true picture of site selection in photochemical hole-burning experiments is more complex than one of simply assuming only a natural linewidth for the molecular homogeneous lineshape. Loss of energy upon absorption of a photon to low-energy lattice vibrations, (usually called phonons even in amorphous matrices by analogy to crystalline solids) allows a guest molecule to absorb photons of higher energy than that corresponding to the zero-point energy difference between the ground and excited states (Figure 3). This produces a broad absorption band to the blue of the pure electronic transition, called the phonon band. Therefore, the much sharper absorption feature due to a homogeneously broadened pure electronic transition is generally referred to as the zero- or no-phonon line. The fraction of the oscillator strength found in the zero-phonon line depends on the nature of the excited states of the guest molecule and, if photochemistry occurs, on the photochemical mechanism. The phonon band results in the appearance, after a hole-burning experiment, of ''phonon

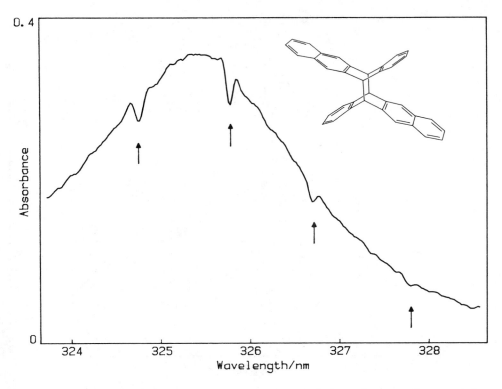

FIGURE 2. Hole burning in the inhomogeneously broadened 0-0 band of *trans*-ditetracene. The arrows indicate the wavelengths of burning. In this example, the widths of the holes are limited by the bandwidth of the laser used for irradiation. (Inset) Structure of the *trans*-ditetracene molecule.

wings'' to both higher and lower energies from the zero-phonon hole (ZPH), as will be discussed in more detail below.

Hole-burning experiments are most readily carried out only on the 0-0 electronic absorption band corresponding to no vibrational excitation in either the ground or excited electronic states. On the other hand, if the chromophore is irradiated into a vibronic band, photochemistry may still occur, but vibrational relaxation to the S_1 (v = 0) level is usually rapid enough to cause significant uncertainty broadening of any photochemical hole produced. Nevertheless, conditions for burning a hole into the 0-0 band may also simultaneously produce ''satellites'' in other vibronic transitions, since destruction of the molecules contributing to the observed absorption at the burn wavelength within the 0-0 band also eliminates the vibronic contribution of these molecules to these other absorption bands. This effect has been utilized to locate vibronic levels in highly broadened spectra in which the vibronic bands would be otherwise unresolvable. The satellite holes are, as a rule, broadened relative to the ZPH in the 0-0 band. Some of this may be homogeneous broadening which results from vibrational relaxation. Inhomogeneous broadening also occurs because a given environment may affect the S_1 (v = 0) level of a molecule in a different way than it affects the S_1 (v = n) levels. In other words, the population selected in the hole-burning process may be homogeneous with respect to the energy of the 0-0 transition, but not homogeneous, for example, with respect to a S_1 (v = 1) $\leftarrow$ S_0 (v = 0) transition.

C. PHOTOCHEMICAL AND NONPHOTOCHEMICAL HOLE-BURNING MECHANISMS

The mechanism of photochemistry may influence the shape of the holes observed. In fact, some photochemical mechanisms will not result in spectral hole burning, effectively

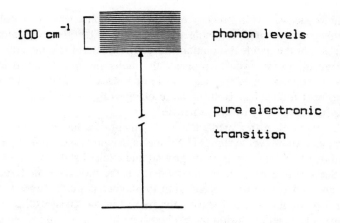

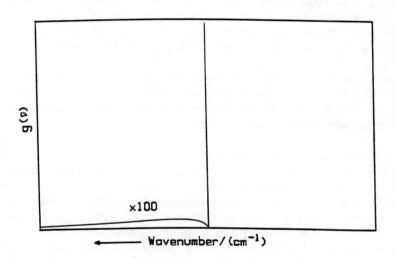

FIGURE 3.   The arrow indicates a pure electronic 0-0 transition ($\nu \geq 15{,}000$ cm^{-1}).
In addition, in a solid matrix simultaneous absorption of a photon and loss of energy
to matrix vibrations is possible, resulting in a nearly continuous band of states at
higher energy than the 0 0. The resulting absorption lineshape for an individual
molecule (or a homogeneous population of molecules) is shown below. In this case,
the Debye-Waller factor, the ratio of the area of the Lorentzian pure electronic feature
to the Lorentzian pure electronic feature to the total area, has been set equal to 0.5.
The phonon band intensity is shown multiplied by 100 for clarity.

eliminating some molecules from consideration for use in optical information storage by
PHB, as discussed by Burland and Haarer.[19] For example, a large geometry change in the
initially excited state will usually result in strong coupling to the lattice phonons, resulting
in an undesirably large contribution to the molecular lineshape from a phonon wing. Furthermore, if the initially excited state is quite short-lived, then the ZPH may also be lifetime
broadened. Systems in which narrow photochemical holes are observed either have some
kind of barrier in the excited state which prevents the photochemistry from occurring immediately on excitation, or else the reaction occurs from some other subsequently populated,
intermediate excited state.

Hole burning may also be observed at low temperatures, even in the spectra of molecules which are not photoreactive. This process is termed "nonphotochemical hole burning" (NPHB).[4,6,20] In this case, although the molecular structure of the guest chromophore is not changed by absorption of the incident radiation, the environment of this molecule is changed by the absorption event. In this type of process, the molecule, when placed in an excited state, may change its orientation in the host matrix or induce a change in the arrangement of the surrounding host molecules. Both of these changes likely result in a change in the zero-phonon transition frequency of the molecule.

In the simpliest theoretical treatment, the environment of a molecule in a glass matrix has been described as a two-level system (TLS): the molecule is modeled as being coupled to two configurations of the glass in both its ground and excited state (e.g., see References 4,6, and 15 and Section II.E below). In this model, for the molecules undergoing NPHB, the barrier to a transition from one configuration to another is much larger than kT in the ground state. For a particular TLS, however, the barrier in the guest excited state may be small enough to permit tunneling during the excited state lifetime. Thus, when the molecule relaxes to the ground state, it finds itself in a different configuration, since a change in the matrix has been induced (Figure 4). Of course, this tunneling process should also contribute to a broadening of the homogeneous linewidth in NPHB by dephasing (see below). Since the change may be subtle, the shift in the peak wavelength of the molecular absorption line may be quite small. Because the "product" (i.e., the same molecule but in a different environment) absorbs at a nearby wavelength, hole burning at that nearby wavelength can produce a filling of the original hole. The existence of the holes depends upon processes with small activation energies, and the holes are thermally unstable, being erased at temperatures at 10 to 20 K. In fact, nonphotochemical holes may not always persist even at 1.4 K (e.g., see Section VII. B.1). For these reasons, NPHB has not generally been considered for optical information storage applications.

D. OTHER SITE-SELECTION EXPERIMENTS

Doppler-free techniques such as Lamb-dip spectroscopy and laser-induced sub-Doppler fluorescence spectroscopy have been used to selectively observe that population within the Doppler profile of a gas, which has zero velocity along the line of observation.[17] An experiment more closely related to PHB experiment is fluorescence line narrowing.[8] In this experiment, excitation into the 0-0 electronic absorption band of a molecule in the condensed phase at low temperature is accomplished with a narrow-band laser. Although difficult to observe, narrow-band fluorescence for the 0-0 transition (i.e., at the excitation wavelength) from the site-selected excited population of molecules may occur. Line narrowing of the vibronic bands in the fluorescence spectrum is also observed to the extent to which the homogeneity in the 0-0 energy difference is preserved for the S_1 (v = 0) $\rightarrow$ S_0 (v = n) vibronic transitions. Phonon side bands are commonly observed in this type of experiment also. However, any molecular photochemical or environmental nonphotochemical change may cause the signal to decrease with time, and therefore, for this technique these processes are undesirable. The difficulty of detecting 0-0 emission line shapes in exact overlap with excitation wavelength and with sufficient resolution makes this technique difficult to apply to studies of homogeneous linewidths.[21]

If the S_1 state is excited, site selection is lost upon intersystem crossing so that the phosphorescence bands are usually broad. However, if $T_1 \leftarrow S_0$ excitation can be accomplished directly, site selection results in narrowed phosphorescence in which the zero-field splitting of the T_1 spin states may be observed.[8]

E. GLASSY MATRICES

PHB has been described as "a spectral probe of the amorphous state in the optical domain."[22] For a review of this aspect, see Reference 5. The importance of developing

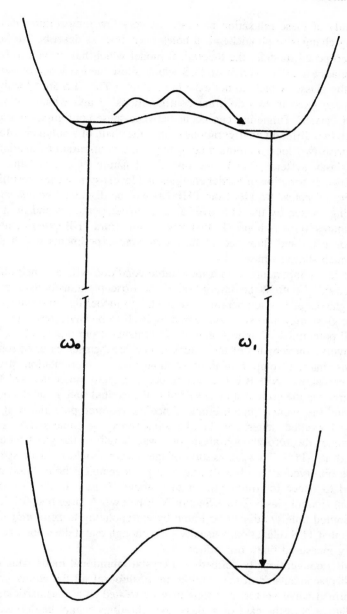

FIGURE 4.   A two-level system which is asymmetric and has a lower barrier in the excited state. Nonphotochemical hole burning occurs when a molecule interacting with the TLS absorbs light at frequency ω_0. The TLS "flips" or tunnels in the excited state, and thus alters the environment of the molecule such that its new absorption wavelength is ω_1.

such probes may be recognized from the fact that measurements of low-temperature heat capacities and thermal conductivities of glasses give temperature dependences which are very different from the behavior observed of crystals. One of the most interesting aspects of these observations is that many different kinds of glasses show similar results.

The glass transition temperature of a polymer is the temperature below which equilibrium cannot be maintained at a given rate of cooling.[23] The approach of a glass to equilibrium must be described by a distribution of rate parameters, since many different processes are involved.[24] Study of these relaxation processes at very low temperatures has been carried out by studying changes in photochemical holes over time, as described in Section VI.B).

Of many proposed models, the theoretical model which has proven to be most useful in describing glassy matrices is that of TLS which allow the existence of very low-energy excitations in the glasses which do not exist in crystals.[25] The TLS are "localized structural defects which can exist in two different configurations"[26] and which are separated by a small potential barrier. Tunneling can occur between these configuration such as, for example, between two different configurations of the side chain of a polymer. More generally,[4] the TLS comprise two local minima among the many configurations available to the nonequilibrium glassy system. Due to the disordered nature of a glass, the TLS must be characterized by a distribution of barrier energies and level asymmetries, resulting in a nearly constant density of states. Results from PHB, as will be discussed below, allow detection of the dephasing caused by the TLS over a range of temperatures and on a time scale of from a few minutes to many hours. In this way, results from PHB experiments complement those of photon echo and fluorescence line narrowing experiments which sample kinetic processes at much shorter times.

A glass at low temperature is in a metastable condition with a nonequilibrium configuration "frozen in".[27] Often, the description of spectroscopic transitions of guest molecules embedded in glasses at low temperature use a TLS to model the environment of the guest molecule. The glass may also be "considered by itself to be characterized by a distribution of double well potentials",[27] as in Figure 4. The change from one level of the TLS to the other could involve movements of the matrix molecules themselves or of segments of host polymer chains, internal rotation of the host molecules, or reorientation of the guest molecule.[28] As noted above, NPHB is believed to occur in guest molecules which interact with TLS whose barrier in the ground state $\gg$ kT but in the excited state is small enough to permit relaxation during the excited state lifetime. Since the required potential well asymmetry of the ground and excited states can be the same only by coincidence, such tunneling in general changes the zero-phonon absorption wavelength of the guest molecule which is interacting with the TLS. Thus, relaxation of the matrix should produce spectral diffusion which may be observed as the broadening of a photochemical hole. Such processes have been observed to occur for times greater than about 10 min in the case of the PHB of quinizarin in an alcohol glass[27,29] (see Section V.B below). A large body of data from which the lifetime-limited hole width can be found by extrapolating to zero temperature (Section V.C) suggests that TLS relaxation is usually slow enough that it does not significantly affect experimentally measured linewidth values.

A phonon is conventionally defined as a crystal vibrational mode which is defined by the relative displacements from the equilibrium position of all the atoms in the crystal. It has a well-defined wave vector, and thus may be treated as a quasiparticle, or as a plane wave in the lattice.[30] In the case of a glass, the vibrations cannot be defined in terms of a regular lattice; however, in the elastic limit of long wavelengths, the concept of phonon is still useful. Evidence from acoustic experiments indicates that phonons do exist in glasses and that they are scattered by the TLS,[25] hence, the term "local phonons". The lifetimes, or coherence lengths, of the excitations in a glass decrease with increasing wave number.[31]

III. INFORMATION STORAGE APPLICATIONS

A. INFORMATION DENSITY

Information storage applications are based on the idea of dividing up an inhomogeneously broadened absorption band into wavelength increments, each of which may or may not have a hole burned in it. This hole or no-hole is interpreted as a 1 or 0 of stored binary information, respectively. This concept allows "multiplexing" many bits of information which may be stored optically at each spatial location on the storage medium. This is in contrast to the way that magnetic storage media are used in which only one bit of inforrmation may be stored at each spatial location. An upper limit on the multiplexing factor, the number of bits that can be stored at a given location, is given by the ratio of the inhomogeneous linewidth to twice the homogeneous linewidth. Various broadening mechanisms may reduce this number in specific systems. However, for example, at least 1600 holes have been burned into a 266 cm^{-1} portion of the 0-0 band of octaethyl porphyn in polystyrene by an interference technique (see Section VI. A).

The area of usable discrete locations for optical storage is ultimately limited by diffraction,[32] but the difficulty of accurately accessing very small spots is a more severe practical limitation. Since addressable storage areas of 10^{-6} cm^2 are considered practically realizable,[33-35] storage densities of 10^9 bits per square centimeter or more is a practical goal. Moerner, who has reviewed the problems that must be resolved in order to apply this idea practically, [35] points out that higher storage densities than this would not necessarily be an advantage so long as sufficiently fast random access to the stored data were possible. This and other engineering problems such as maintaining the storage medium at liquid helium temperatures are "solvable within the current state of the art."[35] The writing and subsequent reading of holes in a 10-μm laser spot at the rate of 30 ns per bit requires the discovery of new optical storage materials, and these materials must fulfill certain well-understood requirements.[33,35,36]

For information storage using PHB, the storage medium must be kept in the temperature range of boiling liquid helium (T ≤4.2 K). Although holes have been observed to persist[37] in an inorganic material even after cycling up to room temperature, in organic materials the holes are significantly broadened if allowed to warm up only to liquid nitrogen temperature (77 K), and these same holes are totally destroyed at somewhat higher temperatures. This destruction occurs because thermal motion of the host causes the environments of the unreacted chromophores to change; these unreacted molecules are thereby reshuffled to fill in the hole. In some systems thermal reversal of the photochemistry may also be a factor in hole filling.

For practical writing and reading of information in this type of storage medium, it would be highly desirable to be able to use diode lasers, which are convenient, reliable, and rapidly tuned.[38] The best candidate currently available is the GaAlAs laser operating from 750 to 850 nm. Holes in a color center and in phthalocyanine in sulfuric acid glass have been burned and detected using a GaAlAs laser.[38,39]

B. PHOTOCHEMICAL MECHANISMS

The nature of the excited state of the chromophore, which has been produced by the site-selecting light, partially determines the suitability of the chromophore for narrow-band hole burning. It has been pointed out[19,32,40] that if the initially excited state is reactive, and/or if it involves a large change in equilibrium geometry from the ground state, then the probability for simultaneous excitation of phonons in the lattice will be higher. Thus, under these conditions, the phonon wings of the observed hole may be much more prominent than the ZPH, a situation which would have an adverse effect on the possibility of using the ZPH for information storage. Furthermore, if the excited state has a short lifetime, the pure electronic transition which produces the ZPH may be significantly broadened. Thus photochemical processes which result in immediate reaction of the excited state are not good candidates for optical information storage.

Burland and Haarer[19] have classified the photochemical processes favorable to the formation of well-resolved ZPHs as either (1) one photon with a barrier or (2) two photon. The two-photon mechanism involves preparation of an intermediate state by the site-selecting light. Photoreaction cannot occur from this intermediate level directly, but absorption of a

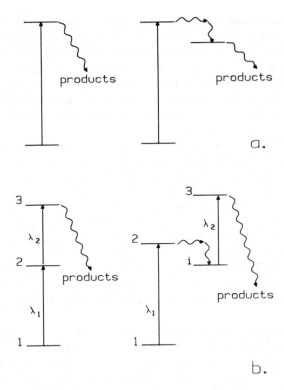

FIGURE 5.   (a) One-photon hole-burning mechanisms. The one on the left involves reaction of the initially excited state; in the other case reaction occurs *via* an intermediate state. (b) Two-photon mechanisms; (Left) the three-level system in which reaction occurs when a photon of wavelength λ_2 is absorbed by the initially excited state; (Right) the four-level system in which an intermediate level absorbs the second photon.

second photon is required to complete the reaction. Two types of two-photon mechanisms have been described.[33] In the three-level mechanism, the intermediate state is the same as the initially excited state; this state lives long enough to absorb the second photon. In the four-level mechanism, the initially excited state relaxes to a second excited intermediate state, such as a triplet state, which then absorbs the second photon. These two mechanisms are shown schematically in Figure 5. "Photon-gated" hole burning, discussed below, is a special case of the two-photon mechanism, in which the second photon is of a different color from the first so that hole burning efficiency may be enhanced by light of a color which does not itself cause photoreaction (see Figure 6). These types of systems possess a distinct advantage for information storage applications over one-photon processes in which the light used for readout of information causes photochemical degradation of the stored information. An additional complication of most one-photon processes reported to date is that they involve a fairly "subtle" chemical reaction such as tautomerization by proton transfer, and thus are generally susceptible to hole filling by thermally activated reversal of the reaction.

C. BOTTLENECKS

For any practical application of PHB in the area of information storage, it is generally required to write the information on the storage medium very rapidly. Thus, for example, a storage medium which uses a photoreactive molecule that reacts in the singlet manifold, but which also has a large intersystem crossing quantum yield, may have a built-in "bot-

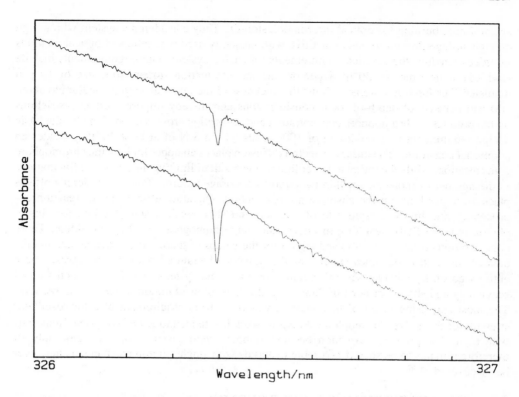

FIGURE 6.   Two photochemical holes, illustrating the gating effect. The upper hole was made by irradiation of the anthracene-tetracene photadduct, AT, with light at 326.4 nm. The second hole was made under the same irradiation conditions and in addition was simultaneously irradiated with light of 440 nm (8-nm bandwidth). The blue light is not absorbed the ground state of AT but is believed to enhance photodecomposition when absorbed by the lowest triplet state (see Section VII.B.5). The spectra are offset for clarity. Full vertical scale equals 0.5 absorbance units.

tleneck'' to the rate with which information may be stored. This comes about because bleaching of the site-selected, ground-state population competes with permanent hole burning. If the quantum yield of intersystem crossing is much larger than the quantum yield of photochemistry, the rate at which information can be written is limited by the relaxation kinetics of the triplet state. For writing times shorter than the lifetime of the triplet state (or other bottleneck state), the rate at which information may be stored depends only on the energy/unit area of the light and the quantum yield of photochemistry.[41]

D. STORAGE MATERIALS: PRACTICAL CONSIDERATIONS

In an information storage application, fast readout of information is generally required and the required signal-to-noise ratio (S/N) for reading accuracy places a lower limit on the intensity of the readout interrogation source. Insofar as the light source used for readout may also cause photochemistry, eventually, degradation of the stored information may occur. These considerations have led to an interest in several recently discovered ''photon-gated'' materials. In these materials the hole burning, caused by the site-selecting light, is enhanced or made more efficient by simultaneous irradiation with light of a second wavelength which is not absorbed by the ground state of the guest molecule. In this type of material, information can be efficiently stored by using both colors simultaneously, but the information readout can be done in the absence of the second wavelength, gating light, thereby causing little or no photochemistry and, hence (ideally), no degradation of the stored information.

Moerner and Levenson[36] have studied the constraints imposed on single-photon-induced spectral hole burning for optical information storage. They considered a system with a high enough storage density to be competitive with presently used magnetic and optical methods in order to deduce the practical requirements for such a system. These requirements include read and write times of 30 ns 4 per bit and an information storage spot size of 10 μm diameter. The latter requirement limits the thickness of the material storage medium to about 100 μm because of depth-of-focus considerations and thereby imposes certain restrictions on the product of chromophore concentration and absorption cross-section. Finally, the basic design requirement of a minimum of 1000 reads with a S/N of at least 26 dB provides an additional constraint. The onset of spectral diffusion places an upper limit on the chromophore concentration, while the ultimate lower limit is the result of limiting the statistical fluctuations in the number of molecules within the addressed storage volume. Power broadening effects place an upper limit on the absorption cross-section. Quantum efficiencies of reaction approaching one imply a rapid rate of reaction and short excited state lifetime for single-photon-induced PHB, resulting in undesirably large homogeneous ZPH linewidths. These considerations establish an allowed region for the values of these various quantities, and the area of this region becomes smaller as the number of required readouts increases. These authors go on to consider several organic and inorganic systems which have been studied previously and show that none of them lie within this allowed region, *even for one readout*. The most surprising result of this analysis was that of the requirement of a low oscillator strength for the active chromophore. Previous work had tended to emphasize use of materials with high absorption cross-section because samples were easier to prepare and study. It therefore remains to be seen if favorable materials with single-photon mechanisms may ever be discovered.[35,36]

E. GATED PHOTOCHEMICAL HOLE BURNING

In the two-photon mechanism of PHB described above, if the second photon is of the same wavelength as the first, site-selecting photon, then the hole-burning rate in the absence of a bottleneck is proportional to the square of incident intensity. This results provides a distinct practical advantage in information storage applications since the source intensity can usually be reduced for readout purposes. However, an even more nearly ideal situation occurs when (1) the second photon required for the reaction is at a different wavelength from the site-selecting first wavelength, (2) the first wavelength is not absorbed by an intermediate state, and (3) the second wavelength is not absorbed by the ground state. In this case, the photochemical reaction (hole burning) can only occur when both wavelengths are present, and light of the second color is often called the gating light. In particular, in this ideal case, the absorption spectrum can be scanned without causing degradation of holes already prepared in the sample spectrum. The possibility of gating by changing other externally accessible factors such as an electric or magnetic field, stress, or microwave irradiation has also been suggested.[35,36]

Several materials have been discovered which show photon-gated photochemical hole burning. These materials are discussed in Section VI.

IV. EXCITED-STATE KINETICS AND HOLE SHAPE

A. GENERAL CONSIDERATIONS

Several papers have been published concerning the shapes of holes observed in spectral hole-burning experiments,[4,42,47] using models which neglect the temperature dependence and spectral diffusion of these shapes (see Section VI. B). In this section we present a simple model based on an extension of work by Friedrich et al.[42]

The rate of absorption of radiation for a molecule whose transition dipole moment makes an angle θ with respect to the electric field vector of the polarized light is given by

$$nB/c \int_0^\infty I(\omega')g(\omega' - \omega)d\omega' = (\pi n/\epsilon_0\hbar^2 c)|\langle\mu\rangle|^2\cos^2\theta \int_0^\infty I(\omega')g(\omega' - \omega)d\omega'$$

$$= \sigma\cos^2\theta \int_0^\infty I(\omega')g(\omega' - \omega)d\omega' \tag{4}$$

in which n = medium index of refraction, B = Einstein absorption coefficient, $I(\omega')d\omega$ = laser intensity between ω' and $\omega + d\omega'$, $g(\omega' - \omega)$ = normalized molecular absorption lineshape, and $\sigma \equiv \pi n|\langle\mu\rangle|\pi^2/(\epsilon_0\hbar^2 c)$.

Here we have neglected attenuation of the laser intensity by the sample, so that intensity does not depend on depth into the sample. This rate of absorption is a function of θ and of the overlap between the laser spectral distribution and the molecular absorption spectrum. A site-selected population of randomly oriented molecules undergoing photochemistry by a one-photon process will decrease as the photochemical reaction proceeds according to the first-order expression

$$dP(\omega,\theta,t)/dt = -\Phi\sigma\cos^2\theta \int_0^\infty I(\omega')g(\omega' - \omega)d\omega' \tag{5}$$

in which Φ is the quantum yield for reaction and $P(\omega,\theta,t)$ is the population of molecules with a zero-phonon absorption between ω and $\omega + d\omega$ and a transition dipole moment that makes an angle θ with the electric field polarization of the light, at time t. Thus,

$$P(\omega'',\theta,t) = P_0\exp[-t\Phi\sigma\cos^2\theta \int_0^\infty I(\omega')g(\omega' - \omega'')d\omega'] \tag{6}$$

in which $P(\omega'',\theta,t=0)$ is taken to be P_0. The sample is assumed to be optically thin, so that P is not a function of depth into the sample. We also assume that P is a function only of burn time, that is, dark processes are neglected (see Section V. B).

In general Φ may be a function of wavelength, offset from the laser line,[42,43] or of the photolysis source intensity in the case of two-photon mechanisms and/or high laser power.[47] When transient ground-state bleaching occurs, for example, by population of a metastable triplet state, it will occur more rapidly for molecules with the largest values of the integral in Equation 5. At a given photolysis dosage the bleaching will be correspondingly less for molecules slightly off resonance. The effect of bleaching depends on the photochemical mechanism; for example, if the molecule is reactive from the initially excited S_1 state without absorption of a second photon, effects include a distortion (and broadening) of the observed ZPH shape and an increase in the intensity of the pseudo-phonon wing (see below).

Readout of the spectral hole with light polarized at an angle α to the burn wavelength polarization gives an absorbance

$$A(\omega,t,\alpha) = 3/4 \int_0^\pi \int_0^\infty P(\omega'',\theta,t)g(\omega - \omega'')(\sin^2\alpha\sin^2\theta + 2\cos^2\alpha\cos^2\theta)d\omega''\sin\theta d\theta \tag{7}$$

By setting $\alpha = 0$ or $\pi/2$ and integrating over θ, it is seen that the sample becomes dichroic at the burn frequency for $t > 0$. To calculate the absorbance measured using unpolarized light, this expression is averaged over all angles α:

$$A(\omega,t) = 3/8 \int_0^\pi \int_0^\infty P(\omega'',\theta,t)g(\omega - \omega'')(1 + \cos^2\theta)d\omega''\sin\theta d\theta \tag{8}$$

Although in some experimental techniques the hole is burned with a laser and read out using a monochromator and arc lamp, the highest resolution is generally obtained if the spectrum is read out using an attenuated laser beam. If the readout laser light is polarized, as is usually the case, this technique also results in greater sensitivity since in the limit of small holes, the hole depth $1 - A(\omega,t)$ detected using light polarized parallel to the burn light source is 1.5 times that detected using unpolarized light, and 3 times that detected using light polarized at right angles to the burn source.

If the photolysis light source intensity (1) is constant over the inhomogeneous bandwidth of the absorber, Equation 6 reduces to

$$P(\theta,t) = P_0 \exp[-t\theta\Phi\cos^2\theta I] \tag{9}$$

with no frequency dependence, and no site selection occurs. In the other limit, if the laser light source bandwidth $<<$ sample homogeneous linewidth, then

$$P(\omega,\theta,t) = P_0 \exp[-t\sigma\Phi\cos^2\theta I_L g(\omega_L - \omega)] \tag{10}$$

where ω_L is the laser frequency and I_L is the laser intensity. In this case, in the limit of small t, the exponential can be approximated by the first two terms of the series expansion,[42] and Equation 7 with $\alpha = 0$ reduces to

$$A(\omega,t) = A_0[1 - (15/8)t\sigma\Phi I_L \int_0^\infty g(\omega_L - \omega)g(\omega - \omega'')d\omega''] \tag{11}$$

in which A_0 is the absorbance due to P_0, the population before photolysis. The zero-phonon part of $g(\omega)$ has a Lorentzian shape. The integral over ω'' thus gives a Lorentzian shape for the ZPH with the hole width equal to *twice* the molecular homogeneous linewidth. Since deeper holes obtained as a result of longer burn times result in broadening of the ZPH, determinations of homogeneous linewidths by hole burning must be carried out in the limit of shallow holes produced at very short burn times (see Section VI. C).

In order to develop the relationship between $P(\omega,\theta,t)$ and the observed absorption spectrum, we shall utilize Equations 10 and 11. If a second simplifying assumption is made, that the molecules are oriented such that their transition dipoles are parallel to the electric field of the hole burning source ($\theta = 0$) the quantity $P(\omega,t)$ has a simple interpretation: the population of molecules remaining at time t with pure electronic transition in the frequency interval between ω and $\omega + d\omega$. Although the situation of oriented molecules with inhomogeneously broadened spectra is artificial, these results are qualitatively similar to those of more lengthy calculations using Equation 7.

The Debye-Waller factor is one measure of the strength of interaction of electronic states with phonons of the lattice, and is given by the probability of a no-phonon transition (pure electronic excited state) divided by the total probability of absorbance. This corresponds to the ratio of the area under the Lorentzian part to the total area of $g(\omega)$. The results of a model calculation based on Equation 7 with $\theta = 0$ and $\alpha = 0$ are shown in Figure 7. In this calculation we assumed a Debye-Waller factor of 0.5 and a molecular absorbance as shown in Figure 4. Using this model we calculated values of $P(\omega,t)$ vs. ω for two different burn times. The results are shown in Figures 7a and c; Figures 7b and d show the corresponding calculated absorption spectra.

The holes in the calculated spectra, Figures 7b and 7d, may be divided into three features: a prominent, relatively narrow ZPH, a "pseudo-phonon wing" to the red of the ZPH, and a "real phonon wing" to the blue of the ZPH. These features are most readily seen in Figure 7d. This descriptive nomenclature is taken from Reference 42. Its significance is as follows:

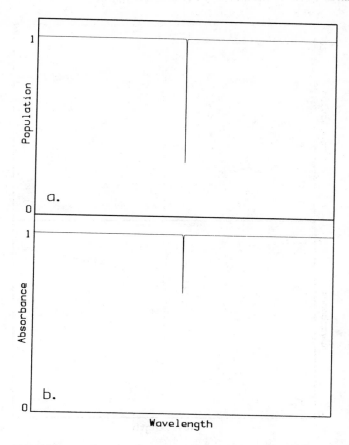

FIGURE 7.   (a) The population of absorbers vs. wavelength after irradiation with a narrow-band laser; (b) the observed absorbance caused by the population shown in (a), obtained from the population by convolution with the molecular lineshape function (Figure 3); (c) the population of absorbers after irradiation for a time 128 times that of (a); (d) the absorbance due to the population shown in (c). The phonon bands are clearly visible in the hole spectrum, and have become asymmetric. In addition, the broadening of the zero-phonon hole is clearly seen by comparison with (b). (Wavelength increases from left to right.)

the ZPH as well as the true phonon band are primarily due to depletion of the population $P(\omega_L,T)$, i.e., to photochemistry as a result of absorption because of a purely electronic (0-0) transition. As seen in Figure 7, hole burning at ω_L produces a negligible effect on the population at $\omega > \omega_L$. The photolyzed molecules with 0-0 band at ω_L also contributed before photolysis to the absorption observed at $\omega > \omega_L$ by absorption of light at frequency ω through their phonon bands. After photolysis this contribution to the absorption spectrum is lost. The real phonon band and the ZPH approximately mirror the molecular line shape as long as the ZPH produced is far from saturation. The pseudo-phonon band, on the other hand, is due to a decrease in population caused when molecules with 0-0 transitions at $\omega < \omega_L$ absorb light of frequency ω_L through their phonon bands. Their contribution to the total absorption at these lower frequencies, chiefly due to their pure electronic transitions, is also lost.

Several characteristics obtained in spectra produced by hole-burning experiments are seen in the calculated spectra. As the fractional ZPH depth approaches the value of the Debye-Waller factor, the growth of the ZPH slows while the pseudo-phonon band continues

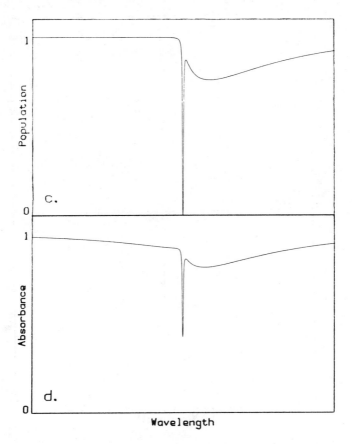

FIGURE 7 (continued)

to grow. In addition the hole shape becomes asymmetric; the pseudo-phonon band grows in relation to the ZPH and the real phonon band. Comparison of Figures 7a and 7 shows that this corresponds to depletion of the population of molecules at ω_L, that is $P(\omega_L, t)$ approaches zero. Absorbance at ω_L causes further reaction in the remaining molecules absorbing through phonon bands or through the wings of the Lorentzian zero-phonon band. This phenonmenon is referred to as photochemical saturation. A second consequence of photochemical saturation is broadening of the zero-phonon feature.

The observed asymmetry between the phonon and pseudo-phonon bands comes about as soon as the population at ω_L is depleted; at this point growth of both the ZPH and the true-phonon band are inhibited. However, the pseudo-phonon band will continue to grow since the population involved is not yet much affected. It has also been suggested that a nonuniform Debye-Waller factor may account for the asymmetry observed at short burn times, as mentioned above.[42]

The ZPH is the feature that is most useful for information storage applications, and the phonon bands are, in this application, a nuisance. Thus, the chromophores chosen as candidates for optical information storage should be picked to have large Debye-Waller factors. The ZPH width obtained depends also upon the temperature, an effect which has been studied extensively (Sections IV. B. 4 and VI. B). For shallow holes, the Lorentzian shape of the zero-phonon feature is preserved. Equations 7 and 8 show that the observed absorption hole is a convolution of the population and the molecular line shape and thus the minimum holewidth, in the limit of low temperatures and shallow holes is *twice* the molecular (zero-phonon) homogeneous linewidth. At low temperatures, under appropriate conditions, the ZPH width has been shown experimentally to approach the lifetime-limited value as discussed below.

B. BROADENING

There are many pitfalls, to use Thijssen and Völker's appropriate description,[48] in the determination of homogeneous linewidths by a spectral hole-burning experiment. Line broadening mechanisms are detrimental to information storage applications as well, since they decrease the value of the multiplexing factor. Those broadening mechanisms, which are the result of properties of the host (e.g., dephasing in a glassy polymer), are of theoretical interest, and have been extensively studied by PHB. Other broadening mechanisms are a result of the experimental method and can be eliminated under appropriate conditions. Various sources of broadening are discussed below.

1. Saturation Broadening

The molecules in a sample which undergo PHB most rapidly are those whose absorption line shape most effectively overlaps the laser line (Equation 5). As the population of such molecules decreases with time due to reaction, other molecules which initially had less spectral overlap absorb proportionately more light, and consequently the hole depth at other wavelengths continues to grow, resulting in a distortion and broadening of the shape of the ZPH. This can be seen by comparing Figures 7b and d. For a variety of reasons, it is possible that only a shallow hole is observed when the population at ω_L has already reached zero. This may be the result of the experimental method, such as attempting to read out the hole with a Xe arc lamp and a monochromator with a bandwidth which was much larger than the ZPH width, or it may result from the ZPH appearing to bottom out when the depth equals the Debye-Waller factor.

With high power, narrow bandwidth lasers, transient bleaching of the resonant population occurs quite readily upon population of metastable levels because a high photon flux is concentrated in a very narrow frequency interval. If PHB is due to a photochemical reaction of the S_1 state, this bleaching represents a bottleneck in the hole-burning process. A bottleneck can also cause broadening of the hole burned in the spectrum of the sample in the same manner as photochemical saturation. Even for photon-gated processes in which a second photon absorbed by an intermediate state is required for reaction, hole broadening will result if the population of the intermediate state is not approximately proportional to the lineshape-intensity overlap (Equation 4) over the whole range of values taken by the integral for resonant and nonresonant molecules in the sample.

2. Power Broadening

Another contribution to the homogeneous linewidth becomes important at high light intensities. This new contribution depends on the square of the electric field strength of the incident light.[11] For a molecule whose transition dipole is parallel to the electric field,

$$\Gamma' = [\Gamma^2 + 2\tau_1\Gamma\omega_R^2]^{1/2} \tag{12}$$

in which Γ' is the observed homogeneous linewidth,

$$\Gamma = (1/\tau_1 + 2/\tau_2)$$

$$\omega_R^2 = |\langle\mu\rangle|^2|E|^2/\hbar^2 \tag{13}$$

and E is the electric field strength. Clearly, power broadening is minimal when $\Gamma \gg 2\tau_1\omega_R^2$. For some molecules, this requirement conflicts with the need to burn holes rapidly, such as may be required for a practical information storage system.

3. Thermal Broadening

PHB experiments are usually carried out at liquid helium temperatures (T ≤4.2 K) in order to reduce contributions to the line shape from thermally populated phonon levels coupled to the ground state of the molecule. Therefore, one source of broadening that can occur is heating of the sample during the burn period, resulting in an increased homogeneous linewidth of the absorbers. This can generally be avoided by using appropriately low laser power levels. When narrow holes that have been burned at 1.5 K are observed at higher temperatures, for example, 10 K, they will appear broadened, and at still higher temperatures they become undetectable due to thermal broadening. However, they may be recovered after such a thermal cycle if again observed at 1.5 K.

Cycling a sample which has undergone PHB at 1.5 K to higher temperatures (such as the technologically important 77 K) usually causes some degradation of the hole spectrum in the sense that the holes will be broadened and shallower when again observed at low temperatures. Both hole broadening and hole filling often occur. Two probable contributions to this phenomenon are suggested in Reference 32. (1) Temperature-activated changes in the matrix can change the environments of some unreacted molecules in such a way that they now adsorb at the photolysis wavelength and fill in the hole; thus their new environment causes them to absorb at the wavelength of the original hole (see Section V. B). This effect is somewhat matrix dependent, and the authors suggest[32] the possibility of increasing the thermal stability of holes by covalently binding the guest to the host polymer or by using harder, cross-linked polymers. (2) Another potential cause of hole filling is thermal reversal of the reaction which was responsible for the PHB.

In experiments conducted on phthalocyanine (H_2Pc) in PMMA, Gutiérrez et al.[32] observed a ZPH width of 0.34 cm^{-1} at 4.2 K, increasing to 0.5 cm^{-1} when measured at 10 K, 0.85 cm^{-1} at 15 K, and approximately 6.5 cm^{-1} at 50 K. However, the hole was found to have decreased in depth by only 10% when cooled again to 4.2 K. A cycle to 80 K resulted in ≃50% decrease in the hole depth, but little difference was seen between cycles to 80 K for 5 min or for over 3 h. At about 100 K, the holes were completely erased; the authors suggest a combination of back-reaction and site destruction as a cause of the hole erasure. Evidence for the role of site destruction was that the thermal stability of the holes was less in a copolymer of styrene and MMA than in pure PMMA.

Although it is not a major factor in designing optical information systems, clearly greater thermal stability of the holes would help prevent loss of stored information in the event of short-term loss of the cryogenic coolant. The site-destruction mechanism of hole filling is dependent on the matrix, while back-reaction depends largely on the characteristics of the photochromic guest. Recently an inorganic system has been discovered which has both a thermally stable matrix and high activation energy for reverse reaction (see Section VII. B. 2 and Reference 37).

4. Spectral Diffusion, Energy Transfer, and Vibrational Relaxation

Spectral diffusion, i.e., broadening of the hole, is known to occur to an observable extent in ≃10 min at 1.35 K for quinizarin in an alcohol glass.[27] Two contributions are likely: reversal of the photochemical reaction and tunneling processes which alter the environments of ground state molecules. Both processes were shown to contribute to an ≃20% reduction in the depth of a narrow hole over 9 h at 1.35 K and a continuous decrease over longer times in a logarithmic manner.

Clearly the system of quinizarin in alcohol glass would be undesirably unstable as an optical information storage medium. However, the photoreaction of quinizarin involves only a change in a hydrogen bond (see Section VI), and other systems should not so readily undergo back-reaction at low temperatures. A large body of results, mostly obtained for polymeric glasses, indicates that for many solutes such relaxation processes do not generally prevent determination of the homogeneous linewidth of the absorber. In other words, this type of broadening does not generally occur to a large extent during the time scale required for the burning-readout process at very low temperatures (Section VI. C) Since the existence of tunneling processes in the TLS is a property of the matrix, elimination of this source of broadening will probably depend on finding guests which are less sensitive to subtle changes in the matrix, as well as using more "rigid" matrices.

If the concentration of the absorbers in the matrix is large enough, Förster-type dipole-dipole energy transfer can occur from an excited molecule to a ground state molecule absorbing at the same or longer wavelength. (Under site-selective conditions the Stokes shift of 0-0 fluorescence is minimal.) In cases of rapid transfer the emission could also be lifetime broadened. Downhill energy transfer could only occur to phonon levels or to low-lying vibronic levels of molecules with zero-phonon absorbance to the red of the initially excited molecule. Energy transfer to a phonon level is unlikely because of the small absorption cross-section of the phonon bands. The rate of energy transfer is given by

$$k_{ET} = KR^{-6}[\cos\theta_{DA} - 3\cos\theta_D\cos\theta_A]^2 g_A(\bar{\nu})\bar{\nu}^{-4} \tag{14}$$

in which K is a factor calculated by Förster[49] and containing the Frank-Condon factor for 0-0 fluorescence of the donor, R is the donor-acceptor distance, θ_{DA} is the angle between transition dipoles of donor and acceptor, θ_D and θ_A are the angles between the vector from D to A and the donor and acceptor transition dipoles, respectively, and $g_A\bar{\nu}$ is the value of the acceptor line shape function at the wavenumber $\bar{\nu}$ of the excitation. Since the value of $g_A(\bar{\nu})$ in the phonon band is very small and in a dilute sample R is large, the rate of transfer to phonon bands of neighboring molecules is unlikely to compete with photochemistry and relaxation of the initially excited molecule. Even for holes burned far from the red edge of the absorption band, it seems unlikely that a molecule satisfying the resonant condition in a vibronic band would be found close enough to the site-selected molecules to undergo energy transfer with high probability.

Vibrational relaxation is another broadening mechanism that was mentioned above. If the inhomogeneously broadened 0-0 absorption band overlaps low-energy vibronic bands, some of the site-selected absorbers may be in excited vibrational states which can rapidly relax to the v = 0 excited vibrational state with resultant lifetime broadening of the absorption loss. This results in a broad component to the hole at the burn wavelength and in addition destroys population to the red of it.

V. EXPERIMENTAL

A. EXPERIMENTAL METHODS

As an example of typical experimental methods used in PHB experiments, we describe below the experimental techniques which we have used in our study of photoadducts.[50] Other methods have been used by other investigators, including additional readout techniques described at the end of this section.

1. Samples

The samples used in our PHB experiments were prepared as thin films in polymers such as PMMA by casting a solution of the polymer and guest in a suitable solvent (e.g., methylene chloride) in a specially prepared quartz cell with a covering to assure slow evaporation at room temperature. The resulting samples are generally clear and range from 0.4- to 0.7-mm thick. The PMMA used was Poly Sciences commercial grade which had been purified before use by precipitation upon addition of methanol to a methylene chloride solution (three times) followed by 250 h of extraction with methanol in a Soxhlet extractor.

Because of the role of the triplet state in the photochemistry of the molecules studied, oxygen was excluded from the samples by placing them under vacuum at room temperature in the liquid helium cryostat, which was to be used for the hole-burning experiments at least 12 h prior to cooling to liquid helium temperatures under a helium atmosphere. The experiments were usually carried out with the samples immersed in liquid helium pumped to below the lambda point.

2. Hole Burning Light Sources

The light source for hole burning was a frequency-doubled CMX-4 flashlamp-pumped dye laser. This laser is equipped with an intracavity Lyot filter, a line-narrowing etalon, and an intracavity doubling crystal. The Lyot filter without the etalon narrows the bandwidth of the frequency-doubled output to approximately 5 cm^{-1} at 325 nm. If the high-finesse intracavity etalon is also used, the measured visible bandwidth is 0.07 cm^{-1}, resulting in a UV bandwidth of 0.09 cm^{-1}. Rhodamine 640 dye was used in most experiments. The laser produces 1-μs duration pulses and 1 to 5 μJ per pulse uv output. A cw He-Cd laser with 50-mW output at 442 nm served as the source of the second color for the majority of the gating experiments carried out on the anthracene-tetracene photoadduct.[50]

3. Optical Setup

Figure 8 shows the optical arrangement used in our experiments.[50] The output of the frequency-doubled dye laser was passed through a Schott UG 1 $\times$ 1 filter to pass the UV and to remove the residual fundamental. A quartz beam splitter sent light to a photodiode in order to monitor the laser output during hole burning. The beam at the sample was larger than the masked sample, to assure burning of the whole portion of the sample which was subsequently interrogated in the experiments to detect the hole spectrum.

4. Detection Method

Two different kinds of hole-burning experiments were routinely carried out in our laboratory. In the first, Figure 10A, after burning with the Lyot-filter-narrowed laser ($\Delta \bar{\nu} \sim$ 5cm^{-1}), the hole was detected in a transmittance mode using a broadband xenon lamp source with a monochromator to scan the spectrum. A 1-m Czerny-Turner scanning monochromator with a dispersion of 0.83 nm/mm of slit width and an f-number of 7 was used to disperse the light from a 150-W, high-pressure xenon lamp. A quartz beam splitter and photomultiplier tube mounted at the exit slit provide a reference signal (I_0) to be ratioed to the signal level produced by light transmitted by the sample (I). The light exiting the monochromator was collimated by a quartz lens, and the image of the exit slit was focused on the masked sample by a second quartz lens. The transmitted light was detected by a photomultiplier tube. The amplified sample and reference signals were fed into a ratiometer whose output was read by a digital multimeter interfaced to a computer. The usual experiment involved a scan over 1 or 2 nm at 213 s/nm using the synchronous motor drive of the monochromator. Typically a spectral bandwidth of 0.02 nm was used.

In the second type of experiment (lower half of Figure 8), the hole was burned at full laser power and then read out with reduced laser intensity in both cases by using the etalon-narrowed laser output. Since 0.09 cm^{-1} is too narrow a bandwidth for detection with a reasonable signal/noise ratio using the xenon lamp per 1-m monochromator combination, the holes were read out by scanning the intracavity, laser-tuning etalon, and measuring the transmittance by the sample of the attenuated laser beam. A quartz beam splitter was used to obtain a reference signal. The sample and reference beams were detected by identical photomultiplier tubes, and the signals were integrated over several laser pulses by a gated integrator circuit, used either as a boxcar or as a straight integrator. Sample and reference signals were ratioed, and the ratio was read by the digital multimeter. The etalon could be

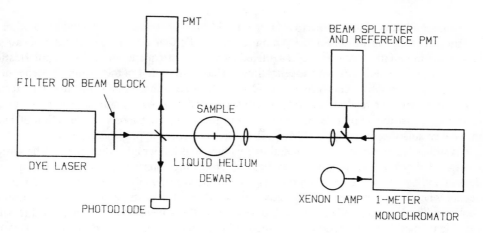

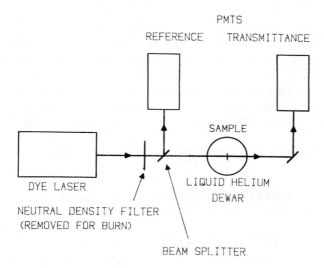

FIGURE 8. Apparatus used for hole-burning experiments.[50] (Upper) Hole burning is carried out using a dye laser for irradiation. A beam splitter diverts part of the laser output to a photodiode in order to monitor the laser power during irradiation. For readout, the laser beam is blocked, the beam splitter is replaced by a mirror, and the spectrum is scanned using a Xe lamp/monochromator with photomultiplier tubes for detectors to measure the incident and transmitted intensity. (Lower) In this case, the hole spectrum is obtained after burning by scanning the laser output using an intracavity etalon. The laser output is attenuated by a neutral density filter, and the incident and transmitted intensities are measured using identical photomultiplier tubes.

scanned across one free spectral range under computer control by a stepper motor. Since the etalon bandpass function is approximately Lorentzian, spectra were deconvolved using the Lorentzian approximation, valid for shallow holes only.

B. ALTERNATIVE READOUT METHODS

In the experiments described above, the holes burned in absorption spectra were detected in transmittance mode. It is also common to detect holes in the fluorescence excitation spectrum using the attenuated burn laser as the excitation source.

A modification of frequency modulation (FM) spectroscopy[51] has been used to detect very small holes with great sensitivity.[41] In this technique, a phase modulator produces weak sidebands on either side of a narrow laser line. If these encounter differential absorption in the sample, this can be detected by monitoring the Fourier component of the transmitted light at the modulation frequency. In a related technique—frequency-modulation polarization spectroscopy[52]—sidebands with polarizations different from that of the carrier are generated. This allows essentially nondestructive readout if the hole is made in an anisotropic material, since the carrier for readout can be polarized perpendicular to the transition dipoles in the material so that the sample is subjected only to the weak sidebands. For detection of holes in weakly anisotropic samples, FM spectroscopy is a more appropriate technique.

Another very sensitive zero-background technique is holographic detection.[53] The sample is irradiated using two crossed laser beams, producing interference fringes at a spacing determined by the angle between the beams and the wavelength of the light. Hole burning occurs where interference is constructive, none where it is destructive. Readout is accomplished with one beam. When the readout beam is at the wavelength of the burn, a diffracted beam is formed in the direction of the second burn beam. Since this is directional, scattered light can be eliminated very effectively. The diffracted light is a zero-background signal which shows the same hole shapes as measurements made in transmittance but is much more sensitive.

Pressure variations are known to cause broadening of spectral holes.[5] This effect has been used in another detection method using ultrasonic modulation.[54] The periodic compression caused by applying 8-MHz ultrasound to the sample through a transducer causes the hole to broaden. The ultrasound was gated at several kHz, and the transmitted light was monitored with a lock-in amplifier so that the kHz modulation in the hole shape was detected. Since only modulation is detected, this is a zero-background technique, and it was possible to scan the whole inhomogeneous band of NaF color centers and observe holes with high sensitivity on a flat baseline.

Readout by scanning an electric field applied to the sample has also been used and is discussed below in Section VI.D.

VI. APPLICATIONS OF PHOTOCHEMICAL SPECTRAL HOLE BURNING TO POLYMER SCIENCE

A. PHOTOCHEMICALLY ACCUMULATED PHOTON ECHOES[55-57]

A time domain approach to PHB was recently demonstrated which allows measurement of the sample dephasing time T_2. It is analogous to accumulated photon echo experiments described by Molenkamp and Wiersma (Reference 58 and references therein). In this case, hole burning was caused by absorption of a series of mutually coherent picosecond pulses from an etalon. Excited molecules with energies within the transform-limited laser bandwidth are thereby produced. If the time between pulses is less than the dephasing time T_2, an interference can occur between a pulse and the previously excited population. For light frequencies within the laser bandwidth where the interference is constructive, the excited state population, and thus hole burning, are enhanced. A lattice of holes is produced, superimposed on the laser bandwidth-limited hole. Since these are produced simultaneously, there is no possibility of storing information by leaving some frequency intervals unburned. However, "exhaustive temporal and spatial information about the signal [used for burning] can be stored in the sample by PHB" provide $T_2 \gg$ time between pulses.[55]

In one experiment,[55] a sample of cryptocyanine in an alcohol glass was irradiated with a series of 3-ps pulses at intervals of 20 ps. The result was a hole with a fine structure of hole modulation at a period of 0.42 cm^{-1}. The width of the modulation is obviously less than the bandwidth of the laser used to produce it. These holes were observed both in transmittance and also by the technique of photochemically accumulated stimulated photon echo. In the latter technique, a single picosecond pulse is used to illuminate the sample, resulting in an emission displaying maxima in intensity at intervals equal to the time between the pulses used for burning. The intensity at the echoes diminishes as the time between burning pulses approaches T_2, which allows the value of T_2 to be measured.

This technique was used to demonstrate the feasibility of storing > 1000 bits at one spot.[57] A lattice of 1600 holes spaced at 0.167 cm^{-1} was burned and detected in a 266 cm^{-1} interval of the spectrum of octaethyl porphin in polystyrene by scanning the dye laser which was producing the picosecond pulses and by burning several times. The contrast (i.e., hole depth) was estimated at 1 to 10% over the burned range, based on the relative intensity of the emission echo.

B. SPECTRAL DIFFUSION AND THE PHYSICS OF GLASSES

As mentioned in Section IV.B.4, spectral diffusion of hole spectra occurs to an observable extent in $\simeq 10$ min at 1.35 K for quinizarin in an alcohol glass.[27] Two contributions are likely: reversal of the photochemical reaction and structural relaxation of the host. The latter should only reduce the hole depth, but not affect the hole area. Both processes were shown to contribute to an $\simeq 20\%$ reduction in the depth of a narrow hole over a period of 9 h at 1.35 K. The two processes are, in principle, distinguishable since tunneling between sites causes hole broadening without filling (for PHB). Back-reaction reduces hole area and also causes broadening, since the change in the guest molecule can cause strains which alter the matrix. The result of both contributions is a broadening of the spectral hole on a logarithmic time scale. Experimental results are summarized in a recent review, Reference 5, in which the decrease in hole area and increase in width are modeled by a distribution of relaxation rates.

Further insight into the nature of the low-energy modes in polymeric glasses is obtained from thermal cycling experiments.[16] Two types of experiments were carried out on phthalocyanine in polyethylene, poly(methylmethacrylate), and polystrene over the range 0.5 to 25 K. One was a measurement of hole width vs. temperature (Section VI. C) and the second was a measurement of broadening after a cycle to a higher temperature. When a hole was raised to a higher temperature, it broadened. Upon cooling to the original temperature, the hole again narrowed, but the original width was not recovered.[16,59] It was found that irreversible broadening occurs only on the first thermal cycle, and that the degree of broadening depends only on the temperature difference (cycle maximum temperature minus the burn temperature). A model was presented[16] in which the hole widths were explained by contributions from TLS modes and local phonon modes.

C. Relation of Hole Width to Homogeneous Linewidths: Dephasing Processes

Spectral diffusion is the result of transitions of the matrix TLS modes. These are known to occur over a wide range of time scales.[5] Dephasing, occurring within the time scale of the experiment, will contribute to the homogeneous linewidth which is observed in the experiment. It is a matter of some controversy whether experiments involving very short times (i.e., photon echo and fluorescence line narrowing) should give the same results as hole burning. It has been suggested[60] that spectral diffusion can result in larger homogeneous linewidths as measured in hole burning experiments. Further, the possibility of dephasing through tunneling processes even at absolute zero has been raised but not confirmed by experiment. A large body of experimental results indicates that lifetime-limited hole widths are obtained for many systems upon extrapolating to 0 K when the experiment is carried out under appropriate conditions. For a thorough review of experimental data, see Reference 21. For recent reviews on the subject of homogeneous line broadening in glasses, see References 5, 13, 20, and 21.

Many hole burning experiments have reported hole widths on the order of 1 cm^{-1}. This large width has been attributed[48] to the use of high burn energies, leading to saturation broadening. In some experiments, this may also be in part an experimental artifact as a result of using a lamp/monochromator combination to read out holes in transmittance. Very narrow holes, e.g., those made with a single-mode laser, are not detectable with an $\simeq 0.1$ cm^{-1} monochromator bandwidth. Thus experimentalists may inadvertently use too much energy to produce an observable but thus broadened hole with a width on the order of the monochromator resolution. Thijssen and Völker[48] point out that, to eliminate these factors, the hole width must be measured as a function of laser power and burn time and detected preferably by scanning with the attenuated laser output rather than a lamp/monochromator. The homogeneous linewidth Γ should then in general be equal to one-half of the hole width extrapolated to zero power.

Theoretical models to explain broadening of homogeneous linewidths in glassy hosts consider interactions between the guest and the low-energy two-level modes of the host, or with a combination of TLS, librational, and phonon modes. Many studies have been carried out which allow the theories to be tested. A fairly general result, with some exceptions, is that the homogeneous linewidth Γ approaches $\Gamma_0 \equiv 1/\tau_1$, the inverse of the natural lifetime, as T approaches 0 K; below 5 K it also is found that

$$\Gamma = \Gamma_o + \beta T^\alpha \qquad (15)$$

Thus, no pure dephasing processes remain at 0 K. Values of α ranging from 1 to 2 are usually reported; $\alpha \simeq 1.3$ is most commonly found. Occasionally a "crossover" to a different value of α is found at some temperature (see below).

A study of hole width vs. temperature from 0.4 to 20 K in S_1 to S_0, 0-0 transitions of dimethyl-*s*-tetrazine (DMST), chlorin, and porphin in 13 different glassy hosts was carried out by Thijssen et al.[61] The polymer hosts investigated were poly(vinyl carbazole), poly(methylmethacrylate), polystyrene, polycarbonate, poly(vinyl alcohol), and poly (ethylene glycol). The results showed that the observed homogeneous linewidth follows Equation 15 with α 1.3 $\pm$ 0.1 for 17 different host-guest combinations (porphin was studied in all 13 hosts). The ratio of Γs for a given solute in two different hosts at the same temperature was found to be independent of the guest and thus should give "information on the relative coupling strength to phonons of the two polymers involved."[61]

In another study,[48] poly(methylmethacrylate), poly(methylacrylate), poly(vinyl acetate), polypropylene, poly(vinyl alcohol), polyethylene, and poly(butyl acrylate) were examined as hosts. A relation was noticed between Γ and the size of the polymer side groups, and the following explanation was suggested. Bulky side groups hold the chains apart, leaving voids between them. The guest is thought to occupy these voids and to have more freedom to move in the larger voids, resulting in faster dephasing. The possible contribution of low-frequency librations of the side groups was also suggested. Thus, for example, at a given temperature porphin in polypropylene or poly(vinyl alcohol) showed smaller values of Γ (and thus β) than it did in PMMA or poly(vinyl acetate). An apparent anomaly in the case of poly(butyl acrylate) was attributed to the plasticizing effect of the butyl side group, which may lead to less void space. The $T^{1.3}$ dependence held for all hosts. Some of the polymers tested are known to be partly crystalline. The authors suggested that the local environment of the guest molecules may be glassy even in these cases since, when the polymer-guest solution is cast, the guest would prevent crystallization in its vicinity. Evidence for this is supported by the fact that, when the guest is diffused into a solid crystalline polymer, the $T^{1.3}$ behavior is not found. In these studies, residual solvent in polymer-guest mixtures led to observed valves of Γ up to 10 times those observed for the same system without solvent. Placing samples in vacuum at 60°C for 72 h was sufficient to eliminate solvents such as p-xylene. The $T^{1.3}$ behavior was not affected by residual solvent.

Porphin, resorufin, crystal violet, pentacene, and DMST in PMMA, poly (vinyl carbazole), polyethylene, poly (ethylene glycol), and ethanol hosts were studied.[62] These represent different PHB mechanisms and, in the case of pentacene, NPHB. Equation 15 is followed with $\alpha = 1.3$ in all cases.

In a study designed to determine the effect of crystallinity in the polymer hosts,[12] the method of sample preparation was varied. Samples of porphin, DMST, and resorufin in PMMA were prepared by dissolving guest and host in solvent and casting a film. DMST was also diffused into solid PMMA. High density ($\simeq 80\%$ crystallinity as determined by DSC) and low-density (58% crystallinity by DSC) polyethylene (HDPE and LDPE, respectively) were used to prepare samples containing porphin and resorufin by mixing ethanol solutions of the guest with the molten polymer, evaporating off the solvent, and cooling at different rates. DMST samples in HDPE and LDPE were made by diffusing DMST into the PE prepared by various annealing procedures. The results confirm the prediction of a previous paper.[62] All samples in PE that were prepared by dissolving and casting, as well as all samples in a PMMA host, follow a $T^{1.3}$ dependence of Γ on T. The samples that were prepared by diffusing DMST into PE showed larger exponents. In all cases Γ extrapolates to $1/\tau_1$ at 0 K.

Detailed comparison of these results with six theoretical models was made.[12] The conclusions are that dephasing in amorphous organic systems must be described by a combination of TLS and librational modes. Semicrystalline polymer hosts show behavior approaching the exponential T dependence of crystals as the degree of crystallinity in the vicinity of the guest increases.

In a study of hole burning in spectra of quinizarin in ethanol-methanol glass,[59] a $T^{1.3}$ temperature dependence of Γ was found above 2 K and a linear temperature dependence below. This implies that the hole width cannot be adequately described by a single process. Extrapolation of the data to $T = 0$, assuming a linear T-dependence, gave the lifetime-limited linewidth. In this case the hole was burned and scanned at the same temperature. If the burn and readout temperatures are not equal, the observed linewidth depends on both, which was explained on the basis of a distribution of barrier heights and hence tunneling rates in the TLS and temperature-induced strain in the matrix. Spectral diffusion in this system is known to be slower if the alcohol glass is deuterated.[5] Measured linewidths were the same in both protonated and deuterated matrices, a result which demonstrated that rapid spectral diffusion in the time between burning and readout was not contributing significantly to measured linewidths.

Octaethyl porphin in polystyrene shows more complex behavior, demonstrating two crossover temperatures: $\alpha = 2.4$ to 2.6 for $T < 0.1$ K, $\alpha = 0.5$ to 0.8 for 0.1 K $<< T$ < 0.3 K, and $\alpha = 1.1$ to 1.2 for $T > 0.3$ K. In PMMA the same compound shows a crossover from $\alpha = 2.3$ below 0.1 K to $\alpha = 1.2$ above 0.1 K. In this case the Γ vs. T behavior also depends upon irradiation wavelength [63-66]

Homogeneous line widths from NPHB of resorufin in ethanol glass werer reported to be four times greater than those found from two-pulse photon echo experiments.[60] A possible reason for the discrepancy is that in photon echo experiments, the dephasing occurring is detected on the timescale of T_2, while in hole burning, a delay of several minutes between burning and scanning is typical. If spectral diffusion occurs within a period of seconds, the hole could thus be broadened before it is detected. The photon echo experiment caused NPHB in the sample, but Γ (from photon echo) was found to be independent of the extent of hole burning. Averaging hole widths of holes $<4\%$ deep gave a limiting low-temperature hole width of 0.05 cm^{-1}, while photon echo results required $\Gamma = 0.011$ cm^{-1}. However, van den Berg and Völker[66] studied the same system and found $\Gamma \simeq .003$ cm^{-1} at 1.5 K and $\Gamma_0 = 0.0007$ cm^{-1}, with a good fit of the data to Equation 15 with $\alpha = 1.3$. These authors found that different results were obtained depending on how the ethanol was cooled; in either case $\alpha = 1.3$ was found, but β is $\simeq 8$ times greater in one phase than in the other.

The fluorescence lifetime of *s*-tetrazine was found by picosecond laser techniques to be 450 ± 50 ps (in benzene at 300 K; cited in Reference 67). Since a dephasing contribution to the linewidth is expected to be <<1/450 ps at T < 1.5 K, an identical value was obtained from the width of a hole in the 0-0 band. In this experiment, the system was an *s*-tetrazine-benzene mixed crystal. "Hole burning" in a vibronic band resulted in a reduction of the entire band; this along with the nearly Lorentzian shape of this band implies that its broadening is mainly homogeneous. This allowed a calculation of the relaxation time of the vibronic level.[67]

PHB has been used to study the primary electron donor in reaction centers of photosynthetic bacteria.[68] Hole widths of $\simeq 400$ cm^{-1} were found. Several possible explanations were suggested, including a very short lifetime ($\simeq 15$ fs), overlapping vibronic levels, and very strong phonon coupling. Theoretical results supported the latter.[69]

D. STARK AND ZEEMAN SPECTROSCOPY

Zeeman shifts large enough to observe by conventional spectroscopic methods are found in metalloporphyrins; however, for the free base the shift is much smaller. Shifts of the order of GHz have been detected by PHB in porphin and chlorin in n-octane, both in the 0-0 transition and in vibronic bands.[70] Stark splitting of the order of 1 GHz, proportional to the applied electric field, of a hole burned at zero field in a single crystal of chlorin in n-octane was also reported. The splitting indicates that the dipoles of the guest molecules in the crystal have two possible orientations. In this experiment the crystal could be rotated to any direction with respect to the applied field. Calculation of the difference between ground and excited state dipole moments from the splitting agreed almost exactly with *ab initio* calculations.[70]

The Stark effect was studied in octaethyl porphin and chlorin in amorphous poly(vinyl butyral) by hole burning with holographic detection (see Section V.C). Although octaethyl porphin has no ground-state dipole moment, a Stark effect resulting in broadening of the hole was observed. In chlorin, the hole both broadened and split. A theory taking into account the polarization anisotropy produced by burning, the difference in excited and ground state dipole moments, and a linear effect due to a matrix-induced dipole moment reproduces the experimental results well.[5,46]

An interesting example of the Stark effect in an amorphous matrix with application to a potential information storage system has been described.[71] The spectrum of a sample containing 9-amino acridine in poly(vinyl butyral) was burned with a HeCd laser. When an electric field E was applied across the sample, the difference in ground and excited state guest molecule dipole moment orientations, and the polarization of the matrix causes a shift in the optical transition of the guest molecules. Since the matrix in this case is amorphous, there is a distribution of frequency shifts for a given electric field. The result is that the hole is filled in when the electric field change is substantial. In this case, when E is returned to its original value, the hole reappears. Thus holes can be burned (1s) or not (0s) at increments of E and fixed wavelength and read out by scanning E over the same range. A system of 19 holes over a range of −50 to +50V was demonstrated.[71] At first, hole burning during readout was a problem, since in this NPHB system, a distribution of burning rates exists, some fast enough to cause burning even with attenuated light. The interesting solution was to preburn a "square" hole by ramping the voltage across the range of interest to eliminate the most susceptible molecules. The experiment was then carried out on the remaining, more cooperative population. In this case, the holes were detected by fluorescence excitation. With regard to information storage applications, the simplicity of scanning the electric field was pointed out. It was suggested that the electrodes could be used to heat the sample locally to erase specific locations of the storage medium.

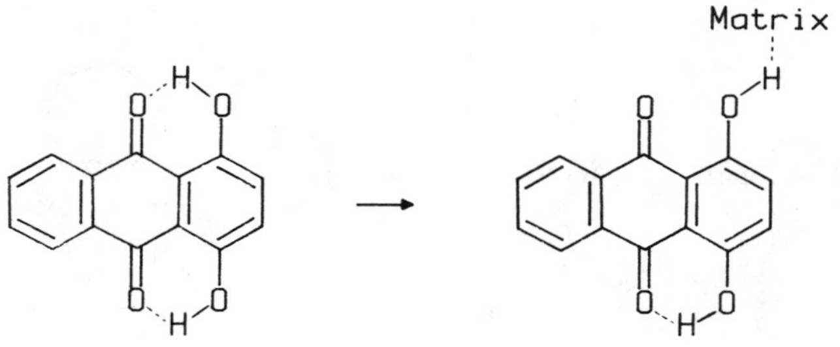

FIGURE 9.   Quinizarin undergoes hole burning only in hydrogen-bonding matrices such as alcohol glasses. The reaction believed to be responsible is shown.

E. STUDIES OF VIBRONIC BANDS AND ENERGY TRANSFER

Lifetime broadening of holes in vibronic bands provides a method of determining relaxation time. One example of this was given in section VI.B above. A systematic study of hole burning in the vibronic bands of porphin[40] revealed no apparent relation of the vibrational relaxation times to the symmetry or the energy of the vibration.

Völker and Macfarlane have also shown[40] that the vibronic bands due to two tautomers of porphin in n-hexane (see Section VII.A.3 below). can be assigned by selectively photolyzing one tautomer, causing the disappearance of one system of peaks.

Biological pigment systems, difficult to study spectroscopically because of inhomogeneous broadening, have been studied by PHB.[4] Hole burning results in a series of "satellites" to lower energies. However, the fact that no vibronic bands are observed in fluorescence suggests that the satellites are caused by energy transfer to lower-lying chromophores in the system. Friedrich and Haarer note three cases.[4] If energy transfer is faster than photochemistry, only the low-energy holes are observed. If energy transfer is very slow, it does not compete with photchemistry, and no satellites are formed. In the intermediate case, both resonant and satellite holes are observed. This is apparently the situation in the pigment phycoerythrin.[4]

VII. REVIEW OF SYSTEMS STUDIED FOR INFORMATION STORAGE USE

A. ONE-PHOTON PHOTOCHEMICAL HOLE-BURNING SYSTEMS

The spectra of many compounds have been studied using PHB experiments with burn sources having wavelengths from the IR to the UV. In this section we indicate the relevant photochemical processes for the compounds most often encountered in the applications which we have considered above—information storage and studies of the properties of glasses. The majority of the one-photon PHB systems reported undergo "subtle" photochemical reactions such as proton transfer resulting in tautomerization or a change in hydrogen bonding.

1. Quinizarin

Quinizarin, dissolved in a hydrogen-bonding matrix, undergoes the reaction shown in Figure 9 when irradiated.[19] No hole burning is observed in non-H-bonding matrices. The 0-0 transition of quinizarin in alcohol glasses has a peak at ≈ 516 nm, making it easily accessible with dye lasers. As noted above, holes in quinizarin undergo filling from reverse reaction and broadening from spectral diffusion processes. Multiple holes have been burned in the band, and, based on their width, an estimated maximum multiplexing factor of 500 was given.[32]

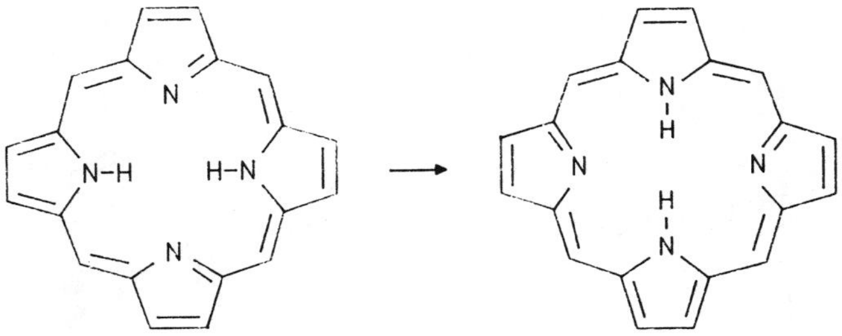

FIGURE 10.   Porphin undergoes a proton tautomerization in the excited state, resulting in hole burning. The two tautomers are degenerate in the free molecule or in liquid solvent, but since the dimensions of the molecule change on tautomerization, they have different energies in a solid host.

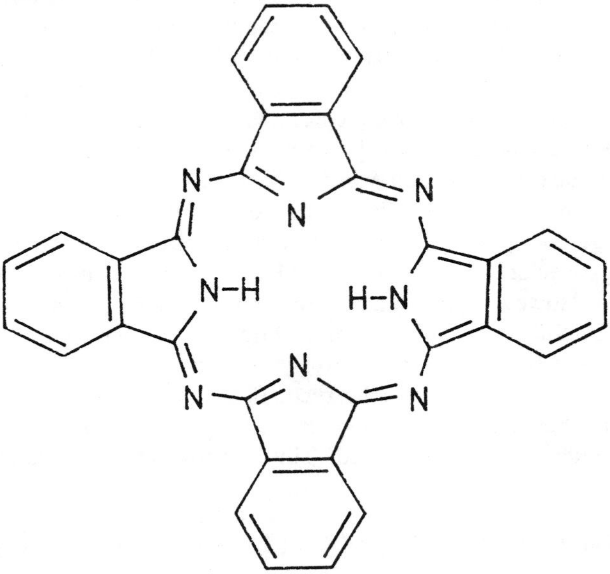

FIGURE 11.   The structure of phthalocyanine. This molecule is believed to undergo hole burning by a tautomerization similar to that shown in Figure 10.

2. Porphin Free Base

This system has been extensively studied.[40] The PHB mechanism, as in phthalocyanine, involves proton transfer resulting in tautomerization (Figure 10). The two tautomers are equivalent in the free molecule and, in solution at room temperature, tautomerization occurs rapidly. In a Shpol'skii matrix (n-hexane) in which only one site is available to the molecules, the two tautomers give rise to two bands in the spectrum, separated by $\simeq 30$ cm^{-1}. In a glass both are merged into one inhomogeneously broadened band.

The reaction is believed to occur during intersystem crossing, with a quantum yield of <1%. Extrapolation of hole width vs. temperature to 0 K gives a lifetime-limited value in many host systems.[40,48,61,62] In n-octane, irreversible hole-filling was found for temperatures $\geq$ 30 K.[40]

3. Phthalocyanine Free Base

Phthalocyanine (Figure 11) was the first molecule for which PHB was reported.[1] PHB is believed to be due to a proton tautomerization analogous to that shown for porphin in

Figure 10. These tautomers are degenerate in a free molecule. In polymer matrices, the inhomogeneous bandwidth is $\simeq 340$ cm^{-1}. The difficulty of poor solubility of this compound is offset by its high extinction coefficient.

Phthalocyanine has a bottleneck (a metastable state with lifetime $\simeq 100\mu$s) and a low reaction quantum yield, so that holes burned for less than 100 μs are extremely shallow.[41]

4. Chlorin

The chlorin molecule is similar to porphin except that one pyrrole ring is reduced; the molecule no longer has the four fold symmetry of porphin (without the N-hydrogens), and thus the two tautomers are energetically inequivalent. Hole burning was observed in the bands due to both tautomers, with a rate 10^3 times greater in the higher energy band.[40]

B. TWO-PHOTON PHOTOCHEMICAL HOLE-BURNING SYSTEMS

Following Moerner et al.[33,72] in this section we will use the following terms and abbreviations: λ_1: wavelength of the initial, site-selecting light; λ_2: Wavelength of the second, gating color; gating ratio: ratio of hole depth with both λ_1 and λ_2 to hole depth with λ_1 alone, at constant λ_1 energy; and TTA: triplet-triplet absorption spectrum. Energy levels (see Figure 5) are (Level 1) ground state; (Level 2) initially excited state; (Level 3) dissociative or reactive state; and (Level i.) intermediate level in the four-level mechanism.

1. Dimethyl-*s*-Tetrazine in Poly(Vinylcarbazole)[67,73-75]

Holes at 567.2 nm in the 0-0 band of DMST have been studied.[73] A satellite hole at $\simeq 550$ nm is observed. These holes have strong phonon bands. Both photochemical and nonphotochemical holes have been observed in this system. The photochemical holes were recovered (but blurred) after heating to 55 K; nonphotochemical holes were destroyed by heating to 12 K, and in fact the nonphotochemical holes decreased to half their original depth over a period of hours, even at 1.8 K. Hole width vs. temperature studies were carried out. For this system, $(\tau_1)^{-1} = 26$ MHz.[61]

DMST has been shown to undergo two photon decomposition into acetonitrile and N_2: a quadratic dependence on light intensity is found when the S_1 0-0 band is selectively excited.[19,74,75] Level 1 is certainly S_1: level i is probably not the T_1 level, since the quantum yield of photolysis is much greater than the quantum yield of intersystem crossing. For DMST in ethanol, an unidentified state with a lifetime of 500 μs and a rise time of 6 μs was observed.[74] A transient absorption was also observed and identified with this state.[19,74,75] Since the singlet state lifetime is 6 ns, other intermediates are probably involved as well.

This compound provides an interesting example of two-photon, one-color hole burning. Thus, one would expect that the hole-burning efficiency would depend quadratically on the excitation intensity. However, as pointed out in Reference 74, for cw excitation the intermediate would absorb λ_1 and undergo decomposition, but excitation with short (<6 μs) pulses would prevent excitation of the intermediate, which would thus allow efficient gating. Furthermore, the excited state absorption spectrum of DMST reported in Reference 75 shows that 570-nm excitation light is inefficiently abosrbed by the intermediate; the maximum in the intermediate spectrum is at $\lambda > 610$ nm. Thus gating should be improved by optimizing the wavelength of λ_2, providing that this intermediate state is indeed the one involved in the photochemistry. We are unaware of any investigation of these aspects of PHB in this compound. It should be noted that this compound also provides one of the few examples of hole burning due to a reaction which gives chemically distinct products, as opposed to more subtle photochemical reactions such as tautomerization, proton transfer, electron ejection, etc.

2. Sm^{+2} in BaClF[37]

In the first reported case of photon-gated hole burning, the two-step photoionization of

Sm^{+2} ions was identified as the responsible process. This is a four-level process, the intermediate levels being identified as metastable ^5D states. λ_1 was 687.9 nm from a cw dye laser (2 W/cm) and λ_2 was 514.5 nm from an Ar^+ laser (20 W/ cm²). A gating ratio of $\simeq 10^4$ was found for these intensities, with the gating ratio increasing to $\simeq 10^7$ when the intensity of gating light was increased.

The role of photoionization was demonstrated in two types of experiments. First, a threshold at $\simeq 500$ nm for gating efficiency vs. λ_2, was observed, indicating that a threshold is involved in the PHB reaction. Second, on one color irradiation at 488 nm, a photocurrent was observed that was 10 times larger than that produced by irradiation at 514.5 nm. The interpretation of this experiment was that the metastable ^5D state is produced in both cases, but only the 488-nm photons have sufficient energy to cause photoionization from that state.[37]

Two kinds of behavior were observed in different samples. In type I, the trap for the ejected electrons appeared to be Sm^{+3} ions, since the reaction could be reversed by irradiation with light, and the action spectrum for this process coincided with an absorption band of Sm^{+2}. Furthermore, when samples were warmed to room temperature for several days and then cooled again to 2 K, the holes were found to have "broadened by less than a factor of 2 with no discernible change in area." This appears to be the only known case of such extreme thermal stability. Type II samples appeared to be dominated by some other electron trap, and they lacked these properties.

3. Carbazole in Boric Acid[76,77]

Gated photochemical hole burning was studied for carbazole in a boric acid glass. This was the first observation of two-color gated PHB in an organic system.[76,77] λ_1 was near the S_1-S_0 origin at 335 nm. Several gating wavelengths from 365 to 514 nm were studied. For $\lambda_2 = 351$ nm, gating efficiencies of 10 to 400 were found, increasing with λ_2 power, but excessive λ_2 power produced broad holes which were attributed to sample heating or to photochemical saturation.

The gating action spectrum was found to correspond in a qualitative way to the TTA spectrum of carbazole measured at 25°C, so that this would be classified as a four-level system where the intermediate level is the lowest triplet state of carbazole. The product spectrum was also determined and appeared to be mainly the carbazole radical cation, indicating that photoionization is the reaction responsible for hole burning. In addition, the quantum yield with respect to λ_2 of hole burning increases exponentially with the energy of the second photon, in agreement with results found by earlier workers for one-color, two-photon photoionization of carbazole in ethanol glass at 77 K. Boric acid is known to facilitate ionization of solutes because of the high dielectric constant. Under similar conditions, but with higher irradiation energies, no PHB was seen in carbazole in PMMA, suggesting that N-H bond clevage, which was observed at room temperature in this system, is not the low-temperature mechanism of PHB for carbazole in boric acid glass.

The holes formed in this system were found to remain, with broadening, after the temperature of the sample was raised to 77 K for several hours, indicating a large barrier for electron tunneling from any matrix traps back to the cation, as well as for any matrix rearrangement.

4. TZT and Chloroform in PMMA[72,79]

Gated PHB was observed in meso-tetra(*p*-tolyl)-Zn-tetrabenzoporphyrin (TZT) in a matrix of PMMA and chloroform. The sample was prepared by dissolving TZT and PMMA in chloroform and evaporating off only part of the chloroform to leave a "wet" sample. The S_1-S_0 transition of TZT is found at 630 nm. Following excitation, the molecule undergoes intersystem crossing with a yield of 0.8. The gating action spectrum is similar to the triplet-triplet absorption spectrum.

In a pulsed experiment in thich the λ_2 pulse was delayed after a λ_1 pulse, the gating effect was found to decay with a lifetime equal to the triplet lifetime. Based on this data, it was concluded that the intermediate level is the T_1 state for this system. Light of $\lambda_2 =$ 350 to 800 nm causes electron transfer from triplet TZT to chloroform. A gating ratio of 30 to 360 was found for this system, depending on hole deptth and other conditions. The gating ratio was only equal to 1 for a sample prepared without chloroform, demonstrating the role of the chloroform as the electron acceptor.

This system is of considerable importance with respect to potential applications in the area of information storage for several reasons. The photochemical system itself offers considerable flexibility, since both donor and acceptor molecules can be varied to optimize performance. The magnesium analog of TZT and other halomethane acceptors have also been studied and shown to exhibit gated PHB.[79] λ_1 is at a convenient wavelength. Detectable holes were made with just one 8-ns, 90-μJ, 632-nm pulse from a N_2-pumped dye laser plus a 200-ms exposure to a 20-mW, 488-nm Ar^+ laser beam. Readout involved averaging 32 to 128 wavelength scans of 0.25 s each of a cw dye laser while measuring the the transmitted intensity.

5. Anthracene-Tetracene Photoadduct in PMMA[50,80]

PHB was observed at 326 nm in the 0-0 absorption band of the anthracene-tetracene photoadduct (AT) with enhancement occurring upon simultaneous irradiation at 442 nm. The gating ratio is approximately 2 under the experimental conditions described in Reference 50. The phonon coupling is fairly strong in this compound: the Debye-Waller factor is approximately 0.5. The reaction responsible for PHB is the cleavage of the adduct, as shown by the product spectrum. The fluorescence spectrum of the products taken after photolysis at 1.5 K, but without warming indicates that the anthracene and tetracene molecules are held in a face-to-face "sandwich" configuration, i.e., in the correct position for the reverse photochemical or thermal reaction. The reverse reaction appears not to proceed at liquid helium temperatures on irradiation with light absorbed by tetracene (442 nm), but it may occur upon irradiation with other wavelengths or at higher temperatures which are still low enough that holes survive in regions of the material which are not irradiated. Holes in this material are broadened, but not eliminated upon cycling to 77 K for short periods of time.[80]

The intermediate state appears to be the T_1 state of AT, so that this is also a four-level system. The gating action spectrum peaks at about 440 nm. It is qualitatively similar to the TTA spectrum, and the metastable triplet state has been shown to be depopulated by irradiation with 442-nm light.[80] The TTA spectrum shows that the T_1 state also absorbs 326-nm light, albeit with a low cross-section. Thus, the one-color PHB appears to be chiefly due to a two-photon process with both photons at 326 nm. Hole burning with enhancement by blue light has also been found to occur in the spectra of the related compounds ditetracene (both isomers) and the adduct of 9-bromoanthracene and tetracene.[80]

6. Co^{+2} in $LiGa_5O_8$[81]

Holes in $LiGa_5O_8:Co^{+2}$ were made at 660.4 nm with 1 W/cm^2 power for 5 s. Gating light at 673.4 nm and 100 W/cm^2 was used. The light sources were cw dye lasers, and the holes were observed in the fluorescence excitation spectrum by attenuating and scanning the shorter wavelength laser. The holes were approximately 100 times broader than expected from lifetimes and laser jitter. A gating ratio of 20 was found. The gating action spectrum was not reported.

The mechanism of hole formation was a two-step photoionization of Co^{+2}. Hole erasure was also possible in this system with an action spectrum similar to Co^{+2} absorption; thus the gating light was chosen at a wavelengtth not strongly absorbed by Co^{+2}.

7. Two-photon Nonphotochemical Hole Burning in Tetracene[82]

NPHB was observed in tetracene in glass (isopropyl alcohol and ethyl ether, 2:5 v/v) and crystal (*n*-hexadecane). The zero-phonon and vibronic emission was studied. The rate of hole burning was determined by measuring the initial slope of zero-phonon fluorescence intensity vs. time, taking into account saturation at $\simeq 50\%$. The authors found that the hole-burning rate varied linearly with intensity in the crystalline matrix but quadratically in glass, indicating a two-photon mechanism in the latter. The intermediate state was not characterized.

ACKNOWLEDGMENTS

Some of the research by the authors which was reviewed in the present work was supported by the Committee on Research of the University of California, Riverside, and by a NASA training grant, NGI-50156. We wish to thank Dr. W. E. Moerner of IBM Almaden Research Center for providing us with preprints describing his recent research results.

REFERENCES

1. **Gorokhovskii, A. A., Kaarli, R. K., and Rebane, L. A.,** Hole Burning in the contour of a pure electronic line in a Shpol'skii system, *JETP Lett.,* 20, 216, 1974.
2. **Shpol'skii, E. V.,** Line fluorescence spectra of organic compounds and their application, *Sov. Phys. Usp.* 3, 372, 1960.
3. **Kharlamov, B. M., Personov, R. I., and Bykovskaya, L. A.,** Stable "gap" in absorption spectra of solid solutions of organic molecules by laser irradiation, *Opt. Commun.,* 12, 191, 1974.
4. **Friedrich, J. and Haarer, D.,** Photochemical hole burning: a spectroscopic study of relaxation processes in polymers and glasses, *Angew. Chem. Int. Ed. Engl.,* 23, 113, 1984.
5. **Friedrich, J. and Haarer, D.,** Structrual relaxation processes in polymers and glasses as studied by high resolution optical spectroscopy, in *Optical Spectroscopy of Glasses,* Zschokke, I., Ed., D. Reidel, Dordrecht, Netherlands, 1986.
6. **Small, G. J.,** Persistent nonphotochemical hole burning and the dephasing of impurity electronic transitions in organic glasses, in *Spectroscopy and Excitation Dynamics of Condensed Molecular Systems,* Agranovich, V. M. and Hochstrasser, R. M., Eds., North-Holland, Amsterdam, 1983.
7. **Rebane, L. A., Gorokhovskii, A. A., and Kikas, J. V.,** Low-temperature spectroscopy of organic molecules in solids by photochemical hole burning, *Appl. Phys. B,* 29, 235, 1982.
8. **Personov, R. I.,** Site selection spectroscopy of complex molecules in solutions and its applications, in *Spectroscopy and Excitation Dynamics of Condensed Molecular Systems,* Agranovich, V. M. and Hochstrasser, R. M., Eds., North-Holland, Amsterdam, 1983.
9. **Steinfeld, J. I.,** *Molecules and Radiation,* MIT Press, Cambridge, MA, 1978, chap. 1.
10. **McQuarrie, D. A.,** *Statistical Mechanics,* Harper & Row, New York, 1976, chap. 21.
11. **Loudon, R.,** *The Quantum Theory of Light,* Oxford University Press, New York, 1983, chap. 2.
12. **Thijssen, H. P. H. and Völker, S.,** Spectral hole burning in semicrystalline polymers between 0.3 and 4.2 K, *J. Chem. Phys.,* 85, 785, 1986.
13. A number of excellent papers on this subject are contained in the issue on optical linewidths in glasses of *J. Lumin.,* 36, 1987.
14. **Lyo, S. K.,** Dynamical theory of optical linewidths in glasses, in *Optical Spectroscopy of Glasses.,* Zschokke, I, Ed., D. Reidel, Dordrecht, Netherlands, 1986.
15. **Reineker, P. and Kassner, K.,** Model calculations of optical dephasing in glasses, in *Optical Spectroscopy of Glasses,* Zschokke, I., Ed., D. Reidel, Dordrecht, Netherlands, 1986.
16. **Schulte, G., Grond, W., Haarer, D., and Silbey, R.,** Photochemical hole burning of phthalocyanine in polymer glasses: thermal cycling and spectral diffusion, *J. Chem. Phys.,* in press.
17. **Hollas, J. M.,** *High Resolution Spectroscopy,* Butterworths, London, 1982.
18. **Selzer, P. M.,** General techniques and experimental methods in laser spectroscopy in *Topics in Applied Physics,* Vol. 49, Yen, W. M. and Selzer, P. M., Eds., Springer-Verlag, Berlin, 1981.
19. **Burland, D. M. and Haarer, D.,** One- and two-photon laser photochemistry in organic solids, *IBM J. Res. Dev.,* 23, 534, 1979.

20. **Jankowiak, R. and Small, G. J.**, Hole-burning spectroscopy and relaxation dynamics of amorphous solids at low temperatures, *Science,* 237, 618, 1987.
21. **Macfarlane, R. M. and Shelby, R. M.**, Homogeneous line broadening of optical transitions of ions and molecules in glasses, *J. Lumin.,* 36, 179, 1987.
22. **Freidrich, J., Wolfrum, H., and Haarer, D.**, Photochemical holes: a spectral probe of the amorphous state in the optical domain, *J. Chem. Phys.,* 77, 2309, 1982.
23. **Heimenz, P. C.**, *Polymer Chemistry: the Basic Concepts,* Marcel Dekker, New York, 1984.
24. **Aklonis, J. J.**, Current problems in understanding the behavior of polymer glasses, in *Photophysical and Photochemical Tools in Polymer Science,* Winnik, M. A., Ed., D. Reidel, Dordrecht, Netherlands, 1986.
25. **Phillips, W. A.**, Introduction, in *Amorphous Solids: Low-Temperature Properties,* Phillips W. A., Ed., Springer-Verlag, Berlin, 1981.
26. **Hunklinger, S. and Schickfus, M. V.**, Acoustic and dielectric properties of glasses at low temperatures, in *Amorphous Solids: Low-Temperature Properties,* Phillips, W. A., Ed., Springer-Verlag, Berlin, 1981.
27. **Breinl, W., Friedrich, J., and Haarer, D.**, Ground state tunneling and optical spectral diffusion in organic glasses, *J. Chem. Phys.,* 80, 3496, 1984.
28. **Hayes, J. M. and Small, G. J.**, Non-photochemical hole burning and impurity site relaxation processes in organic glasses, *Chem. Phys.,* 27, 151, 1978.
29. **Breinl, W., Friedrich, J., and Haarer, D.**, Spectral diffusion of a photochemical proton transfer system in an amorphous organic host: quinizarin in an alcohol glass, *J. Chem. Phys.,* 81, 3915, 1984.
30. **Kittel, C.**, *Introduction to Solid State Physics,* John Wiley & Sons, New York, 1986.
31. **Wong, J. and Angell, C. A.**, *Glass Structure by Spectroscopy,* Marcel Dekker, New York, 1976, 413.
32. **Gutiérrez, A. R., Friedrich, J., Haarer, D., and Wolfrum, H.**, Multiple photochemical hole burning in organic glasses and polymers: spectroscopy and storage aspects, *IBM J. Res. Dev.,* 26, 198, 1982.
33. **Lenth, W. and Moerner, W. E.**, Gated spectral hole-burning for frequency domain optical recording, *Opt. Commun.,* 58, 249, 1986.
34. **Castro, G.**, Recent progress in optical storage by photochemical hole burning, in *Photochemistry and Photobiology,* Vol. 2. Zewail A., Ed., Harwood Academic Publishers, Chur, Switzerland, 1983.
35. **Moerner, W. E.**, Molecular electronics for frequency domain optical storage: persistent spectral hole burning, *J. Mol. Electron.,* 1, 55, 1985.
36. **Moerner, W. E. and Levenson, M. D.**, Can single-photon processes provide useful materials for frequency-domain optical storage?, *J. Opt. Soc. Am. B,* 2, 915, 1985.
37. **Winnacker, A., Shelby, R. M., and Macfarlane, R. M.**, Photon-gated hole burning: a new mechanism using two-step photoionization, *Opt. Lett.,* 10, 350, 1985.
38. **Pokrowsky, P., Moerner, W. E., Chu, F., and Bjorklund, G. C.**, Reading and writing of photochemical holes using GaAlAs-diode lasers, *Opt. Lett.,* 8, 280, 1983.
39. **Lee, H. W. H., Huston, A. L., Gehrtz, M., and Moerner, W. E.**, Photochemical hole-burning in a protonated phthalocyanine with GaAlAs diode lasers, *Chem. Phys. Lett.,* 114, 491, 1985.
40. **Völker, S. and Macfarlane, R. M.**, Photochemical hole burning in free-base porphyrin and chlorin in n-alkane matrices, *IBM J. Res. Dev.,* 23, 547, 1979.
41. **Romagnoli, M., Moerner, W. E., Schellenberg, F. M., Levenson, M. D., and Bjorklund, G. C.**, Beyond the bottleneck: submicrosecond hole burning in phthalocyanine, *J. Opt. Soc. Am. B,* 1, 341, 1984.
42. **Friedrich, J., Swalen, J. D., and Haarer, D.**, Electron-phonon coupling in amorphous organic host materials as investigated by photochemical hole burning, *J. Chem. Phys.,* 73, 705, 1980.
43. **Friedrich, J. and Haarer, D.**, Transient features of optical bleaching as studied by photochemical hole burning and fluorescence line narrowing, *J. Chem. Phys.,* 76, 61, 1982.
44. **Köhler, W., Breinl, W., and Friedrich, J.**, Laser photochemistry with polarized light in low temperature glasses, *J. Phys. Chem.,* 89, 2473, 1985.
45. **Kador, L., Schulte G., and Haarer, D.**, Relation between hole-burning parameters: free-base phthalocyanine in polymer hosts, *J. Phys. Chem.,* 90, 1264, 1986.
46. **Meixner, A. J., Renn, A., Bucher, S. E., and Wild, U. P.**, Spectral hole burning in glasses and polymer films: the Stark effect, *J. Phys. Chem.,* 90, 6777, 1986.
47. **Jalmukhambetov, A. U. and Osad'ko, I. S.**, Dependence of photochemical and photophysical hole burning on laser intensity, *Chem. Phys.,* 77, 247, 1983.
48. **Thijssen, H. P. H. and Völker, S.**, Pitfalls in the determination of optical homogeneous linewidths in amorphous systems by hole-burning. Influence of the structure of the host, *Chem. Phys. Lett.,* 120, 496, 1985.
49. **Förster, Th.**, Transfer mechanisms of electronic excitation, *Discuss. Faraday Soc.,* 27, 7, 1959.
50. **Iannone, M., Scott, G. W., Brinza, D., and Coulter, D.**, Gated photochemical hole burning in photoadducts of polyacenes, *J. Chem. Phys.,* 85, 4863, 1986.
51. **Bjorklund, G. C. and Levenson, M. D.**, Frequency modulation (FM) spectroscopy, *Appl. Phys. B,* 32, 145, 1983.

52. **Romagnoli, M., Levenson, M. D., and Bjorklund, G. C.,** Frequency-modulation polarization-spectroscopy detection of persistent spectral holes, *J. Opt. Soc. Am. B,* 1, 571, 1984.
53. **Renn, A., Meixner, A. J., Wild, U. P., and Burkhalter, F. A.,** Holographic detection of photochemical holes, *Chem. Phys.,* 93, 157, 1985.
54. **Huston, A. L. and Moerner, W. E.,** Detection of persistent spectral holes using ultrasonic modulation, *J. Opt. Soc. Am. B,* 1, 349, 1984.
55. **Rebane, K. K.,** Laser study of no-phonon lines in the inhomogeneously broadened spectra via photochemical hole burning, *J. Lumin.,* 32, 744, 1984.
56. **Rebane, A. K., Kaarli, R. K., and Saari, P. M.,** Burning out a complex-shaped hole by a coherent series of picosecond pulses, *Opt. Spectrosc. (USSR),* 55, 238, 1983.
57. **Kikas, Ya. V., Kaarli, R. K., and Rebane, A. K.,** Limiting number of photochemical holes in an inhomogeneously broadened spectrum, *Opt. Spectrosc. (USSR),* 56, 238, 1984.
58. **Molenkamp, L. W. and Wiersma, D. A.,** Optical dephasing in organic amorphous systems. A photon echo and hole-burning study of pentacene in polymethacrylate, *J. Chem. Phys.,* 83, 1, 1985.
59. **Breinl, W., Friedrich, J., and Haarer, D.,** Linewidth study of the dye molecule quinizarin in solid alcohol glasses, *Phys. Rev. B,* 34, 7271, 1986.
60. **Walsh, C. A., Berg, M., Narasimhan, L. R., and Fayer, M. D.,** A picosecond photon echo study of a chromophore in an organic glass: temperature dependence and comparison to nonphotochemical hole burning, *J. Chem. Phys.,* 86, 77, 1987.
61. **Thijssen, H. P. H., van den Berg, R., and Völker, S.,** Thermal broadening of optical homogeneous linewidths in organic glasses and polymers studied by photochemical hole burning, *Chem. Phys. Letts.,* 97, 295, 1983.
62. **Thijssen, H. P. H., van den Berg, R., and Völker, S.,** Optical relaxation in organic disordered systems submitted to photochemical and non-photochemical hole burning, *Chem. Phys. Letts.,* 120, 503, 1985.
63. **Grokhovskii, A. A. and Rebane, K. K.,** Hole burning in organic glasses: linewidth study and some applications, Abst. 3rd Int. Conf. Unconventional Photoactive Solids, Schloss Elmau, West Germany, 1987.
64. **Gorokhovskii, A., Korrovits, V., Pal'm, V., and Trummal, M.,** Temperature broadening of a photochemical hole in the spectrum of H_2-octaethylporphin in polystyrene between 0.05 and 1.5 K, *Chem. Phys. Letts.,* 125, 355, 1986.
65. **Gorokhovskii, A., Korrovits, V., Pal'm, V., and Trummal, M.,** Photochemical burning of a gap in the spectrum of an impurity in an amorphous polymer at 0.05-1.5 K, *JETP Lett.,* 42, 307, 1986.
66. **van den Berg, R. and Völker S.,** Optical homogeneous linewidths of resorufin in ethanol glass: an apparent contradiction between hole-burning and photon-echo results?, *Chem. Phys. Lett.,* 137, 201, 1987.
67. **De Vries, H. and Wiersma, D. A.,** Lifetime-limited photochemical hole-burning in *s*-tetrazine-benzene mixed crystals at 2 K, *Chem, Phys. Letts.,* 51, 565, 1977.
68. **Boxer, S. G., Lockhart, D. J., and Middendorf, T. R.,** Photochemical hole-burning in photosynthetic reaction centers, *Chem. Phys. Lett.,* 123, 476, 1986.
69. **Hayes, J. M. and Small, G. J.,** Photochemical hole burning and strong electron-phonon coupling: primary donor states of reaction centers of photosynthetic bacteria, *J. Phys.,* 90, 4928, 1986.
70. **Dicker, A. I. M., Johnson, L. W., Noort, M., Thijssen, H. P. H., Völker, S., and van der Waals, J. H.,** MHz resolution Zeeman and Stark spectroscopy of porphins via photochemical hole burning, in *Photochemistry and Photobiology,* Vol. 2. Zewail, A. H., Ed., Harwood Academic, Chur, Switzerland, 1983.
71. **Bogner, U., Beck, K., and Maier, M.,** Electric field selective optical data storage using persistent spectral hole burning, *Appl. Phys. Letts.,* 46, 534, 1985.
72. **Moerner, W. E., Carter, T. P., and Bräuchle, C.,** Fast burning of persistent spectral holes in small laser spots using photon-gated materials, IBM Research report, 1986.
73. **Cuellar, E. and Castro, G.,** Photochemical and nonphotochemical hole burning in dimethyl-s-tetrazine in a polyvinyl carbazole film, *Chem. Phys.,* 54, 217, 1981.
74. **Dellinger, B., Paczkowski, M. A., Hochstrasser, R. M., and Smith, A. B., III,** Observation of transient intermediates in the photochemical decomposition of substituted *s*-tetrazines, *J. Am. Chem. Soc.,* 100, 3242, 1978.
75. **Burland, D. M. and Carmona, F.,** Photodissociation in molecular crystals, *Mol. Cryst. Liq. Cryst.,* 50, 279, 1979.
76. **Lee, H. W. H., Gehrtz, M., Marinero, E. E., and Moerner, W. E.,** Two-color, photon-gated spectral hole-burning in an organic material, *Chem. Phys. Letts.,* 118, 611, 1985.
77. **Lee, H. W. H., Gehrtz, M., Marinero, E. E., and Moerner, W. E.,** Observation of two-color, photon-gated spectral hole-burning in an organic systems, *J. Chem. Phys.,* submitted.
78. **Carter, T. P., Bräuchle, C., Lee, V. Y., Manavi, M., and Moerner, W. E.,** Mechanism of photon-gated persistent spectral hole burning in metal-tetrabenzoporphyrin/halomethane systems: donor-acceptor electron transfer, *J. Phys. Chem.,* 91, 3998, 1987.

79. **Carter, T. P., Bräuchle, C., Lee, V. Y., Manavi, M., and Moerner, W. E.,** Photon-gated spectral hole burning via donor-acceptor electron transfer, *Solid State Phys.,* submitted.
80. **Iannone, M. and Scott, G. W.,** unpublished results.
81. **Macfarlane, R. M. and Vial, J. -C.,** Photon-gated spectral hole burning in $LiGa_5O_8Co^{+2}$, *Phys. Rev. B,* 34, 1, 1986.
82. **Gorokhovskii, A. A., Kikas, Ya. V., Pal'm, V. V., and Rebane, L. A.,** Characteristics of hole burning in the spectra of organic molecules in glassy matrices, *Sov. Phys. Solid State,* 23, 602, 1981.

INDEX

A

Ablated matter, see UV laser ablation
Ablation of polymers by lasers, 56—57, see also UV laser ablation
Ablative photodecomposition of polymers, see UV laser ablation
Absorbance, 188, 241—242
Absorbing defects, 101, 104, 112—115, 117
Absorbing impurities, 108, 112, 115, 127
Absorption grating, 182
Accumulated photon echoes, 250—251
Acetals, 5
Acrylic monomers, 5, 30
Active Q-switches, 99
Ambient temperature, 109, 111
Amplitude hologram, 183, 190, 199, 202
Anthracene-tetracene photoadduct in PMMA, 259
bis-Anthrylmethylether, 220
Argon-ion laser, 8—11, 13, 43—44, 47
  charge transfer complexes, 50—51
  photosensitive systems, 49
Aromatic ketones, 4, 39
Arterial blockages, 61
Arthroscopic surgical techniques, 61, 70
Aryldiazonium salts, 5
Azido compounds, 42
Azobenzene, 219
Azobisisobutyronitrile, 115, 125

B

BA/CAc, temporal grating process, 204
Beam profile, 104
Beer-Lambert's law, 76, 81
Beer's law, 165
Benzoin isobutyl ether, 41
Benzophenone, 41
Benzophenone in polymethylmethacrylate matrix (BP/PMMA), 187—189
Biphotonic absorption, 75—80
Bleaching quantum efficiencies, 122, 124
Bottlenecks in information storage, 238—239
BP/PMMA, see Benzophenone in polymethylmethacrylate matrix
Breaking stress, 114
Brittle fracture limit, 114, 116—117, 127
Broadening mechanisms, 245—247
Bulk components, high-power laser systems, 98—99
Bulk optical strength, 101, 110, 117, 127
Bulk-to-surface peak fluence damage threshold ratio, 105

C

CAD, see Computer-aided design
CAM, see Computer-aided manufacturing
Cancerous tumor reduction, 61

Carbazole in boric acid, 258
Carbazole in poly-methyl-methacrylate matrix, holography with, 202
Carbon formation, 109
Carbonic acid ether, 126
Cationic photoinitiators, 5, 43
Cationic photopolymerization, 22—23
Cellulose acetate butyrate, 106
Cellulose triacetate, 119—120
Charge transfer complexes, 25, 50—51
Chlorin, 257
Chloroform in PMMA, 258—259
Coating industry, 2
Coatings, 38
Coating technology, 22
CO_2 lasers, 44—45
  hardening behavior, 52
  photosensitive systems, 49
  softening of thermoset polymers, 57—59
CO^{+2} in $LiGa_5O_8$, 259
Complexity considerations, 69
Computer-aided design (CAD), 74, 91—92
Computer-aided manufacturing (CAM), 74
Continuous wave holographic gratings, 182—187
Continuous-wave lasers, 104
Contrast, 10
Conventional PIH, detection sensitivity limitations of, 195
Copolymerization, 117
Copolymers, 124
Copper phthalocyanine (CvPc), 100
Copper vapor lasers, 44
Cost factors, 68—69
Crosslinking, 5, 8, 11, 15, 38, 203
  azido compounds, 42
  cycloaddition reactions, 42
  epoxides, 42
  thermal effects of lasers in, 51—56
  thio-ene systems, 41—42
Crosslinking polymerization, photoresist applications, 45—49
Cryptocyanine, 99
Crystallinity in polymer hosts, 253
Curing of polymers, thermal effects of lasers in, 51—56
Curing of powders, 67
Curing resins, potential applications of lasers in, 59—68
CvPc, see Copper phthalocyanine
Cycloaddition reactions, 42

D

Damage probability, 104, 112
Debris shields, 98—99
Debye-Waller factor, 242—245, 259
Dentistry, laser hardening, applications of, 59—61
Dephasing, 230, 245, 251—254

I

Waveguide directional coupler, 206
Wavelength-dependence measurements, 112
Wavelength range, 44

X

Xanthene dye, 126
Xanthene dye 11B, 103, 126
XPS analysis, 153—154
X-ray photoelectron spectroscopy, 149

Y

Young' modulus, 114, 116, 127

Z

Zeeman shifts, 254
Zero-phonon band, 244
Zero-phonon hole (ZPH), 232—233, 237, 241—245
ZPH, see Zero-phonon hole